I0891183

ROUTLEDGE LIBRARY EDITIONS: HISTORY AND PHILOSOPHY OF SCIENCE

SOVIET MARXISM AND NATURAL SCIENCE

SOVIET MARXISM AND NATURAL SCIENCE

1917–1932

By

DAVID JORAVSKY

Volume 17

Routledge
Taylor & Francis Group
LONDON AND NEW YORK

First published in 1961

This edition first published in 2009 by Routledge
2 Park Square, Milton Park, Abingdon, Oxon, OX14 4RN

Simultaneously published in the USA and Canada
by Routledge
52 Vanderbilt Avenue, New York, NY 10017

Routledge is an imprint of the Taylor & Francis Group, an informa business

British Library Cataloguing in Publication Data
A catalogue record for this book is available from the British Library

Library of Congress Cataloging in Publication Data
A catalog record for this book has been requested

First issued in paperback 2013

Publisher's Note
The Publisher has gone to great lengths to ensure the quality of this reprint but points out that some imperfections in the original copies may be apparent.

Disclaimer
The Publishers have made every effort to trace copyright holders and would welcome correspondence from those they have been unable to contact.

ISBN 13: 978-0-415-85078-0 (pbk)
ISBN 13: 978-0-415-42029-7 (Set)
ISBN 13: 978-0-415-47486-3 (Volume 17)

SOVIET MARXISM AND NATURAL SCIENCE

1917 – 1932

by

DAVID JORAVSKY

Routledge and Kegan Paul

LONDON

First published 1961
by Routledge and Kegan Paul Ltd
Broadway House, 68–74 Carter Lane
London, E.C.4

Printed in Great Britain
by Cox and Wyman, Ltd
London, Fakenham and Reading

To
DORIS

CONTENTS

PART FOUR
THE GREAT BREAK, 1929–1932

PART FIVE
PHYSICS AND BIOLOGY IN THE FIRST PHASE 1917–1932

PREFACE

THIS book did not grow out of an interest in the capability of Soviet technologists, which has had a tendency to dominate recent discussions of Soviet science. It grew out of an interest in the intellectual history of the Russian Revolution, out of a desire to understand the modern analogues to Marat and Lavoisier in an earlier revolution, or to Calvin and Servetus in another. It may be naïve for a contemporary of Lysenko and N. I. Vavilov, who enter the last pages of this history and will be major figures in the sequel, to believe that interest in such a subject can be different from animus or enthusiasm, that a desire to understand such people need not be a desire to expose or vindicate. Naïve or not, that is my belief. I hope that this history will be a contribution, however limited, to the satisfaction of a similar interest and desire in others.

The focus here is on Soviet Marxist philosophy of natural science, as it developed in its first phase, from 1917 to 1932. 'Natural science' is used in the conventional sense to mean the systematized knowledge of nature, with the exception—equally conventional—of human nature. The 'philosophy of natural science' is, of course, a controversial concept that may be used to indicate such diverse things as emotions, *Weltanschauungen*, or methodologies that claim derivation from and application to natural science. Fortunately, it has not been necessary to attempt a definition of my own in order to review the arguments that Soviet Marxists fell into when trying to define theirs. But it has been necessary to make extended excursions into their philosophical heritage from pre-revolutionary times, into certain policies and dissensions of the Soviet Communist Party from the Revolution to the early 'thirties, into the changing nature of the Soviet scholarly community during the same period, and into some contemporaneous controversies in biology and physics. The reader should not expect to find these collateral subjects fully treated. For example, he will not find a complete survey of pre-revolutionary Marxist views on natural science. Nor will he find a full history of natural science and higher education in the Soviet Union during the period under review. He will find only certain parts of such collateral histories—those parts that are necessary for an understanding of the first phase in the interaction between Soviet Marxism and natural science. I have tried to draw lines of demarcation narrow enough to avoid superficiality, yet broad enough to avoid distortion by artificial isolation.

I hope that this history will help, if only slightly, to lessen the widespread confusion concerning Soviet Marxism and its relations with Soviet natural science. Soviet Marxists have contributed to this confusion by interpreting their intellectual history in accordance with their shibboleths, which change rather frequently, and by their general reluctance so far to write substantial intellectual histories of the Soviet period. Non-Communist authors have contributed to the confusion by an excess of boldness, by the proclivity that many have shown towards magisterial judgments on the basis of insufficient evidence. One author, for example, writing in a scholarly journal, based a history of the theory of relativity in Soviet physics and philosophy on three 'sources', two of which were merely ambiguous passing references to Einstein's theory in Soviet articles on other subjects.[1] Such hasty generalizations help to explain why non-Communist authors have pinned such diverse labels on Soviet Marxism as mechanistic materialism, a curious sort of positivism, voluntaristic idealism, pragmatism, an abortive version of 'ideal realism', a perverted twin of Thomism, or merely a philosophically meaningless pistol for shooting the opponents of Lenin and Stalin. Estimates of this philosophy's relations with Soviet natural science have been hardly less diverse; weighed down by few facts, such estimates have tended to drift with the tides of public attitudes towards the Soviet Union.

There are of course other reasons than hasty generalization for the diversity of views on the relations between Soviet Marxism and natural science. Not the least of them is the lack of consistency and clarity with respect to the crucial terms used to discuss Marxist philosophy, both by Marxists themselves and by philosophers at large. Difficulties peculiar to the various types of Marxists will be examined in Chapter 1, but here it seems appropriate to review some crucial ambiguities of a broader provenience, in the hope of averting as many obscurities as possible.

'Metaphysics' will be used below in the conventional sense to describe a type of philosophy that seeks knowledge of universal reality beyond that given by the empirical sciences, and by other methods than those used in the empirical sciences. 'Positivism' will be used to describe a type of thought that would limit knowledge to the methods and results of the empirical sciences. But serious reservations must be made in these definitions. The tortuous course of 'positivism' since Comte shows the difficulty—some would say the impossibility—of escaping from 'metaphysics', of excluding from philosophy knowledge and methods that are not appropriate to the empirical sciences. Accordingly, the distinction just given between 'metaphysics' and 'positivism' may be artificial from a philosophical point of view; it is used here without philosophical commitment to describe an historically observable cleavage among philosophers. Moreover, in applying the terms to Marxists, most of whom claimed to transcend both 'metaphysics' and 'positivism', I am not judging them fools or liars. In the first place, I am not using these terms in the special pejorative senses that they have for Marxists. (These meanings will be described in Chapter 1.) In the second place, I know of no other terms to describe the historically observable cleavage in Marxist thought that will be traced through the Chapters below.

'Materialism' will be used below to indicate a type of philosophy that regards 'matter' as the universal substance. But one must bear in mind that 'materialists'' definitions of 'matter' have ranged from the Democritean—e.g., 'extended, impenetrable, eternally existent [particles that are] susceptible of movement or change of relative position'[2]—to the nearly agnostic—e.g., 'the extra-mental cause of sense experience'.[3] At one extreme 'materialism' is 'metaphysical', even though the term itself may be proscribed; at the other, 'materialism' is presented almost solely as an epistemology and becomes rather 'positivistic'. As the reader no doubt realizes, and as he will find time and again in what follows, this ambivalence characterizes not only 'materialists' as a group but often the mind of an individual 'materialist'.

'Mechanism' has a history of even more diverse meanings than 'materialism'. It will be used here to indicate a type of philosophy that considers 'reduction' the chief, or perhaps even the only method of scientific explanation. The meaning of 'mechanism' therefore depends in each case on the meaning given to 'reduction', and on the boundaries, if any, that are set to its applicability. 'Reduction' may be defined most comprehensively as the attempt 'to explain complicated . . . phenomena as the result of combinations of simpler ones'.[4] In further definition the 'mechanist' reveals his tendency towards 'metaphysics' or towards 'positivism', since he may or may not try to assign ontological as well as epistemological significance to 'reduction'[5]. And whatever his views on the ontological significance of 'reduction' may be, he may restrict the applicability of 'mechanism' to one or another area of study. A 'vitalist', for example, may believe that the physical sciences will ultimately be 'reduced' to the mechanics of elementary particles, but he remains a 'vitalist' by denying that the same can happen to the biological or social sciences. A 'mechanist' social theorist 'reduces' human behaviour to such determinants as geographical conditions, race, means of production, or *id* and *ego*, but he may be indifferent or even hostile to 'mechanism' in other fields.

'Dialectics' is an ancient philosophical term richly overgrown with various species of meaning. The species of interest to the present study is a very comprehensive one: a type of philosophy that perceives a universal process of change or transformation to be the essence of reality, whether mental or external. Some will object that this definition is too comprehensive. Under it, what some Marxists have scorned as 'vulgar evolutionism' would be 'dialectics'. Under it, Cratylus, who was so overwhelmed by the sense of endless change that he would not assume the stability of concepts long enough to sustain conversation, and Hegel, who was comparatively comforting with his vision of recurring triads, would both be 'dialectical' philosophers. Similar objections could be made to the broad definitions of other philosophical terms given above. But the historical method that I have tried to follow prohibits judgments of what, once and for all, 'dialectics' or 'materialism' should mean; it allows only descriptions of what, in different contexts, such terms have meant. Virtually all of the authors whose views are analysed below claimed to be upholding 'dialectical materialism'. I have tried to characterize their divergent views accurately, and to discover the circumstances that helped shape their divergencies. I leave it to the philosopher to decide which came closest to genuine 'dialectical materialism'.

Some of the confusion concerning Soviet Marxism is due to difficulties of translation. The translator must constantly be on guard both against the carelessness that smudges what is precise, and against the officiousness that tidies up obscurities. But even when he consciously aspires to this golden mean, he finds in Russian some elements both of clarity and of confusion that cannot be reproduced in English. The Russian *ideinyi*, to take a simple example, would have an exact equivalent in the adjective 'ideal', if 'ideal' were understood in the sense, 'pertaining to an idea or belief'. But the adjective 'ideal' is generally understood in its other senses, and one must translate *ideinyi* either as 'ideological' or as 'ideational', each of which carries implications lacking in *ideinyi*. Similarly, the Russian *nauka* like the German *Wissenschaft*, acquires new connotations when it becomes 'science'. A number of Russian philosophical terms, to take a case that is particularly troublesome, are loan translations of German words that have no felicitous equivalents in English, especially when the terms have Hegelian connotations. *Predstavlenie* (*Vorstellung*), *sniat'* (*aufheben*), *zakonomernyi* (*gesetmässig*)—these examples may suggest to the reader the translator's perplexities. If the reader will bear in mind also that grammatical differences can make exact translation impossible (English articles and verb tenses, for example, force unwonted niceties on Russian, which has no articles and fewer tenses than English), he will understand why the readability of some translations in the text below has been marred by the insertion of the Russian original, or even a German equivalent, in parentheses or brackets. I made such insertions whenever there seemed to be danger of significant misunderstanding.

I want to thank Columbia University, the Social Science Research Council, the Research Progam on the History of the CPSU, the American Academy of Arts and Sciences, and Brown University, for fellowships and grants that enabled me to work up this book. I am also grateful to the following teachers and colleagues, who were kind enough to read various parts of the manuscript, and contributed their informed criticism: Robert S. Cohen, A. Erlich, Donald H. Fleming, Philipp Frank, George L. Kline, Walter Landauer, Mary Mothersill, Hermann J. Muller, Henry L. Roberts, Ernest J. Simmons. I am indebted most of all, for advice and criticism, to Professor G. T. Robinson. It hardly needs saying that he and the others are not responsible for the interpretations I have arrived at, or for the errors I may have committed.

Many thanks to Jane Blanshard for doing the Index.

Two chapters, 16 and 19, have appeared previously, the former in *Daedalus*, Summer, 1960 and the latter in *Journal of the History of Ideas*, Jan., 1959.

Brown University
April, 1960

DAVID JORAVSKY

PART ONE

THE PRE-REVOLUTIONARY HERITAGE

I

ORTHODOX MARXISM AND NATURAL SCIENCE

RECENT intellectual history is considerably simplified by writers who tick off the philosophy of Marx and Engels as a typical nineteenth-century materialism comparable, say, to Taine's[1], but history of this kind hardly enables one to understand the powerful yet diverse appeal of Marxism. Fortunately such writers are becoming rare; it is getting to be a generally recognized fact that the philosophy of Marx and Engels contained contradictory elements. Many critics use this fact to dismiss Marxism as 'a philosophical farrago', to quote one of the most recent and well-informed of them,[2] and logic may be on their side. It may, for example, be an appropriate and deadly stroke to comment, as Rudolph Stammler did in 1896, that astronomers do not form a political party to bring on a predicted eclipse.[3] But the historian must wrestle with phenomena that the logician may dismiss as absurd. The very combination of revolutionary and scientific claims that Stammler and others have found inherently ridiculous has made a quite different impression on many of Marx's followers. 'The irresistible power of attraction that draws the socialists of all countries to this theory [Marxism],' Lenin wrote in 1894, in his first published work,

> lies precisely in the fact that it unites a rigorous and most lofty scientism [*nauchnost'*] (being the last word in social science) with revolutionism, and unites them not by chance, not only because the founder of the doctrine combined in his own person the qualities of a

scientist and a revolutionary, but unites them in the theory itself intrinsically and inseparably.[4]

However harshly a philosopher may judge this characterization of Marx's theory, an historian can hardly fail to agree that Marx's claim to give scientific guidance to those who would transform society has been one of the chief reasons for his doctrine's enormous influence.[5]

In this central aspect of Marx's thought—and of Engels', one must add, for their views are virtually inseparable[6]—lie the chief determinants of their philosophy of science. Social theory was their predominant concern, but, if they were to prove that theirs was a scientific social theory, they had to assume or demonstrate some continuity between it and natural science. A similar problem is of course faced by all social theorists who would be social scientists,[7] but Marx and Engels had the special problems of those who would also be revolutionaries, not on a separate level of their existence, but in their very work as scientists.[8] The result was that they claimed to have discovered not merely a scientific method that is applicable both to natural and social science, but one that

> includes in its comprehension and affirmative recognition of the existing state of things, at the same time also the recognition of the negation of that state, of its inevitable breaking up; . . . [one that] lets nothing impose on it, and is in its essence critical and revolutionary.[9]

To accomplish these divergent purposes Marx and Engels placed great stress on 'contradiction', a term that has many meanings in their writings. Sometimes they used it in its everyday senses, and then, of course, they did not claim it as a virtue of their philosophy but charged it as a sin against their opponents. In their own system they claimed to have incorporated '*dialectical* contradiction', something they found both in thought and in material processes. At times 'dialectical contradiction' appears to mean the simultaneous truth of opposed statements (Marx and Engels not bothering to distinguish among the traditional types of logical opposition); sometimes the opposition of forces moving in different directions; sometimes the tensions or simply the differences among aspects or parts of a mental or material process, the tensions or differences being regarded as the cause of the process. And very likely other meanings of 'contradiction' can be discovered in the writings of Marx and Engels, for they were intent on thinking in 'dialectically contradictory' fashion.

For the purposes of the present study it is not necessary to review the major theses of Marx and Engels' contradictory or dialectical philosophy, which are already widely known in any case. For present purposes it is necessary to review only one: the contradiction in their philosophy of science between metaphysics and postivism—in the sense, one must stress, that was given these terms in the Preface. For Marx scornfully rejected Comte's '*Scheisspositivismus*' ('Positivist rot' is the bowdlerized version in the standard English translation),[10] and he and Engels heaped so much ridicule on 'metaphysics' that it became a special pejorative in the Marxist vocabulary, indicating theorizing with insufficient evidence in search of the chimera of changeless substance. Obviously therefore Marx and Engels did not try to join 'positivism' with 'metaphysics', as they understood these terms. Indeed it is significant that there are no terms in their philosophy to describe what was called metaphysics and positivism in the Preface: the viewpoints of those who respectively do and do not intend philosophy to be a separate science that characterizes reality as a whole. Apparently Marx and Engels felt no necessity of choosing between these two viewpoints; their own philosophy seemed to them to be and at the same time not to be a separate science characterizing reality as a whole:

> As soon as each separate science is required to get clarity as to its position in the great totality of things and of our knowledge of things, a special science dealing with this totality is superfluous. What still independently survives of all former philosophy is the science of thought and its laws—formal logic and dialectics. Everything else is merged in the positive science of nature and history.[11]

This and similar statements[12] have been ceaselessly quoted by Marxists with positivist inclinations, but it has been possible for Marxists with metaphysical inclinations to turn the same passages to account. For the quoted passage exempted dialectics from the predicted dissolution of philosophy into 'positive science', and Marx and Engels' customary definition of dialectics was 'the science of the general laws of motion and development of nature, human society and thought.'[13] In effect, one can argue, this picture of dialectics as the queen science restores what Engels rejected: 'a special science dealing with the great totality of things and of our knowledge of things', provided 'the great totality' be conceived as processes rather than fixed things.

It would be most interesting, if this were a philosophical analysis of Marxism, to inquire whether Marx and Engels achieved a logically defensible synthesis of these contradictory attitudes, or whether they were inclined more towards one than the other. For the purposes of the present work it is sufficient to note that both attitudes may readily be found in their writings. The best place to observe this, with respect to the philosophy of *natural* science, is not in the works that Marx and Engels published during their lifetime. In 1885 Engels explained that

> a knowledge of mathematics and natural science is essential to a conception of nature which is dialectical and at the same time materialist. Marx was well versed in mathematics, but we could only partially, intermittently and sporadically keep up with the natural sciences.[14]

It is mainly in the voluminous correspondence and notebooks of the two men that the record of their intermittent and sporadic study is to be found.[15]

The correspondence is especially intersting because it reveals the groundlessness of the notion that Marx did not share Engels' views on the philosophy of natural science, that Marx held to positivism, Engels to metaphysics. If Marx had generally less to say on the subject of natural science, the reason was his deference to Engels in this field. Once, for example, when he read several books on physiology because influenza had made him 'quite incapable of work', he wrote to Engels that a certain criticism of phrenology reminded him favourably of Hegel, but he added apologetically: 'You know that everything 1. comes late with me, and 2. I always follow in your footsteps.'[16] Actually Marx was exaggerating when he said 'always', for there was at least one occasion on which he disagreed with Engels on a question of natural science, and the incident is worth recounting, for it shows the contradictory nature of Marx and Engels' philosophy of natural science very clearly, with Engels in the role of positivist and Marx in that of metaphysician.

Marx took a fancy to the theories of one Trémaux, who had ventured to explain evolution in terms of the properties of various soils, and, as usual in such matters, he turned to Engels for an opinion.[17] Engels dismissed Trémaux's theories as a ridiculous construct, guilty of terrible blunders in geology. 'By the way', he concluded, 'let no one tell me any more that the Germans alone know how to "construct" systems, the french [*sic*] beat them hollow at that.'[18] Marx, somewhat piqued, reminded Engels of the argument between Cuvier and 'German *Naturphantasten*' (a play on *Naturphilosophen* [philosophers of nature] and *Phantasten* [Visionaries])

> who were quite able to *express* Darwin's fundamental idea, however little they were able to *prove* it. Nevertheless this did not obviate the fact that Cuvier, who was a great geologist . . . was wrong and the people who expressed the new idea were right. Trémaux's fundamental idea concerning *the influence of the soil* . . . is, in my opinion, an idea that needs only to be *expressed* in order to acquire once and for all rights of citizenship in science, and this quite independently of Trémaux's exposition.[19]

Here, it would seem, Marx was implying that philosophy, even the 'drunken speculation' that he and Engels had renounced in 1845 in favour of 'matter-of-fact philosophy',[20] could establish truths in science 'once and for all', presumably impervious to empirical refutation. Yet this was the same author who, in a famous passage in *Capital*, disdained the method of '*a priori* construction' that was characteristic of Hegel's dialectics, and described his own dialectical method as 'not only different from the Hegelian, but . . . its exact opposite'.[21]

It was not merely in sentimental declarations of the last sort that Marx showed his respect for empirical science in contrast to '*a priori* construction' and 'drunken speculation'. Perhaps the most convincing demonstration of this respect was the painstaking research that was evident in the work he published 'as a scientific man', and his abstention from publishing any extensive statement of his views on natural science.[22] In private he sometimes used periods of illness or recreation to study certain areas of natural science (chiefly those dealing with change or development, such as biological evolution or the differential calculus), and on occasion he used his sense of philosophical superiority to pass judgment on the works he studied. Trémaux's book, he wrote to a friend *after* Engels had denounced it, is full of blunders, but 'it is still—for all that and all that—an advance on Darwin'.[23] In one letter to Engels he indicated a desire, which he did not find time to satisfy, to publish a pamphlet explaining what was rational (*das Rationelle*) in Hegel's universal method.[24] But it is noteworthy that he did not publish, and expressed no desire to publish, an extensive treatise on the philosophy of natural science. The incidental references to natural science in his published works were generally arguments by analogy to prove the scientific nature of his social theory, rather than efforts to show natural scientists what to think about their subject.[25]

For a long time Engels was similarly diffident in public, even though Marx generally deferred to him in matters concerning natural science. But he was disturbed by the spread of rival philosophies among the radicals to whom he and Marx looked for followers, and in 1873 he jotted down a plan for a polemic against the 'vulgar materialism' of Vogt, Moleschott, and Büchner:

> One could let them alone and leave them to their not unpraiseworthy if narrow occupation of teaching atheism, etc., to the German Philistine, but for 1) the abuse directed against philosophy (quote passages) which in spite of everything is the glory of Germany, and 2) the pretension to apply the theories of natural science to society, and to reform socialism—all this compels us to turn our attention to them.[26]

These characteristically polemical considerations were the beginning of Engels' project of writing an extended treatise on the philosophy of natural science. But in 1876 and 1877 political considerations required him to lay aside this project and write a series of polemics against Dühring, whose criticisms of Marxism had been gaining a considerable following in the German Social Democratic Party.[27] Engels incorporated in the famous *Anti-Dühring* some of his views on natural science, writing to Marx that his recent studies enabled him to 'move with a certain amount of freedom and security' in this field, but he also confessed that he still felt somewhat insecure and in need of 'great caution'.[28] Perhaps this caution is one of the main reasons, along with Engels' constant ridicule of Dühring's claim to have established 'final truths', that *Anti-Dühring*, which was for two generations the chief Marxist text on the philosophy of natural science, seems more positivist than metaphysical.

In the Preface to the second edition of 1885 Engels acknowledged 'a certain clumsiness', and boasted of caution, in treating natural science.[29] To be sure, he also restated his metaphysical vision:

It goes without saying that my recapitulation of mathematics and the natural sciences was undertaken in order to convince myself in detail—of which in general I was not in doubt—that amid the welter of innumerable changes taking place in nature, the same dialectical laws of motion are in operation as those which in history govern the apparent fortuitousness of events; the same laws as those which similarly form the thread running through the history of the development of human thought and gradually rise to consciousness in the mind of man; the laws which Hegel first developed in all-embracing but mystical form, and which we made it our aim to strip of this mystic form and to bring clearly before the mind in their complete simplicity and universality.[30]

But following this confident assertion of his metaphysics was an equally categorical declaration that lends itself to positivist interpretation:

. . . to me there could be no question of building the laws of dialectics into nature, but of discovering them in it and evolving them from it. . . . To do this systematically and in each separate department is a gigantic task.[31]

The impression of positivism was strengthened by Engels' cheerful remark, after mentioning his projected treatise on the dialectics of nature: 'It may be, however, that the advance of theoretical natural science will make my work to a great extent or even altogether superfluous',[32] for natural science was allegedly turning dialectical spontaneously. But then Engels cautioned that natural scientists work with concepts, and the art of working with concepts, he said, is not inborn but requires 'real thought'. This 'real thought', philosophy in short, developing for two and a half thousand years, had allegedly culminated in dialectical materialism. If natural scientists would assimilate the results of this development, Engels concluded, natural science would rid itself both 'of any isolated natural philosophy standing apart from it, outside it, and above it', and 'of its own limited method of thought, which was its inheritance from English empiricism',[33] In Engels' view, dialectical materialism had 'transcended' (*aufgehoben*) both: destroyed them and at the same time absorbed their best elements.

This Preface to the second edition of *Anti-Dühring* was in a literal sense the summing-up of Engels' philosophy of natural science, for it was written in 1885, when he had just set aside his work on the subject—this time for good, as it turned out—in order to complete *Capital*.[34] (Marx had died in 1883 with only the first volume published.) The mass of study notes, comments, and first drafts that Engels had accumulated in approximately eight years of work on the philosophy of natural science (1873–1876 and 1878–1883), posthumously published as *Dialectics of Nature*, may therefore be expected to display a similar mixture of metaphysical and positivist elements. And so they do, but the emphasis is on the metaphysical element, for Engels' projected adversary in this unfinished work was not Dühring's philosophy of 'final truths', but 'narrow-minded empiricism', 'mechanistic materialism', 'vulgar evolutionism', and 'pan-inductionism'. The continuing hero through the fragmentary manuscript was 'Hegel, in whose [word indecipherable because of an ink blot] the synthesis of the sciences of nature and their rational grouping are a greater achievement than all the materialistic stupidities put together'.[35] Towards natural scientists who 'believe that they free themselves from philosophy by ignoring it or abusing it',[36] Engels was didactic:

> Whatever pose natural scientists adopt, philosophy rules over them. The question is only whether they want to be ruled by some vile fashionable philosophy, or whether they want to be guided by a form of theoretical thought that is grounded on acquaintance with the history of thought and its achievements.[37]

But the comment immediately following this admonition shows that the positivist element was present even in the *Dialectics of Nature*:

> Physics, beware of metaphysics!—that's quite right, but in a different sense [from the one usually intended].
>
> By contenting themselves with scraps of the old metaphysics

natural scientists still continue to leave philosophy some semblance of life. Only when natural and historical science have absorbed dialectics, only then will all the philosophical rubbish—with the exception of the pure theory of thought—become superfluous and disappear in positive science.[38]

It is noteworthy that Engels showed a readiness not only to criticize natural scientists on philosophical issues—Newton, for example, is tagged an '*Induktionsesel*' (inductive ass)[39]—but also to overstep the hazy boundaries that separate philosophy from science and to speak with authority in the latter field. He was perhaps just on the boundary between philosophy and physics in rebutting Clausius' theory of the 'heat death' of the universe,[40] but he steped fairly into physiology in charging the Faulhorn experiment of Fick and Wislicenus with ignoring important aspects of metabolism.[41] J. B. S. Haldane, Engels' English editor, argues that Engels has been vindicated in both cases by the development of natural science since his time.[42] Whether or not Haldane is right is beside the present point, which is simply that Engels felt confident enough both of the power of dialectical materialism, and of his own grasp of natural science, to criticize not merely the philosophical attitudes but also the scientific theories of contemporary natural scientists. It is easy to take this as evidence of the metaphysical overweening that presumes to tell what is true about nature with little or no regard to facts and empirically established theories, the kind that Hegel made notorious.[43] But such an interpretation of Engels' attitude towards natural science ignores the fact that the *Dialectics of Nature* is a fragmentary, unfinished work. (At the end of his criticism of the Faulhorn experiment, for example, Engels noted: 'All this has to be greatly revised.'[44]) What is more, such an interpretation overlooks Engels' repeated acknowledgment of an empirical criterion of truth: 'We are all agreed that . . . in theoretical natural science one must not construct connections into the facts, one must disclose them in the latter and, upon disclosing, prove them, so far as possible, empirically.'[45]

Indeed, it is possible to argue that this empirical criterion of truth was more nearly characteristic of Marx and Engels' attitude towards the natural sciences than their occasional readiness to derive theories from philosophical principles and prescribe them to specific sciences The chief evidence for this argument would be the absence of such prescriptions from the works they published in their lifetime, and the consequent fact that their followers have never been able to agree on what specific theories of natural science, if any, are incumbent on Marxists by virtue of their philosophy. To be sure, Lysenko and his supporters have been responsible for the widespread assumption that Marx and Engels did commit their followers to belief in the inheritance of acquired characters, but many other Marxists, as the reader will see below, have read the masters quite differently. Nor is this divergence hard to understand, if one reads Marx's and Engels' comments on biology with an open mind.

As the case of Trémaux indicates, their interest in biology was focused on the theory of evolution. For one thing, they were amused by a parallel between Darwin's theory and political economy. 'It is noteworthy,' Marx wrote to Engels,

> that among beasts and plants Darwin recognizes his English society with its division of labor, competition, opening of new markets, 'inventions,' and the Malthusian 'struggle for existence.' It is Hobbes' *bellum omnium contra omnes*, and it reminds one of Hegel, in the *Phenomenology*, where bourgeois society is represented as a 'spiritual animal kingdom' ['*geistiges Tierreich*'], while in Darwin the animal kingdom is represented as a bourgeois society.[46]

More significantly, Marx and Engels were convinced that Darwin had delivered 'the mortal blow' to teleology in natural science by providing a rational explanation of functional adaptation in living things and by proving his explanation empirically.[47] On the most general level, they welcomed Darwin's theory, and complementary theories of geological and cosmic evolution, as confirmation of their belief that throughout nature (the human variety included) present reality continually 'negates' itself, continually gives rise to a different reality in accordance with natural laws that can be established scientifically.[48] These were presumably the reasons for their repeated statements to the effect that Darwin's work 'contains the basis in natural history for our view [of human history]'.[49] Indeed, Marx wanted to dedicate parts of *Capital* to Darwin, but Darwin declined the honour because, he wrote Marx, he did not know the work, because he did not believe that direct attacks on religion advanced the cause of free thought, and finally because he did not want to upset 'some members of my family'.[50]

In short, Marx and Engels, who were not directly involved either in biological research or in agriculture, had an ideological or philosophical interest in biology, rather than a biologist's interest *qua* biologist, or a farmer's *qua* farmer. *Anti-Dühring*, the chief source for Marxist philosophy of natural science in the half century that separated its appearance from the publication of *Dialectics of Nature*, revealed Engels' intense interest in such issues as 'spontaneous generation' or 'the struggle for existence' but showed little concern for the biologist's problems that did not point clearly towards *Weltanschauung*. He knew, for example, that the causes of continuity and change in the heredity of the individual organism were unknown to biologists, but he appears to have regarded this as a technical problem whose solution was of little moment to Marxism.[51] A few incidental remarks revealed his acceptance of the nearly universal belief in somatically induced alterations of heredity, but his followers hardly noticed these remarks in the period before the Bolshevik Revolution.[52]

Those who would deny this view of Marx and Engels as empiricists in their attitude towards biology and the other natural sciences must place chief reliance on the posthumous *Correspondence* and *Dialectics of Nature*. Analysis of these works could be used to argue that the philosophy of Marx and Engels was inherently metaphysical and dogmatic, but did not clearly reveal these traits in respect to natural science simply because the two authors were not directly involved in the natural sciences whether as scientists or as technologists. For the purposes of the present work it is not necessary to decide what Marx and Engels 'really meant'. It is sufficient to note that their writings could easily inspire among their followers diverse attitudes towards natural science.

As Marxism grew to be the main ideological force in the European socialist movement,[53] internal diversity grew with it, until Marx made his famous remark, concerning a group of Marxists in France, '*Tout ce que je sais, c'est que je ne suis pas marxiste, moi!*'[54] ('All that I know is that *I* am not a Marxist!') When the deaths of Marx (1883) and Engels (1895) removed the possibility of appealing to the 'founders' for the elucidation of disputed points, this diversity grew greater still, until it erupted at the end of the nineteenth century in the famous controversy between revisionism and orthodox Marxism. The issues tended to differ from one country to another, but the common central problem in theory was the attitude that socialists should take towards the heritage of Marx and Engels. The term 'revisionist' was first applied to Eduard Bernstein and his followers, who considered themselves Marxists, though they insisted on the necessity of altering or abandoning many of Marx's doctrines;[55] but it has come also to describe the wide variety of socialists who have not considered themselves Marxists but have felt free to adopt parts of Marxism and reject others. To describe 'orthodox' Marxists as those who would not tolerate any alteration of Marx's doctrine is to shut one's eyes to the fact that they often have altered it, though, when they have done so consciously, they have represented themselves as changing the 'form' rather than the 'essence' of Marxism, or as 'developing Marxism further' rather than 'revising' it.[56] Thus, orthodox Marxism is better described as a temper, an attitude of strong conviction that the fundamental teachings of the 'founders' are true, and may be altered to suit changing circumstances only with the greatest caution, lest revolutionary socialism be turned into opportunism.

One of the characteristic, though not universal,[57] ideas of revisionism was that ethical beliefs rather than scientific knowledge were the proper theoretical basis of socialism. The orthodox could not combat this idea without some effort to demonstrate a continuity between ethical and scientific judgments, and since they were orthodox they generally claimed to base such demonstrations on dialectical materialism. But the inherent contradictions in the philosophical heritage of Marx and Engels, and the diversity of personalities, intellectual traditions, and current problems in various countries produced remarkably different combinations of ideas under the rubric of dialectical materialism. It is not within the compass of the present work to attempt a thorough survey of these combinations, but some foreshadowings of trends of thought in post-revolutionary Russia need to be examined.

The case of Paul Lafargue, for example, is noteworthy for his startling union of mechanistic materialism with Cartesian rationalism, which had considerable influence on the Soviet discussions of the 'twenties.[58] Lafargue was concerned to answer the argument of Jaurès and others that ideas of justice are the theoretical basis of socialism, and that they are fundamentally different from scientific ideas. He thought to accomplish this by a sweeping reduction of all ideas, ethical and scientific, to the influence of sensations on the brain. The brain, he argued, transforms sensations into ideas as 'a dynamo transforms motion into electricity'. In a quaintly Victorian version of Diderot's analogy between the mind and a harpsichord, Lafargue said that ideas are placed in the head 'as new airs are added to a barrel organ'.[59] But he reasoned further that certain fundamental ideas, placed in the head by generations of similar experiences, have become hereditary characteristics: 'We must admit that it [the brain] possesses the molecular arrangement destined to give birth to a considerable number of ideas and concepts'.[60] Thus Lafargue used mechanistic materialism to support his belief in innate ideas as a significant source of true knowledge, whether in mathematics or in ethics. And thus he believed that he had overcome the disjunction between ethical and scientific arguments for socialism: both were ultimately empirical in nature, but the scientific was preferable since it was consciously and systematically empirical.[61]

Lafargue was unusual not only in the nature of his philosophy but also in the attention that he gave to natural science. The German leaders of Marxist orthodoxy were largely indifferent to the philosophy of natural science. 'Historical materialism,' wrote Franz Mehring, 'is a self-contained theory, designed for the cognition of the historical development of human society, a theory that draws its justification from itself alone and allows itself to be amalgamated with the methods of natural science just as little as it raises natural scientific claims for itself.'[62] Indeed Karl Kautsky, who was probably *the* leading theorist of German, and therefore of European, orthodox Marxism in the period before the First World War,[63] was not especially interested in epistemology even within the hotly disputed field of social theory. In 1898 he wrote to Plekhanov, who had been prodding him to take a stand against the neo-Kantian philosophy that was then the vogue among revisionists:

In any case I must openly declare that neo-Kantianism disturbs me least of all [the views of the revisionists]. I have never been strong in philosophy, and, although I stand entirely on the point of view of dialectical materialism, still I think that the economic and historical viewpoint of Marx and Engels is in the last resort compatible with neo-Kantianism. . . . If Bernstein was moulting [*gehäutet*] only in this respect, it would not disturb me in the least.[64]

And in 1909, when Kautsky was asked for an opinion of the Russian Marxists' quarrel over Machism, he wrote that he deplored it. He was himself a dialectical materialist, but Marxian social theory could be united with Machist philosophy too, for 'Marx proclaimed no philosophy, but the end of all philosophy'.[65] The basic doctrine of Marxism was that social consciousness is determined by social being.

Whether this conception is based on 18th-century materialism, or on Machism, or on Dietzgen's dialectical materialism, is not all the same for the clarity and unity of our *thought*; but it is a question that is entirely inconsequential for the clarity and the unity of our *party*. Individual comrades may study this as private people, as they may the question of electrons or Weissmann's law of heredity; the *party* should be spared this.[66]

Thus Kautsky and Mehring were not far from the notion that Marxism is not a universal but only a social philosophy; they usually declared themselves opposed to this revisionist view, but for the most part they ignored the implications of such declarations. If, as the reader will see, a similarly paradoxical position was widespread among orthodox Soviet Marxists of the 'twenties, the influence of German orthodoxy must have been a considerable cause.[67]

There is probably broad significance in the fact that *Russian* Marxists prodded Kautsky to make a major issue of dialectical materialism. As G. D. H. Cole has shown in his monumental *History of Socialist Thought*, Russian Marxism was sharply different in many important respects from Marxism in Western Europe. The repressive nature of the Russian government made Russian Marxists, revisionists included, almost uniformly revolutionary, and, by cutting them off to a very large extent from the everyday parliamentary politics that absorbed the energy of West European Marxists, repression turned Russian Marxists in on themselves and greatly magnified the importance of theoretical arguments in their eyes.[68] But whatever the causes may have been, the fact is clear that arguments over general philosophy (*Weltanschauung*, epistemology, the philosophy of science) were a prominent feature of Russian Marxism. For a time they were largely disputes between an ephemeral group of 'Legal Marxists', who rejected dialectical materialism,[69] and the bulk of Russian Marxists, who were generally agreed on orthodoxy in philosophy, though they were increasingly divided on political issues into the Menshevik and Bolshevik factions. But then, as the two factions grew into two separate parties, disputes between philosophical revisionism and orthodoxy appeared within both.

Plekhanov, the chief Menshevik theorist, and Lenin, the chief Bolshevik *tout court*, did not differ to any great extent in general philosophy. Both were orthodox. In fact, orthodox Bolsheviks, overlooking their political hostility to Plekhanov, regarded his philosophical works as the chief 'classics' after those of Marx and Engels. The point bears stressing, for the past thirty-five years of Bolshevik writing have quite obscured it. After Lenin's death in 1924 his philosophical writings became equally authoritative 'classics', and after 1930 a retrospective equation of philosophical and political positions turned the historical record upside down. Lenin was made the philosophical teacher, Plekhanov an errant pupil, no less Menshevik in his philosophy than in his politics.[70] Thus, if we are to understand the philosophical heritage that orthodox Soviet Marxists began with in 1917, it is necessary to set aside their own accounts and study the original sources in their original context.

Of course there were connections between the philosophical and political differences of pre-revolutionary Russian Marxists (they will be reviewed in the following section), and there were philosophical differences between Lenin and Plekhanov. In epistemology Plekhanov showed a greater sensitivity than Lenin to the sceptical trend of thought that stems from Locke, Berkeley, Hume, and Kant. Plekhanov conceded that preceptions do not 'resemble' or 'reflect' the external objects or processes that arouse them. He preferred to say that perceptions 'correspond' to things-in-themselves, and on one occasion he went so far in stressing the differences between perceptions and objects[71] that the Machists were able to ridicule his 'luckle attempt to reconcile Engels with Kant by the aid of a compromise—a thing-in-itself that is just a wee bit knowable'.[72] In *Materialism and Empirio-criticism* Lenin was indigant with the Machists for taking undue advantage of Plekhanov's lapses, as Lenin viewed them. To shut the door firmly against Humean arguments, Lenin shouted that perceptions are 'copies, photographs, pictures, mirror-reflections of things'.[73] Plekhanov did not bother to reply to Lenin's criticism, but Aksel'rod the Orthodox, an admiring pupil of Plekhanov's and a fellow Menshevik, did. She argued that Lenin espoused not materialism but 'naïve realism', which identifies objects and our perceptions of them, and is fundamentally akin to *Machism*. 'But materialism,' she declared flatly, 'takes the point of view that sensations, which are aroused by the action of various forms of the motion of matter, are not like the objective processes that gave rise to them'.[74]

Aksel'rod was not the only one to assert a kinship between Lenin's theory of reflection and the Machist epistemology that he flayed. Three of Lenin's Machist opponents condescendingly instructed him on the difference between materialism, which they pictured as a transcendentalist philosophy that requires sensations to be regarded as symbols of things-in-themselves, and common-sense realism, which shrugs its shoulders at metaphysical efforts to find the substance of things beyond our perceptions of them. Machism was such realism made systematic, Iushkevich wrote, and Lenin would see this if he were not bewitched by Engels' terminology.[75] And Bazarov commented that Lenin's 'realistic "common sense" is not yet entirely stifled, but only slightly knocked off course by metaphysicians of one school or another.'[76]

In order to see this disagreement between Lenin and Plekhanov in its proper dimensions, one must bear in mind that Aksel'rod and the Bolshevik Machists had political reason to stress it. As the reader will see below, Aksel'rod was intent on correlating Bolshevism with Machism, and the Bolshevik Machists wanted to win Lenin from his temporary bloc with Plekhanov and the 'Party Mensheviks'. As far as the philosophical issues are concerned, it is noteworthy that Plekhanov withdrew his comparison of perceptions with 'hieroglyphs' or 'conventional symbols', which stand for objects but do not resemble them. Plekhanov called this comparison an unnecessary 'terminological concession to our philosophical opponents'.[77] Moreover, it must be noted that neither Plekhanov nor Lenin was nicely uniform in his choice of terms to describe the relation between perceptions and the objects of perceptions. Plekhanov tended to favour the term 'correspondence', while Lenin, forewarned by the Machist use of this term as an opening for attack, tended to favour the term 'reflection'; but neither took pains to use only his favoured term, and neither developed a full epistemology clearly opposed to the other's. Both allowed the difference between them to remain embryonic —and therein lies the chief significance of the episode. For both were chary of deep involvement in epistemology, since both held to Marx's famous 'Thesis on Feuerbach':

> The question whether objective truth is an attribute of human thought—is not a theoretical but a *practical* question. Man must prove the truth, i.e. the reality and power, the 'this-sidedness' of his thinking, in practice. The dispute over the reality or non-reality of thinking that is isolated from practice is a purely scholastic dispute.[78]

Thus both Plekhanov and Lenin were inclined to shrug off, with a gesture towards successful human 'practice',* the epistemological problems raised by their commitment to a materialist philosophy.[79]

Of course polemics cannot be conducted entirely with gestures. Some argument is needed, and it is probably true to say that Plekhanov's argument against the so-called Machists showed a greater sensitivity than Lenin's to the force of the Humean and Kantian tradition. The reason for this difference may be contained in a letter that Lenin wrote to Gorky as he was beginning *Materialism and Empirio-criticism*:

> Plekhanov is in essence *altogether* right against them [the Machists],

* In Marxist philosophical writings the term 'practice' has any or all of the following meanings: (1) sense data, (2) workable techniques in agriculture and industry, (3) effective policies in social and political affairs, and (4) theories that are ultimately verifiable by any of the foregoing, singly or in combination. Accordingly, the statement often found in Marxist writings, that 'practice is the criterion of truth', is susceptible of many different interpretations both as a general dictum and as applied in particular cases.

only he is unable, or unwilling, or too lazy to say it *concretely*, circumstantially, simply, without superfluous intimidation of the public with philosophical niceties. And I at any price will say it *my way*.[80]

But one must also note that both Plekhanov and Lenin had an interest in philosophy beyond the immediate needs of polemics. It is significant that Plekhanov's interest took him to the materialists of the seventeenth and eighteenth centuries,[81] while Lenin's studies concentrated in the end on Hegel. It is furthermore worth noting that, as Lenin studied Hegel, his epistemological disagreement with Plekhanov recurred to him and the comments that he left in his notebooks are tantalizingly suggestive of a new turn in his thought, towards a greater emphasis on dialectics than one can find in *Materialism and Empirio-criticism*.[82] If Lenin had lived long enough or been less busy with politics, one wonders whether he would have persevered in his efforts 'to apply dialectics to the *Bildertheorie* [the theory of reflection], to the process and development of knowledge'.[83] It may be that his commitment to the priority of 'practice' over theory, which is reaffirmed in these same notes,[84] would have made him grow impatient with such an enterprise and drop it as 'scholasticism' before he had made much more than the little progress one can find in his fragmentary *Philosophical Notebooks*. But this is idle speculation. For present purposes it is sufficient to note that there were embryonic epistemological differences between Plekhanov and Lenin, which played a subordinate role in the philosophical discussions of the 'twenties, when the works of Plekhanov and Lenin as well as those of Marx and Engels were considered the 'classics' of orthodoxy.

If Plekhanov was more sensitive than Lenin to the Humean tradition in epistemology, the situation was reversed in regard to anti-materialist arguments based upon new developments in physics. The only really harsh words that *Materialism and Empirio-criticism* had for Plekhanov as a philosopher concerned his neglect of

> the connection between the new physics . . . and Machism. . . . To analyse Machism and at the same time to ignore this connection—as Plekhanov does—is to scoff at the spirit of dialectical materialism, i.e., to sacrifice the method of Engels to the letter of Engels. Engels says explicitly that 'with each epoch-making discovery even in the sphere of natural science' (not to speak of the history of mankind),

'materialism inevitably must change its form'. . . . Hence, a revision of his propositions in natural philosophy is not only not 'revisionism' in the accepted meaning of the term, but, on the contrary, is necessarily demanded by Marxism.[85]

Following these brave words Lenin undertook not an explicit revision of Engels but an extended rebuttal of a number of authors who had been arguing that recent discoveries—such as the atom's divisibility, the 'decay' of radioactive matter, or the variability of mass in elementary particles—made materialism untenable, since they destroyed the conceptions that lay at the basis of materialism.[86] The essence of Lenin's counter-arguments was that 'the sole "property" of matter with whose recognition philosophical materialism is bound up is the property of *being an objective reality*, of existing outside our mind'.[87] All other characteristics of matter, Lenin said, were for the specific sciences to describe in their endless approach to the asymptote of absolute truth. This argument was not so much a revision of the form of Engels' materialism as it was a special emphasis on the positivist element in Engels, the distaste for an ontological philosophy.

Yet even in this discussion (Chapter V) one can find some evidence of Lenin's belief in the ontological significance of his philosophy. Stressing 'the temporary, relative, approximate character' of such physical concepts as the atom or electron, picturing them as '*landmarks* of the cognition of nature by man's progressing science',[88] Lenin concluded that the electron would prove no less 'inexhaustible' than the atom. This has been interpreted by some Soviet Marxists as a brilliant forecast of the discovery of other sub-atomic particles than the electron.[89] It may indeed be that, though one must note that other parts of the same Chapter were used to a contrary purpose by many Marxists during the 'twenties, who argued that in the electron modern science had found the ultimate building block of the universe.[90] It is not germane to the present point to ask which of these views on physical particles was closer to Lenin's actual meaning. The point here is simply that Lenin's famous book could be—and was—not only an inspiration to those Marxists who regarded the philosophy of science as pure epistemology, but also a supporting text for those who wanted a philosophy that would assert ontological truths.

Another example of this paradox is Lenin's denunciation of 'energeticism'. Arguing, after Boltzmann, that motion cannot exist without matter, he rejected as idealist Ostwald's theory that energy rather than matter is the universal substance.[91] To be sure, this may be regarded as a philosophical argument, but it can have an influence on the acceptance of physical theories; Lenin's authority would be used by some participants in the Soviet discussions of physics to discredit the relativistic denial of a fluid-like ether acting as a medium for physical interaction. But their opponents would be able to match quotation against quotation. A field of force in 'empty' space may be regarded as a material substance, if Lenin was right in defining materialism as 'recognition of nature's objective regularity [*zakonomernost'*, *Gesetzmässigkeit*] and of the approximately true reflection of this regularity in man's head'.[92] Moreover, Plekhanov, raising his head from the history of philosophy to make one of his rare comments on the new physics, said that 'energeticism' *was* a form of materialism, as Joseph Priestley's acceptance of Boscovich's atoms proved.[93] On the whole, one must stress, there was very little in Lenin's *Materialism and Empirio-criticism* that would commit a dialectical materialist to one physical theory rather than another, for it was clearly Lenin's desire to avoid the examination of physical theories as such. 'It is far from our intention,' he wrote at the outset, 'to deal with special physical theories. What interests us exclusively is the epistemological conclusions that follow from certain definite propositions and widely known discoveries.'[94]

Though Lenin did not comment on new propositions and discoveries in biology, other orthodox Marxists showed a readiness to take them for granted too, hastening only to put a dialectical materialist gloss on them. The connection between Darwinism and Marxism was one of their favourite topics, but it was seen as a problem of justifying socialism against the attacks of rugged individualists, of proving that the law of tooth and nail in the animal kingdom at large must by a natural evolution give way to a collectivist ethics within the human enclave.[95] They were also concerned to refute the argument that Darwinism proved gradual evolution to be the natural pattern of change rather than sudden revolution. Thus, when De Vries put forward his theory of mutation as the basis of evolution, both Kautsky and Plekhanov seized upon it as proof that natural development proceeds in accordance with the laws of materialist dialectics, by leaps as well as gradually.[96] And, as far as the present writer has been able to find out, none of their 'orthodox' comrades, Lenin included, condemned this argument as tainted in any way with 'revisionism'.

It may be argued that the orthodox Marxists did, nevertheless, regard dialectical materialism as an ontology no less than an epistemology. Lenin's comments on the 'inexhaustibility' of the electron and on the impossibility of energy without matter show, one may argue, what metaphysical arrogance he might have assumed towards physical and biological theories if he had been more concerned with them. But this argument assumes that a purely epistemological assessment of natural science is possible; some would protest that Lenin could not have avoided ontological commitments, even if he had been an out-and-out positivist. Indeed, an analogous argument was repeatedly used by Lenin against a variety of positivists. They pretend to avoid ontology, Lenin said, but in fact they cannot escape the choice between materialism and idealism. Plekhanov had been making the same point against avowed positivists, and the result was that orthodox Marxists came to use the term 'positivism' as a pejorative, indicating a renunciation of metaphysics that is unsuccessful, or even disingenuous, and generally results in idealism. Without attempting to decide the philosophical question whether the goal of positivism (the avoidance of metaphysics) is attainable, one may note without possibility of doubt that many thinkers have felt drawn to that goal. And the paradoxical fact is that Lenin himself exhibited fairly strong tendencies in that direction. 'From the point of view of Marx and Engels,' he once wrote, 'philosophy has no right to a separate independent existence, and its material is dissolved among various branches of positive science.'[97] Even in his posthumous *Philosophical Notebooks* (published in 1925–1929), which were largely a study of Hegel's *Logic* and were therefore markedly metaphysical in tendency, one can still detect the conflicting element of positivism.[98] Lenin, no less than Plekhanov, Marx, and Engels, left a contradictory philosophical legacy to his followers.

2

LENIN AND THE PARTYNESS OF PHILOSOPHY

SOVIET Communists and some of their most militant opponents are agreed in attributing to Lenin the current Soviet Marxist doctrine of 'partyness' (*partiinost'*), which is generally understood to mean simply the ideological control of philosophy (and of art and scholarship generally) by the Communist Party's Central Committee. N. V. Valentinov, for example, reaching back nearly fifty years to recapture the philosophical arguments he had with Lenin, attributes to his deceased opponent the belief that 'knowledge of the laws of social life . . . —precisely because epistemology, the theory of knowledge, is a party science—can be only the privilege of the Party headed by Lenin'.[1] Valentinov concludes that 'from this book [Lenin's *Materialism and Empirio-criticism*] the road goes straight, well smoothed by bulldozers, to a state philosophy, resting on the GPU–NKVD–MGB'.[2] Except for the final reference to the political police, this is very like the doctrine of partyness that M. B. Mitin inferred from *Materialism and Empirio-criticism* in 1930, when Valentinov was about to emigrate and Mitin was establishing himself as one of the leading philosophers in the Soviet Union:

> The philosophy of dialectical materialism is the official point of view, the world view of the Communist Party. . . . There is not and cannot be a philosophy that wants to be considered Marxist-Leninist philosophy while denying the necessity of ideational-political and theoretical leadership on the part of the Communist Party and its leading staff.[3]

But neither Valentinov nor Mitin could adduce an explicit statement of this version of partyness from *Materialism and Empirio-criticism*, for it is not there. Indeed, in all of Lenin's life there were only two or three occasions on which he approached this version of the partyness of philosophy. The doctrine of partyness recurs frequently in his writings, but generally in a different sense.

Usually Lenin's doctrine of partyness was merely a new name for the standard Marxist sociology of knowledge. If being determines consciousness, and the essence of social being is class conflict, then social theory cannot be disinterested, as professors hypocritically claim it to be; willy-nilly, consistently or confusedly, every social theorist serves the interest of one contending class or another.[4] At the age of twenty-four, in his first published work, Lenin used this argument against those who called themselves abstractly the Friends of the People and claimed to speak for universal justice.[5] A year later, in 1895, he turned the same type of argument against Struve, a fellow Marxist who had described the rise of capitalism in Russia without denouncing it or urging its revolutionary overthrow. Lenin rejected such dispassionate objectivity in favour of the 'more consistent, more profound, fuller . . . objectivity' of the materialist, and coined the term *partiinost'* or partyness to describe it: 'Materialism includes, so to speak, partyness, enjoining one in any judgment of an event to take directly and openly the standpoint of a definite social group.'[6] Thus, the Marxist sociology of knowledge, which Lenin chose to call 'partyness', was a prescription as well as an analysis, a programme of action for the proletarian theorist as well as a sociological dissection of all theorists. And it is clear that in both respects partyness was a figure of speech: social theories were correlated with the interests of various *classes*, and the Marxist was exhorted to take the side of the working *class*. In 1895 Lenin could hardly have correlated social theories with genuine political parties or demanded that the Marxist subordinate his theory to the control of the one true party of the proletariat, for in 1895 there were no genuine political parties in Russia, but only trends of thought, local 'circles', and study groups. To be sure, only a few years later genuine parties were organized, and Lenin announced his famous theory of the 'vanguard party', according to which the standpoint of the working class is taken clearly and consistently only by the party of the working class—that is, by Lenin's Bolsheviks.[7] Logically then, one may conclude, he must have insisted on a Bolshevik monopoly of proletarian truth in the field of general philosophy no less than politics. But the evidence indicates that, more often than not, he continued to understand the partyness of philosophy in the broad sense of commitment to the standpoint of the working class, without specification as to party. To understand the reasons for this anomaly, one must trace the connections between Lenin's philosophical development and the political history of Russian Marxism.

In the late 'nineties, against Mikhailovskii and the 'subjectivist' school of sociology Lenin held up Marxism as objective social science; against Struve's 'objectivist' interpretation of Marxism he held up Marxism as a call to action. He seems to have been indifferent to the epistemological difficulties of this demand for thinking that would be wilful at the same time that it was realistic, that would be volition as well as cognition. Struve offended his orthodox soul by arguing that Marxism lacked a solid philosophical foundation, but Lenin's initial reply was a positivistic shrug. 'From the point of view of Marx and Engels', he declared, 'philosophy has no right to a separate independent existence, and its material is dissolved among the various branches of positive science'.[8] When Struve and the 'Legal Marxists' began to argue that the philosophical foundation for Marxist social theory was to be supplied by Kantianism, Lenin protested indignantly that Engels had rejected Kantianism and that Plekhanov had proved dialectical materialism to be the legitimate and inevitable product of the whole development of modern philosophy.[9] He wanted to go beyond such argument from authority to the analysis of issues, but he was stayed by diffidence. 'I am very conscious of my philosophical ignorance,' he wrote to a friend in 1899, 'and do not intend to write on these themes until I study up.'[10] In Siberian exile he began to study up, meanwhile writing to friends that Plekhanov should defend orthodox Marxism against the neo-Kantians.[11]

He could recognize the neo-Kantians' revisionism, for they explicitly rejected dialectical materialism, but he was initially taken in by a new trend of philosophical revisionism that did not openly criticize Marx and Engels. In 1899 A. A. Bogdanov published his first philosophical work, and Lenin, carefully studying it in Siberia, was kindled with enthusiasm. He thought 'Bogdanov' a pseudonym for Plekhanov, and was so incensed by a hostile review of the book that he planned to write a reply.[12] But then he went abroad to work with Plekhanov on *Iskra* (The Spark), Russia's first Social Democratic newspaper, and 'Plekhanov', he later confided to a friend, 'explained to me the erroneousness of Bogdanov's views'.[13] L. I. Aksel'rod ('the Orthodox', as she called herself), another disciple of Plekhanov's, who had earned a Swiss doctorate in philosophy, was asked to write a critique of Bogdanov's philosophy for the newspaper.[14]

The political split of the Russian Social Democrats in 1903 had no initial effect on this incipient division in philosophy, especially since Plekhanov at first stood with Lenin on the Bolshevik side of the political split. By 1904, however, a rearrangement occurred that was to have a profound influence on Lenin's understanding of the partyness of philosophy: Plekhanov, his mentor in philosophy, shifted to support of the Mensheviks, while Bogdanov, coming to join the *émigrés* with two more philosophical works, each more openly revisionist than the one before, was eager to join the Bolshevik faction.[15] Pulled in opposite directions by philosophical and political convictions, Lenin made a characteristic choice. He made a deal with Bogdanov: the Bolshevik faction was to be philosophically neutral; philosophy was to be considered a private matter and was not be be discussed in the faction's publications.[16] The major difficulty of this arrangement was that the *Mensheviks* were not bound by it. Three of the seven editors of the Bolshevik newspaper (Bogdanov, Bazarov, and Lunacharsky) were revisionists in philosophy, and the Mensheviks could hardly pass up such an opportunity to impugn the Bolshevik claim to orthodoxy. *Iskra*, which had become a Menshevik paper, finally published Aksel'rod's long-delayed critique of Bogdanov in November, 1904. She opened with the sarcastic remark that *Lenin* had asked her to write the critique, obviously hoping to embarrass the Bolsheviks and perhaps to instigate internecine philosophical polemics among them.[17]

But they ignored the provocation, holding to their pact of philosophical neutrality.

In 1905 philosophical issues were pushed into the background by the mounting revolution in Russia, but they were not entirely forgotten. Plekhanov, accusing the Bolsheviks of deviating towards Blanquism in politics, suggested that 'Machist' philosophical revisionism was a contributing cause.[18] Lenin used the forum of the Third Party Congress in May, 1905, to reply, sneering at Plekhanov for

> dragging in Mach and Avenarius by the ears. I certainly don't understand what relation these writers, for whom I have not the slightest sympathy, have to the question of social revolution. They have written of individual and social organization of experience, or something of that sort, but, truly, they have not speculated on social revolution.[19]

And Lenin dismissed the issue with a *tu quoque*: 'Doesn't Plekhanov know that Parvus [whom Lenin considered a Menshevik] has become an adherent of Mach and Avenarius?'[20] Plekhanov, however, was not willing to drop the matter. He was well aware, he wrote, that Lenin regarded Mach and Avenarius as 'extraneous subjects. But then, for him all other philosophical "subjects" are extraneous, for he has always been quite indifferent in regard to philosophy'.[21] He speculated that Lenin's simple mind was being perverted by his 'Machist' entourage; they should prove their denial of a connection between 'Machist' philosophy and Bolshevik politics by showing what consequences 'Machism' did have for Social Democratic politics.[22] But the Bolsheviks still turned down the challenge, repeating the assertion that philosophy was an extraneous issue, and the denial that the Bolshevik faction was predominantly 'Machist'.[23] In October, when the Revolution reached its climax, Lenin sent Plekhanov an urgent plea to join the Bolshevik newspaper. He recognized that Plekhanov might feel disinclined to work with the three 'Machist' editors, but he swore to Plekhanov that they had never tried and were not trying to connect their philosophical views with any kind of official Party business, and assured him that they were sincere and friendly in their desire to work with him.[24] The plea failed in its purpose, in part, one supposes, because Plekhanov could not stomach Lenin's refusal to set a Party line for philosophy.

Yet the paradoxical fact is that shortly after Lenin wrote this plea to Plekhanov, he struck off an article that was to be endlessly quoted after 1930 to prove that his doctrine of partyness meant continuous Party control of philosophy.[25] In October, 1905, the Tsar was forced to promise civil liberties, and Lenin hastened to warn his exhilarated comrades of the dangers involved with the opportunities. The Asiatic censorship was being struck down, he wrote, but it could be replaced by the European mode of publishing, in which commercialism, careerism, and anarchic individualism determine literary production. To avoid this calamity, newspapers, publishing houses, book stores should become party enterprises; and, whatever the other parties might do or fail to do, the Social Democratic Party must discipline its authors. Many inconsistent intellectuals would try to use the Party's publishing facilities now that it was safe to do so; they must be taught that literary activity is '"a wheel and screw"' of the great Social Democratic mechanism. With malicious delight Lenin pictured the outcries of consternation at such a suggestion.

> What! You desire the subjection to collectivism of such a delicate, individual business as literary creation! You want workmen to decide questions of science, philosophy, esthetics by a majority of votes! You deny the absolute freedom of absolutely individual creation of ideas.[26]

Lenin had a double reply. He was not denying the writer's freedom, for anyone who feared Social Democratic discipline could stay out of the Party and 'enjoy' the hypocritical freedom of bourgeois publishing, which was actually a masked enslavement to capitalism. For intellectuals who were bold enough to seek their real freedom in service to the proletariat, Lenin had a few words of reassurance:

> It is self-evident that literary work is least of all submissive to mechanical uniformity, to levelling, to the rule of the majority over the minority. It is self-evident that, in this work it is absolutely necessary to guarantee great scope to individual initiative, to individual propensities, scope for thought and fantasy, form and content. All this is indisputable, but all this proves merely that the literary

part of the proletariat's Party work cannot be mechanically identified with other parts of the proletariat's Party work.[27]

And Lenin returned to the main theme of his article: the need for discipline of the writer and for Party control of publishing.

It is fairly clear that 'bourgeois' freedom for the writer was linked in Lenin's mind with the apparent birth of 'bourgeois' democracy, and there was nothing in this essay that suggested the slightest qualms at the prospect of victorious Social Democracy eliminating the writer's 'bourgeois' freedom along with the bourgeois publisher. The two crucial questions that rise in this connection were not answered. If a writer's submission to the Party's discipline was a free action, since it was a voluntary choice between proletarian partyness and 'bourgeois-anarchic individualism', would it still be free when the proletarian cause triumphed and the 'bourgeois' Press was eliminated? Presumably Lenin regarded such a question as one of the many utopian demands for 'the recipes of the future', which he put off writing until the future should arrive. The second question, however, concerning Lenin's assurance of full scope to individual creativity within the limits of partyness, was an immediate issue, for the Party already existed. For Social Democratic writers where was the line to be drawn between disciplined uniformity and creative originality, between mandatory Party doctrine and permissible private belief? Lenin's essay of November, 1905, did not answer this question, but, as far as general philosophy was concerned, his record for the previous decade was a clear answer. He had supported the drive that had pushed neo-Kantianism outside the pale, but he had been the chief promoter of official neutrality towards 'Machism', even though he had belatedly learned from Plekhanov to regard it as a revisionist philosophy.

In the course of 1906 and 1907, as the Revolution expired, divisive recriminations sprang to new life among the Russian Marxists. For those who identified self-realization and realization of the cause it was bad form to parade individual feelings, with the result that Lenin's letters and the reminiscences of him are singularly unrevealing of his inner mood. But from that very identification of the self and the cause we can easily infer how Lenin must have felt as, from a *dacha* in Finland, he watched the final unco-ordinated outbursts of peasant rebels go down before Stolypin's punitive expeditions to the countryside. This was the time, the summer of 1906, when Bogdanov emerged from jail with the third instalment of his latest philosophical work, *Empiriomonism*, a copy of which he presented to Lenin.[28] Reading it and falling into acrimonious disputes with another Bolshevik 'Machist' who was a regular visitor at the *dacha*, Lenin filled three notebooks with a critical letter to Bogdanov.[29] 'An explanation in love', he called it subsequently, though Bogdanov found it so insulting, if we can accept a secondhand report, that he returned it, commenting that he would regard it as unopened and unread in order to maintain his friendship and political collaboration with Lenin.[30] Lenin toyed with the idea of publishing the long letter as 'Notes of a Rank-and-File Marxist on Philosophy', but refrained.[31] Philosophical neutrality was still a condition for unity of the Bolshevik faction, and he had still to overcome his diffidence in philosophy. In 1907 a further strain was put on Bolshevik unity by the government's call to elect the Third Duma on the basis of an arbitrarily revised electoral law. Bogdanov, arguing for a boycott of the election, again clashed with Lenin, but we have Lenin's word for it that there was no connection between this political disagreement and the longstanding one in philosophy, especially since Bazarov, one of the foremost 'Machists', took the Leninist side on the issue of the boycott.[32] In any case, Lenin carried his point, and Bogdanov acquiesced in the decision to participate in the election, until a similar issue was raised in the summer of 1908.

In the interim a great intensification of the philosophical discord brought the Bolshevik faction to the verge of a split. With the Revolution unmistakably dead, at least for the time being, a fever of abstract discussion laid hold of the Russian intelligentsia, the Marxists included. Bogdanov at last saw fit to reply to Plekhanov's taunts; in a legal publication of the Bolsheviks, shortly before the government cut short its existence, he threw down a challenge. He demanded that Plekhanov cease his casual dismissals of 'Machism' as though it were a dead dog and publish a full-fledged analysis of the issues.[33] It may well have been this challenge that decided Lenin to enter the philosophical quarrel, with the urgent motive of forestalling a war of orthodoxy and revisionism in which the Bolsheviks would be revisionists, the Mensheviks orthodox. He must have indicated his decision to Gorky, for already in January, 1908, Gorky, recovering from tuberculosis on Capri, was calling the Bolshevik leaders to conference.[34] 'Either a congress,' he wrote to Lunacharsky, 'or we enter the swamp of new splits, misunderstandings, and so on. Il'ich [Lenin], it seems, has already entered!'[35]

Apparently Gorky was simultaneously pleading the Machists' case to Lenin, in letters that have not been published, for Lenin's letters of early February, which have been published, replied to Gorky on that subject. He was determined, he wrote, to save Bolshevism from charges of philosophical revisionism. This would mean a public dispute with fellow Bolsheviks, but if the dispute were conducted outside the faction's illegal newspaper, a split could be avoided, for there was no essential connection between the philosophical issues and the political programme on which the faction had maintained almost perfect unity.[36] He had hardly written this when three nearly simultaneous events all but precipitated an immediate split. The 'Machists' published their most audacious work to date, *Essays on the Philosophy of Marxism*, which caused Lenin to 'rage with indignation'.[37]

> To assure the reader that 'faith' in the reality of the external world is 'mysticism' (Bazarov), to confound materialism and Kantianism in the most hideous fashion (Bazarov and Bogdanov), . . . to teach the workers 'religious atheism' and the 'deification' of the higher human powers (Lunacharsky) . . .! No, this is too much! . . . I would sooner be quartered than agree to participate in an organ or a group preaching such things.[38]

About the same time that this infuriating book appeared, *Neue Zeit*, the most important organ of Marxist orthodoxy, whose softness towards 'Machism' had already caused Lenin to grumble in private, printed a translation of Bogdanov's article, 'Ernst Mach and Revolution'.[39] In a little preface the translator informed the German audience that the Russians had unfortunately made a political issue of Mach: the Bolsheviks had made his philosophy the basis of their faction, while the Mensheviks defended the materialism of Spinoza and Holbach.[40]

The most divisive event was the arrival at the Bolshevik editorial office of a 'Machist' article by Gorky, the faction's most prized literary figure, who had come to regard the philosophical dispute as a clash between 'Philistine materialism', preaching 'historical fatalism', and 'a philosophy of activism'.[41]

At a meeting of the editorial board on February 24, 1908, Lenin berated the 'Machists plainly, coarsely'.[42] But an immediate split was still avoided by a reaffirmation of the agreement of 1904, which was however made more explicit, and apparently even put down in writing, the first Bolshevik resolution on philosophy.[43] Gorky's article was to be rejected, and in the future *The Proletarian*, the faction's illegal, official newspaper, was to publish no articles on philosophy of any sort. In legal publications philosophical articles could be printed, on condition that orthodox dialectical materialism and 'Machism' were to have equal space.[44] Finally, notices were to be placed both in *Neue Zeit* and in *The Proletarian* to the effect that philosophy was *not* a factional issue within Russian Social Democracy, since 'Machists' and dialectical materialists were to be found within both the Menshevik and the Bolshevik factions.[45] In this way, Lenin wrote to Gorky the next day, the Bolsheviks could engage in a public philosophical debate, which was necessary to refute the widespread notion that the Bolsheviks were uniformly 'Machist', without disrupting the tactical unity of the faction.[46] Thereupon Lenin plunged feverishly into a study of philosophy, from which he emerged, in the fall of 1908, with the manuscript of *Materialism and Empirio-criticism*.[47]

It has been necessary to retrace the origins of this famous book in such unduly minute detail because of the widespread, erroneous view of it that stems from the Soviet revaluation of partyness in 1930. In that revaluation a one-to-one linkage of political and philosophical groups was asserted to be virtually a universal law, and retrospectively Lenin was given credit for striking down the 'Machists' as a political no less than philosophical deviation.[48] A variant of this interpretation has been given currency outside the Soviet Union by Bertram Wolfe, who cautions against the excessive stress of Soviet scholars on the linkage of political and philosophical groups, but reads the political significance of Lenin's chief philosophical work largely as they do: as a stroke at political dissidents within the Bolshevik faction.[49] Such an interpretation is achieved by telescoping the philosophical disagreements between Lenin and the 'Machists', which had their origin in 1901 and reached the breaking point in February, 1908, with the political disagreements within the Bolshevik faction that began to emerge clearly only in the summer of 1908. In order to achieve this interpretation one must also disregard the fact that the original sources, including *Materialism and Empirio-criticism* itself, never suggest the political motives that Wolfe and the Soviet scholars attribute to Lenin. The sources show that he *had* a political aim in writing the book, but it was not to join the philosophical and political issues that Russian Marxists were arguing about; it was to separate them. The Mensheviks were trying to join them, in order to picture the Bolsheviks as philosophical no less than political revisionists, and Lenin's chief political hope in working up his book was, as he wrote to Gorky, that 'the Mensheviks will be reduced to politics, and that is death for them'.[50]

As he set to work, the correlation of Bolshevism with 'Machism' was restated at greater length and with more plausibility than ever before. In the Menshevik newspaper A. M. Deborin, a 27-year-old disciple of Plekhanov's in Geneva, published a lengthy analysis of 'Machism' that built up to a double correlation between it and Bolshevism:

> The stamp of 'subjectivism' and 'voluntarism' lies on all the tactics of so-called Bolshevism, the philosophical expression of which is Machism. . . . Our Machist-shaped Marxists are conscious Bolsheviks, who 'give meaning' to the practice and tactics of the latter. And Bolshevik tacticians and practical people are unwitting [*bessoznatel'nye*] Machists and idealists.[51]

The last sentence was apparently Deborin's method of accounting for the fact that Lenin, the chief 'practical person' of Bolshevism, had been silent in public—with the one fleeting exception noted above—concerning the controversy over 'Machism'. About the same time that Deborin's article appeared, in April, 1908, Lenin proclaimed his entry into the philosophical arena—in defence of Plekhanov. He described Plekhanov as 'the sole Marxist in international Social Democracy who, from the point of view of consistent dialectical materialism, has given a critique of the improbable banalities that the revisionists have effused', and he decried 'the profoundly mistaken attempts to palm off old and reactionary philosophical trash in the guise of criticism of Plekhanov's tactical opportunism.'[52] Having thus made clear that he approved of Plekhanov's philosophy as strongly as he disapproved of his politics, Lenin promised a series of articles or a special pamphlet in criticism of the *Essays on the Philosophy of Marxism*, 'by Bogdanov, Bazarov and others'.[53] (By his special mention of the Bolshevik contributors he underscored his intention of raising the philosophical discussion above the factional division of Russian Marxists.)

While this proclamation was dragging through the press, Bogdanov gave a lecture on philosophy to the *émigré* colony in Geneva, and Lenin chose this opportunity to make another public display of the contervailing orthodox tendency within Bolshevism.[54] He worked up ten questions for one of his followers to address to Bogdanov. (He himself was away in the British Museum digging up material for his book.) The three questions selected for special stress were designed to demonstrate that the only possible choice in philosophy was between idealism and materialism; that an external world existed independently of the mind; and, finally, that 'Machism has nothing in common with Bolshevism, that Lenin has repeatedly protested against Machism', that the Mensheviks were guilty of 'imposture' in picturing their faction as the home of philosophical orthodoxy.[55]

These public announcements did not, however, put an end to such Menshevik claims. Aksel'rod the Orthodox, writing another criticism of 'Machism', granted that there were empirio-critics in the Menshevik faction and dialectical materialists in the Bolshevik, but she denied any significance to the fact.

> The question concerns not the views of one or another individual person, who can always be inconsistent, but a tendency as a whole, which, on closer and more attentive examination, always discloses a logical regularity [*zakonomernost'*] that gives a basis for philosophical generalization. But here another question can be raised, to wit, whether it is possible to seek a logical connection between two such incommensurable quantities as factional politics, on the one hand,

and a general philosophical trend, on the other hand. In other words, what can there be in common between epistemology, which has to do with the primary conditions of scientific experience, and a comparatively narrow, limited, local and ephemeral political tendency? In spite of the important and imposing appearance of this question, it does not withstand the slightest criticism.[56]

This, with the long analysis that followed, was the most striking anticipation in pre-revolutionary Russian Marxist literature of the linkage of politics and philosophy that was to triumph in Soviet Russia. The bewildering convolutions of Russia's revolutionary history would, in the late 'twenties, join Aksel'rod the Orthodox with Bogdanov (the 'Machist') in a bloc opposing the first major step towards a conscious linkage of epistemological issues and factional politics. In 1908–1909 Lenin was the chief opponent of such a linkage.

It must not be imagined that Lenin treated philosophy with academic detachment in *Materialism and Empirio-criticism.* On the contrary. Still a bit diffident in philosophical matters, yet passionately certain that he was defending elementary truths against opponents who were not honestly disputing but slyly undermining them, Lenin fluctuated between dreary stretches of inordinate quotation and violent explosions of invective. In both respects Engels' *Anti-Dühring* was obviously his model, but there were significant differences. Long quotations, which had been Engels' method of setting up the targets for his polemical fire, were also, in Lenin's book, a frequent substitute for independent exposition and analysis; and Engels' sarcastic laughter was transformed into furious scolding.[57] In both respects *Materialism and Empirio-criticism* was clearly a step towards the frightful style of quote and club that would dominate Soviet philosophical writing for a generation following 1930.[58] But, stylistic considerations aside, the book's concept of partyness was hardly an anticipation of the rule that the Bolshevik leaders are the areopagus of philosophical truth. Partyness was one of the book's main themes, but always in the broad sense. The only possible parties in philosophy, Lenin argued, were idealism, which had been serving the interests of priests and exploiters since the Hellenic age, and materialism, which for an equally long time had been the banner of science and revolution. His reply to the Menshevik correlation of Bolshevism and

'Machism' was contemptuously to ignore it, and to show by example how philosophical issues should be separated from factional politics. Though he demurred from one aspect of Plekhanov's philosophy, he defended him and Aksel'rod the Orthodox against the Bolsheviks Bogdanov and Bazarov.[59] Of course he did not specify the factional affiliations of those he was defending or attacking, for his whole point on the partyness of philosophy was that it transcended factional differences among revolutionaries; willy-nilly the comrades who had embraced 'Machism' were supporting the ancient party of reaction. Driving the point home at the very end of the book, he called attention to the fact that Mach, though proclaiming himself a freethinker, had dedicated a book to a philosopher who supported the Kaiser politically as well as ideologically. And to this concluding 'demonstration' of the true partyness of philosophy, Lenin appended a supercilious sneer at the Mensheviks' misuse of the concept:

> Plekhanov, in his remarks against Machism, has been concerned less with refuting Mach than with dealing a factional blow at Bolshevism. For this petty and wretched exploitation of fundamental theoretical differences he has already been deservedly punished—with two books by Machist Mensheviks.[60]

The political discord among the Bolsheviks, which arose while *Materialism and Empirio-criticism* was being written, reinforced Lenin's commitment to the supra-factional understanding of the partyness of philosophy. In the spring of 1908 some of the Bolshevik organizations in Russia reverted to the mood that had, a year earlier, produced the demand for a boycott of the Duma elections. If Bolshevism was truly revolutionary, they felt, it should not traffic in the parliamentary politics of a pseudo-constitutional régime. They demanded that the faction recall its delegates in the Duma, or at least oblige them to make such revolutionary speeches there as to precipitate their ejection. When the news of this *otzovism* (from the Russian for 'recall') reached the Bolshevik leaders in emigration, their unity, already severely strained by public disputes over philosophy, was pushed towards destruction. Lenin wanted the faction's newspaper to condemn *otzovism* as a semi-anarchist deviation, while Bogdanov thought it should be discussed as a permissible variant of Bolshevism.[61] At the same time the opposite of *otzovism* was pushing the Menshevik faction towards a split of its own. Some Mensheviks, dismissing the illegal organizations as virtually dead, stressed legal organizations as the basis of a renovated Party. Other Mensheviks, including Plekhanov, not only opposed this 'liquidationism', but were moving towards a break with its supporters, which was accomplished towards the end of 1908, when Plekhanov withdrew from the Menshevik newspaper and set about establishing his own.[62] It was probably at this point that Lenin, seeing the possibility of a bloc with Plekhanov and his 'Party-Mensheviks', decided to press for a complete split with the Bolshevik 'Machists' and *otzovists.*

As late as December 19, 1908, he showed a lingering reluctance to make the break, by permitting his sister to tone down the denunciation of the Bolshevik 'Machists' in *Materialism and Empirio-criticism,* which she was seeing through the press.[63] But a month later he seems to have changed his mind; he plunged for a split by wilfully violating the resolution of February, 1908, concerning the Bolshevik newspaper's philosophical neutrality.[64] Overriding Bogdanov's protest, he got *The Proletarian* to publish a blast at Lunacharsky's philosophy of 'god-building'.[65] (His rationalization was that the article concerned religion rather than philosophy.) At the end of February he sent out a call for a plenary session of *The Proletarian's* editors, and wrote his sister to abstain from toning down the invective against the Bolshevik 'Machists': 'Our relations with them are *completely broken.* There is no point in toning down, it isn't worth it.'[66]

At this point, one might think, philosophical and political issues would become fused, and the Conference would lay down a Bolshevik line for philosophy as well as politics. Yet 'Machism' was not on the agenda, when the editors of *The Proletarian* gathered in the upstairs room of a Parisian café late in June, 1909; and through several days of debate the chief epithets that Lenin and Bogdanov flung at each other were not 'idealist' and 'dogmatist'—though each had such an opinion of the other—but 'semi-anarchist' and 'centrist'.[67] ('Semi-anarchist', because Bogdanov wanted the faction, tolerating *otzovists,* to erect itself into a separate, ardently revolutionary party that would have no truck with Mensheviks of any kind; and 'centrist', because Lenin wanted to purge the faction of *otzovists* and form a bloc with Plekhanov's 'Party-Mensheviks'.)[68] And at the end, though Lenin had his way completely and the split with Bogdanov was formally consummated, the Conference's published resolutions contained no reference to 'Machism'. One zealous Leninist (Skrypnik) moved that empirio-criticism be condemned along with 'god-building', but the motion was defeated.[69]

The reasons for this apparent anomaly are not hard to find. Lenin would not have put 'Machism' on the agenda in any case, for he would not admit that he was calling in question the resolution on philosophical neutrality. And, aside from such tactical considerations, he had just brought out a whole book demonstrating that philosophical issues transcended the factional politics of Russian Social-Democracy. (*Materialism and Empirio-criticism* finally came off the press about two months before the Conference met.)[70] He was hardly ready to turn about and urge the establishment of a philosophical line mandatory for all Bolsheviks. Indeed, he had followed up *The Proletarian's* denunciation of Lunacharsky's 'god-building' with an article defending the right of religious believers, even ordained priests, to be Bolsheviks as long as they did not proselytize within the faction.[71] Bogdanov agreed that religious propaganda must be proscribed, and the debate on the newspaper's denunciation of Lunacharky therefore turned on the question whether his 'god-building' was actually a religion or a philosophy, like Feuerbach's, that used religious terms but was genuinely atheistic.

Lunacharsky, who was not present but submitted a statement, charged the Leninists with hypocrisy for insisting that 'god-building' was a religion, and, half mockingly, half seriously, demanded the establishment of a clearly defined philosophical line:

> Bebel declared once that we have no dogma and therefore cannot have heretics. Let us grant that he was mistaken, that both heresy and its condemnation are a possible phenomenon in Social Democracy. But don't we have a right to expect that in our Party free thought will have at least the same guarantees as are given to it by the Catholic Church, which prides itself on its intolerance?[72]

And he described with ironic admiration the Church's careful thoroughness in the examination of suspected heresies. Bogdanov forthrightly stated a similar view. If the editors' meeting were an authentic Bolshevik conference, he said, he would introduce a resolution recognizing the departure from the faction's philosophical neutrality and 'making clear the limits of aberrations', which would include a condemnation of religious terminology and of idealism.[73] Voicing this demand for a philosophical line without Lunacharsky's irony, he was more obviously inconsistent when he added that he was against 'restricting the great proletarian movement within authoritarian limits'.[74]

These efforts to provoke a clear definition of the Bolshevik line in philosophy were in vain; the Leninists condemned 'god-building' but held to their position that the resolution on the faction's philosophical neutrality had not been upset. But then Bogdanov walked out of the conference, and the Leninists were released from the necessity of diplomacy. Zinoviev, presumably speaking for Lenin, introduced a resolution on philosophy, but three substitute motions were immediately offered. Tomsky condemned philosophical discussion altogether, as a flight from the painful tasks of revolutionary agitation, and proposed that all philosophical articles be banned in the Bolshevik press. Rykov wanted a simple repeal of the resolution on philosophical neutrality, arguing that any effort to define the Bolshevik position in philosophy was fruitless, since no two Bolshevik philosophers could agree with each other. (The last remark provoked Lenin to cry 'Slander!') Kamenev, sharing Rykov's liberalism but somewhat more hopeful of finding a common philosophical denominator, proposed that the Bolshevik newspaper be allowed to carry philosophical articles written from 'the point of view that the core of West-European revolutionary Social Democracy holds to'.[75] To which Lenin remarked, somewhat cryptically, that it was impossible to determine beforehand what course the philosophical discussions would take. Thereupon Zinoviev's resolution was adopted by a vote of six to one, with two abstentions. It amended the old resolution of February, 1908, by striking the requirement that the faction's legal publications must give equal space to 'various philosophical tendencies', and implicitly left in force the ban on all philosophical articles in the illegal newspaper. But this ban was not to be interpreted as a sign of the editors' neutrality: 'If philosophical questions should come up in the [Party's] Central Organ, the representatives of [*The Proletarian's*] extended editorial board in the Central Organ should take the definite [*opredelennuiu*] position of Marx-Engels' dialectical materialism.'[76] Thus the editors were formally committed to philosophical orthodoxy, however laconically defined; Lenin had made sure that the Mensheviks could not link philosophical revisionism with his faction.

This resolution was the second occasion on which Lenin approached the narrow conception of the partyness of philosophy—or the first, if the article on Party literature, in which he had suggested the possibility of deciding philosophical questions by a majority of votes, is considered too vague to be relevant. It would be a mistake to emphasize the nearness of the approach. For one thing, the resolution was not made public (it was first published in the 'thirties), and thus could not become a part of the faction's platform, binding on all who joined. When Bogdanov and Lunacharsky, in the manifesto announcing their new faction, the *Vperedists*, followed the advice they had given Lenin and stated their philosophical position, Lenin exclaimed with amazement that they were the first 'of all the groups and factions of our Party' to make philosophy part of their platform.[77] Very likely the amazement was genuine, for in 1912 Lenin readmitted the *Vperedists*, even though they continued to publish 'Machist' works; he required only that they abstain from publishing their philosophical views in the Bolshevik press.[78]

Nor did Lenin ever make a serious effort, as Aksel'rod and Plekhanov had done, to demonstrate an organic connection between 'Machism' and a specific political deviation. On one occasion he seemed on the verge of doing so. In reply to an argument that the dispute over philosophy was a bagatelle without relevance to the political cause, Lenin asserted that the defence of materialism 'is *always* connected "by an organic, real bond" with "the Marxist socio-political movement" —otherwise the latter would be neither Marxist, nor socio-political, nor a movement'.[79] But this comment, which Soviet authors have quoted time and again in their repeated efforts to demonstrate the philosophical roots of political deviations and the political offshoots of philosophical deviations, was in context merely another example of Lenin's understanding of partyness in the broad sense. For it was the introduction to an argument that different historical periods demand attention to different aspects of Marxism, making philosophy a relatively unimportant issue at one time and a crucial one at others.[80] He did not try to demonstrate an organic connection between dialectical materialism and his own political line, or between 'Machism' and *otzovism*. Indeed, when the re-established unity with Bogdanov broke down in 1914, and Lenin had to explain both why the new unity had been possible and why it had now broken down, he stressed the *lack* of an organic connection between politics and philosophy: 'The point is that the *Vperedists* were pasted together of heterogeneous anti-Marxist elements, . . . [of] "Machism" and "otzovism". . . . Against [Machism] both the Menshevik G. V. Plekhanov and the Bolshevik V. Il'in [Lenin] struggled energetically.'[81]

If Lenin had tried to show an organic connection between dialectical materialism and his own political line, he would have ruled out the bloc that he was seeking with the 'Party-Mensheviks'. He therefore had political reasons for his repeated insistence that dialectical materialism was defended by Marxists of all factions.[82] Even so, Aksel'rod the Orthodox and Plekhanov still grumbled that he was easier on the Bolshevik 'Machists' than on the Menshevik, and they explained the minor difference between his interpretation of dialectical materialism and Plekhanov's as evidence of 'Machist' influence on Lenin's 'prosaic and unphilosophical nature'.[83] But, however condescendingly, they took the hand that Lenin held out to them. Plekhanov expressed joy that the departure of the revisionists surrounding Lenin had again made visible his own head, 'with its more or less narrowly and woodenly understood Marxism'.[84] If Lenin did not reply in kind, the reason was not simply his desire for a political bloc with Plekhanov. Even when the bloc was failing, and Lenin was accusing Plekhanov of political 'idiocy', he wrote in an encyclopedia[5] 'On the question of the philosophy of Marxism and of historical materialism the best exposition is G. V. Plekhanov's . . .'; and farther along in the same bibliographical article, he listed

Aksel'rod's writings ahead of his own.[85] Evidently he had a genuine admiration for their philosophical work.

Thus Lenin approached the narrow concept of partyness in one respect—the official Bolshevik newspaper was to be kept pure of revisionist philosophy—but he lacked the narrow concept in four other respects. He did not seek out organic correlations between philosophical and political deviations; he did not assert a Bolshevik monopoly of philosophical orthodoxy; he did not require philosophical conformity among Bolsheviks; and he did not try to maintain continuous supervision of philosophical discussion by the Central Committee. Ironically, some of the Marxists who accused him of authoritarianism came somewhat closer to the narrow concept of partyness than he did: Plekhanov and his disciples by arguing a one-to-one correlation between philosophical and political trends, Bogdanov and the *Vperedists* by demanding a clearly defined Bolshevik line in philosophy.[86]

Lenin's broad concept of partyness survived the Bolshevik Revolution and helped shape the intellectual milieu of Soviet Marxists until the end of the 'twenties, that is, until conditions arising after Lenin's death engendered the narrow concept of partyness. It would be a mistake to attribute to Lenin either a streak of conscious liberalism, inhibiting the full development of the narrow concept of partyness, or a cunning omniscience that anticipated this full development but concealed its prescience until the time for revelation should arrive. The second view is contradicted by too many facts to be seriously maintained, even if its underlying assumption could be accepted. The first view overlooks the fact that Lenin could be liberal only by accident, for he was illiberal on principle. Liberalism, the willingness on principle to give free play even to ideas one loathes, is founded on doubt that any one idea is absolutely true, and Lenin hated such doubt as one of the chief hindrances to revolutionary action. Absolutely sure that he and his Party knew the way through the capitalist present of blind necessity to the socialist future of conscious freedom, he believed that his Party *enhanced* freedom by extending its control wherever refractory forces blocked that way. If he did not draw the inference that his Party had to establish its undisputed, rigid control all over, the reason was that obstacles did not rise up all over, and that other people than Bolsheviks supported one or another of his chief ideas, philosophical orthodoxy included. Audaciously and wilfully as he steered his revolutionary course, still he ran before a mighty wind of popular, spontaneous revolution. When fatal obstacles would loom on every side, and contrary winds and intractable calms would beset his Party at every turn, in the time of Stalin's 'revolution from above', his illiberal legacy would be developed to the ultimate degree.

PART TWO

THE SOVIET SETTING 1917-1929

3

INTRA-PARTY POLITICS AND PHILOSOPHY

QUITE abruptly at the end of 1929 Soviet Marxists discovered that their philosophical discussions of the preceding twelve years had been determined by the clash of political factions within the Party. By 1931 they worked out the formula that is still standard in the Soviet Union. The Party had been fighting a war on two fronts in philosophy as in politics. There had been a group of mechanistic authors who had provided the philosophical justification of the right deviation; there had been a group of Deborinite or Menshevizing idealist authors who had done the same service for the 'left' (always in quotes) deviation; and there had been 'the Party' (never in quotes), now triumphant over all deviants.[1] Fortunately, no one recalls that M. B. Mitin, the chief author of this standard formula, as late as April, 1929, reviewed the controversy over mechanistic materialism without a word about the right deviation, and in June attacked the right deviation without a word about mechanism. Certainly he did not connect the Deborinites with the left deviation, for he was then a Deborinite himself.[2]

Outside observers of Soviet Marxism have reacted to this formula in different ways. When it was first announced, *émigré* Russians who had followed the philosophical discussions laughed sarcastically, supposing that nothing more was involved than an abrupt decision of Stalin and his place-hunting young men to control philosophical discussion.[3] They had seen no evidence that might, even with hindsight, support the newfound formula; and they had overlooked the actual political significance of the preceding philosophical discussions, in part because they denied the very possibility of finding class or party interests in abstract philosophical views. A similar interpretation echoes in recent Thomist accounts of Soviet philosophy, which, in their historical sections, rely heavily on the writings of the Russian *émigrés.*[4] Somerville's book on Soviet philosophy skirts the issue of intra-Party politics, largely ignoring it in the account of developments before 1929, and translating the retrospective political allegations made since then into vague and non-committal terms.[5] R. A. Bauer's book on Soviet psychology, taking these allegations very seriously, imaginatively constructs a covert manœuvre by Stalin in the philosophical discussions of the 'twenties.[6]

A good deal of the confusion on this question could be avoided by introducing a few elementary distinctions. Soviet Marxist discussions of the philosophy of science during the 'twenties may have had connections with politics, understood as factional dissension and manœuvring within the Bolshevik Party, or with politics, understood as state policy. And in either case, one can seek either consciously formed connections, which may be presumed to have left some clear evidence of themselves, or an unintended cross-fertilization of ideas, which is difficult to prove conclusively. The possibility of such unintended cross-fertilization will be examined in a later chapter, and connections between the philosophy of science and state policy that was not the object of factional discord will be a recurrent theme all along. But first one must set aside the legend that already in the 'twenties there were consciously formed connections between the factional struggle within the Party and the discussions of the philosophy of science. There is no direct evidence of overt connections between the two fields, but the legend thrives nevertheless, on the assertion of indirect evidence of covert connections.

Without the possibility of archival research one can probe for covert efforts to connect the philosophical and political dissensions of the 'twenties only by searching for truly distinctive ideological affinities and repeated chronological coincidences between developments within the two fields. The search usually begins with the assignment of determinist presuppositions to

Bukharin's political thought, voluntarist ones to Trotsky's, and an opportunistic mixture of both to Stalin's—a highly simplified formula that has some substance but can hardly be regarded as a great truth from which lesser facts may be logically deduced.[7] Moreover it seems doubtful, even if these standard labels did adequately and accurately describe the philosophical presuppositions of the three political factions headed by these men, that the same lables might with just precision be pasted on corresponding groups in the philosophical discussion. Intellectual history would indeed be easy to write—and hardly worth writing—if there were an unexceptionable logic by which the political and philosophical views of men and their parties could be inferred from each other. How, to take a concrete example, could the following facts be fitted into the scheme that makes the opposition of determinism and voluntarism the essential element both of the conflict between Bukharin and Trotsky and of the conflict between the mechanists and the Deborinites? In a little controversy of 1922–1923 between Bukharin and Sarab'ianov, each accused the other of unwittingly adopting a philosophical basis for Menshevism. In Sarab'ianov's view that society's 'superstructure' can change only when the 'base' has already changed, Bukharin saw the kind of determinist argument that the Mensheviks used to deny the Marxist legitimacy of the Bolshevik Revolution.[8] And Sarab'ianov perceived the seeds of a similar determinism in Bukharin's view 'that man's will is not at all free, that it is determined by the external conditions of man's existence'. Against this Sarab'ianov cried out spiritedly: 'No, Comrade Bukharin, not at all! The limits of the human will are determined *not only* by the external conditions of man's existence but also by man himself and his own will.'[9] After 1923 Bukharin and Sarab'ianov ignored each other, and Sarab'ianov, who became a leader of the *mechanist* faction, defended his voluntaristic philosophy against the attacks of the *Deborinites*.[10]

But it is quite beside the purpose of the present work to attempt a definition of the philosophies implicit in the political and social theories of Bukharin, Trotsky and Stalin, and it would be premature at this point to attempt a definition of the basic differences between the mechanist and Deborinite factions in the philosophical controversy. Let it be assumed then, for the moment, that the mechanist faction in philosophy and the right deviation in politics did exhibit a truly distinctive ideological affinity, and that the same may be said of Trotskyism and the Deborinite faction, while Stalin may be described as co-operating with Bukharin to defeat Trotsky and then appropriating both the Trotskyite political programme and its alleged Deborinite correlate to defeat Bukharin. Does *chronology* indicate that conscious though surreptitious efforts were made by either the political or the philosophical factionalists to exploit these presumed correlations?

From the autumn of 1923, when the 'Platform of the Forty-six' appeared, until the Fifteenth Party Congress in December, 1927, the political controversy was between a left opposition, originally led by Trotsky and ultimately by Zinoviev and Kamenev as well, and the dominant Party leadership, in which Stalin and Bukharin appeared to be indissolubly joined. If this dominant leadership had been covertly trying to exploit its alleged affinity with the mechanist group in philosophy, one would expect to find some evidences of official favour being extended to the mechanists. But the contrary is true. As the reader will see below, almost from the start of the philosophical controversy, and certainly from 1926, there were rather clear signs of high-placed sympathy with the Deborinites. By the end of 1927, when the Fifteenth Party Congress pronounced a final anathema on the left opposition, all but two (or four at most) of the Marxist-Leninist journals, societies, and institutions of higher learning were either active supporters of Deborin's philosophy or benevolently neutral. If the dominant Party leadership had been conscious of a distinctive affinity between the Deborinites and the Trotskyites, it is difficult to understand how the former could have moved from triumph to triumph as the latter were being pushed towards destruction. Apparently the dominant leadership regarded the Deborinites favourably for other reasons than those of political factionalism.

In 1928, as the First Five Year Plan got under way, there was a split within the dominant group that had defeated the left opposition. In the autumn of 1928 Stalin called publicly for a struggle on *two* fronts, against a right as well as a left deviation. In January, 1929, the Political Bureau heard Stalin denounce

Bukharin as the leader of the right deviation, and in April, 1929, the full Central Committee and the general public heard the same.[11] During the same month of April, 1929, the Deborinites achieved their ultimate triumph over the mechanists at the Second All-Union Conference of Marxist-Leninist Scientific Institutions, which adopted resolutions formally condemning mechanism as anti-Marxist. It would seem that mechanism and its alleged political correlate, the rightist opposition, were suffering defeat together, and from this approximate coincidence Bauer has inferred a covert manœuvre by Stalin. Bauer pictures Stalin as sponsoring the victory of the Deborinites in order to justify a break with determinism and to gain 'a powerful weapon against the theories of . . . Bukharin'.[12]

This reasoning is plausible until one examines the facts a little more closely than Bauer apparently has done. In the first place, the victory of the Deborinites had already been assured before the split occurred between Stalin and the right deviation In the second place, if one is to imagine Stalin deciding in 192.8 or early in 1929 to make the Deborinite ascendancy official and complete in order to gain 'a powerful weapon' in the gathering conflict with the right deviation, one would think that he would have got the Deborinites to give a political, anti-rightist emphasis to the propaganda against the mechanists. But in 1928 and through much of 1929 allusions to intra-Party politics in the philosophical controversy were just as rare and haphazard as they had been previously. At the Conference of April, 1929, where the mechanist group was formally condemned, the right deviation was not mentioned in the main speeches by Deborin and O. Iu. Shmidt. During the debates some *mechanists* raised the issue by describing the *Deborinites* as philosophers of the right deviation, and some Deborinites, including Deborin himself in his 'Concluding Remarks', threw the charge back at the mechanists.[13]

When the philosophical portion of the Conference had ended, V. P. Miliutin, a prominent economist and a member of the Party's Central Control Committee, gave an administrative report on the work of the Communist Academy, in the course of which he remarked:

Comrade Deborin said here that we, the economists, historians, legal scholars, have not taken an active part in the struggle that the philosophers have been obliged to carry on with the mechanists. But we can present another bill: in the struggle with the right deviation, which the historians and economists have been obliged to wage, the philosophers have unfortunately participated insufficiently. . . . In Comrade Deborin's speech the line of dialectical materialism's further penetration into the field of natural science was correctly projected, but I should say that a greater penetration of dialectical materialism is also needed in the field of the social sciences, and also a more active struggle on the front of bourgeois philosophy and revisionist views.[14]

It will be noted that Miliutin demanded the philosophers' help in the Party's factional conflict, but he did not specifically picture mechanistic materialism as the philosophical basis of the right deviation. Evidently neither he nor the philosophers had been informed of Stalin's alleged intention to use Deborinite philosophical arguments against the right deviation. Similar ignorance was displayed in a subsequent decree of the Central Committee elaborating on the 'organizational questions' that had been raised at the Conference of April, 1929; the decree contained no explicit mention of the philosophical controversy, with or without reference to the drive against the right deviation.[15]

Furthermore, Stalin's speeches against the right deviation show no distinctive signs of an intention to use the Deborinite version of dialectical materialism. Three days after the philosophical Conference dispersed, a plenary meeting of the Central Committee gathered to hear Stalin's major polemic against Bukharin and the right deviation. Stalin had almost nothing to say about dialectical or mechanistic materialism; in his criticism of Bukharin as a theorist, he concentrated on some of the criticisms that Lenin had once made of Bukharin.[16] Characteristically, Stalin ignored the most philosophical criticism that Lenin had made of Bukharin, the considerable lecture on the difference between dialectics and eclecticism that occurs in Lenin's well-known speech, 'Once Again on the Trade Unions'.[17] Instead Stalin quoted inordinately long passages to demonstrate Bukharin's alleged divergence from Lenin in political theory. As an introduction to this attack on Bukharin's orthodoxy, he quoted the famous passage from 'Lenin's Will' that describes Bukharin as something of a '"scholastic"' who "'has never studied, and . . . never fully understood dialectics'".[18] Only in Stalin's incidental observation, as he commented on this quotation, that 'dialectics is the soul of Marxism' can one find the slightest evidence of a possible borrowing from the Deborinites, but it is much more likely that he was simply repeating a cliché common to many more Marxists than the Deborinites.[19] Certainly Sten, a leading Deborinite who was in Stalin's audience, did not behave as if he were witnessing the application of his principles to politics. He interrupted Stalin twice to defend Bukharin.[20]

There is, finally, a fourth reason that it seems wrong to read the coincidence of the defeat of mechanism and the drive against the right deviation as evidence of a conscious though surreptitious effort to use the Deborinite philosophy as a weapon against the right deviation. Even after Miliutin had plainly indicated the Central Committee's desire for the philosophers' aid in the campaign against the right deviation, and after Stalin had publicly criticized Bukharin as a theorist, there was a gap of some months before the Deborinites complied with Miliutin's request and followed Stalin's example.[21] Very likely the leading Deborinites were reluctant to abandon the safety of their previous separation from intra-Party politics, and were, furthermore, perplexed by the novelty of their assigned task. Moreover, when the leading Deborinites did come out with philosophical criticisms of the right deviation, Stalin appears to have been dissatisfied rather than pleased.

In December, 1929, there was another coincidence of meetings. At a series of meetings in the Communist Academy the Deborinites pictured mechanistic materialism as the philosophical basis of the right deviation, and just after these meetings were wound up, a Conference of Marxist Agrarian Economists assembled in the same building to hear Stalin deliver another major polemic against the right.[22] Stalin once again argued against the right without mentioning mechanistic materialism, while the Deborinites had dwelt on machanism as the heart of the right deviation. Moreover, they pictured themselves as showing the Party leaders the way to criticism of the right deviation, while Stalin complained that 'the theorists' were lagging behind the Party's practical leaders. If Stalin may be presumed to have included the Deborinite philosophers among these lagging theorists, it would appear that his first public comment that could possibly be interpreted as showing an awareness of the Deborinite philosophy was a complaint of its uselessness to the Party. Thus, if Stalin consciously intended to use the arguments of the Deborinites as 'a powerful weapon' against Bukharin, then he must be presumed also to have intended to keep the borrowing quite secret, so much so that he removed the evidence from his polemics against Bukharin, and censured rather than praised the Deborinites when they finally came out with a philosophical critique of the right deviation. This seems an excessively long chain of inferences to hang on the lone fact of a coincidence of meetings in April, 1929.

For all these reasons it is highly improbable that the Party leaders made conscious efforts to gain their ends in the Party's factional struggles by covert manœuvres in the philosophical discussions. Indeed, the direct evidence, the actual record of the philosophical discussions, suggests something quite different: a widespread, conscious effort to keep the philosophical discussions separate from the Party's factional quarrels. Consider the case of I. I. Stepanov (or Skvortsov-Stepanov), who was a leading mechanist in the philosophical discussions and a leading opponent of Trotskyism. When one recalls that the first attack on Stepanov's philosophy came from a Deborinite philosopher who belonged to the left deviation, as did one other prominent Deborinite (Sten and Karev),[23] it seems strange that Stepanov did not allege a connection between his philosophical and political opponents. Yet the fact is that he did not. He was not above the injection of *ad hominem* arguments in his philosophical polemics. He once described the Deborinite Gredeskul

> a mature, honoured Cadet [i.e., Constitutional Democrat] in the past; a solid professor, an irreproachable idealist in the past and the present. If in pre-revolutionary times Deborin, half Zionist and half Menshevik of the extreme right wing, exhorted us to vote for the Cadet, why should he not form a bloc with Professor Gredeskul now against the materialist danger?[24]

The Deborinites countered this sort of attack with arguments that mechanist philosophy was logically a justification of Menshevism or Social Democracy, but they did not mention factional deviations within the Bolshevik Party.[25]

There were only four exceptions to this mutual abstention from allusions to the Party's internal conflict. In 1927 the Deborinite A. A. Maksimov submitted an article to *Under the Banner of Marxism*, in which he accused Stepanov of supporting a philosophy that justified Trotskyism. News of the accusation reached Bukharin, who wrote the editors that it would be 'ridiculous, and politically—how shall I put it?—not mighty clever' to link *Stepanov* with Trotskyism. 'The [philosophical] discussion can and should go on, but not with such methods as in essence hardly correspond to reality and in politics are plainly harmful. This is my *personal* opinion.'[26] Deborin, the chief editor, wrote back that he shared the opinion, and had killed the offensive passage a month before Bukharin's letter arrived.[27] But Deborin did allow Maksimov and two other authors to drop fleeting assertions, without reference to Stepanov, that mechanism in philosophy and Trotskyism in politics were natural correlates.[28]

That is the whole record—outside the archives—of efforts to link philosophical discussion and the Party's factional struggle before the end of 1929. The fact that these premature efforts were made by over-zealous Deborinites suggests the probable reason for Stepanov's failure to charge the Deborinites with Trotskyism. With the exception of Karev and Sten, they were as firmly opposed to political deviation as he.[29] Otherwise one cannot understand his role in the Communist Academy early in 1928. He was sent by the Party leadership to guard the Academy 'against each and every school and non-school deviation [*shkol'nykh i neshkol'nykh uklonov*].'[30] Very likely the expulsion of Trotsky and other leaders of the left deviation from membership in the Communist Academy, which was carried out by a plenary meeting towards the end of March, 1928, was the result of Stepanov's instigation. But the same plenary meeting extended membership in the Academy to two leading Deborinites.[31] Apparently Stepanov saw no discrepancy in this simultaneous acceptance of Deborinites and expulsion of Trotskyites, for his last philosophical polemic, published about the time he died in October, 1928, taunted leading Deborinites with their pre-revolutionary membership in non-Bolshevik parties, but still made no effort to connect them with Trotskyism.[32]

Another noteworthy instance of an apparent tendency to keep the factional struggles in politics and philosophy separate was Bukharin's anomalous position in the philosophical discussions of the 'twenties. When his *Theory of Historical Materialism* appeared in 1921, Soviet Marxist philosophers felt free to receive the book either hostilely or favourably.[33] But during 1924, when the controversy between the Deborinites and the mechanists was taking shape, hostile judgments of Bukharin's book disappeared almost entirely from the Soviet press, and remained conspicuous by their absence until the Sixteenth Party Conference in April, 1929, made criticism of Bukharin an element of the Party line. During that period, 1924–1929, no extensive discussions of Bukharin's philosophy, whether *pro* or *contra*, were published. A number of passing references to his views were made by philosophers of both blocs; these references were almost uniformly favourable, and were often accompanied by awkward silences in regard to major issues on which Bukharin's views were sharply at variance with those of the author citing him. One senses the existence of a gentleman's agreement to keep Bukharin out of the philosophical controversy, for the similarity between his views and those of some mechanists fairly clamoured for comment. The silence is especially striking in the numerous Deborinite attacks on Bogdanov; never, during the period 1924–1929, did these attacks mention Bukharin, the most famous of Bogdanov's disciples.[34]

It is tempting to turn to the simple explanation that the Deborinites were afraid of being called Trotskyites if they attacked Bukharin during the years 1924 to 1929. That was indeed the implication of some remarks made by one of the first Deborinites to break the silence on Bukharin in 1929:

> . . . Comrade Bukharin has occupied an altogether different position [from that of the much-criticized mechanists]. For people working in the field of Marxist theory, it has been no secret that Comrade Bukharin in some fundamental, philosophical questions of a general theoretical nature, has not kept to the positions of Marx and Lenin, but has tried to 'develop' their views 'further'. But in the eyes of the broad masses of readers, he appeared to be a true follower of Marx and Lenin. This happened because for the broad masses Comrade Bukharin's political role, which was deservedly his until a

very recent time (until he became the ideologist of the right deviation), served as a guarantee of theoretical orthodoxy.[35]

Still, as the reader has already seen, two of the leading Deborinites *were* Trotskyites, for a time at least, during the period 1924–1929; yet they too observed the truce on Bukharin's philosophy when participating in the philosophical controversy.[36] The obvious supposition is that they did so in order not to embarrass all the other Deborinites who were opposed to Trotskyism. In other words, the Deborinites found it possible to ignore their own political differences in the service of their philosophical cause.

But how is one to interpret the mechanists' failure to make capital of the similarity between their own and Bukharin's views? In retrospect it may seem odd that they neglected the argument that all opponents of Trotskyism must share Bukharin's mechanistic philosophy, and, conversely, that opposition to the mechanistic philosophy was the equivalent of support for Trotskyism. Yet they did not advance this argument. Stepanov, for example, ignored Bukharin when writing polemics against the Deborinites. Stepanov, to be sure, had the same problem as the anti-Trotskyite Deborinites: one of his collaborators in the mechanist faction was a leading Left deviationist in the political struggles.[37] But the present writer feels, as has already been indicated, that Stepanov's main reason for refraining from intruding political factionalism in the philosophical controversy was that most of his opponents in philosophy were his supporters in the political wars within the Bolshevik Party. Moreover, not all the leading mechanists approved of Bukharin's views on Marxist philosophy; in the period before 1924 the future mechanist Sarab'ianov, as well as the future Deborinite Gonikman, had exchanged polemics with Bukharin regarding the philosophy expressed in the latter's *Theory of Historical Materialism*.[38]

One can only speculate about the reasons for Bukharin's disregard of the philosophical controversy and the controversy's disregard of Bukharin. The assumption of a sort of gentleman's agreement to keep intra-Party politics out of the philosophical controversy does not explain why Bukharin did not, like Stepanov, put aside his role of Party leader for occasional participation in the philosophical controversy. Perhaps he was simply too busy with other matters. Perhaps he was sulking. In 1923 he publicly expressed his anger at the editors of *Under the Banner of Marxism* for sheltering an alleged enemy of the Party, who had criticized Bukharin's philosophy. Bukharin announced his intention of having nothing more to do with *Under the Banner of Marxism*, which was the chief journal of Soviet Marxist philosophy.[39] Or perhaps the philosophers considered Bukharin a popularizer of historical materialism rather than a significant figure in the elaboration of dialectical materialism.[40] Whatever the causes may have been, the fact of mutual avoidance between 1924 and 1929, with occasional nods of polite respect towards Bukharin from philosophers of both blocs, is undeniable.

Other major figures in the Party's political factions also remained aloof from the clash of mechanists and Deborinites. Zinoviev, for example, in 1925 celebrated the publication of some of Lenin's philosophical notes by a leading article in *Bol'shevik* that was silent concerning the philosophical controversy, in spite of the fact that Lenin's notes were an object of special contention in it.[41] Nor did Trotsky, the chief leader of the Left deviation, take a position in the philosophical controversy. In 1925, while exhorting a meeting of chemists to support the Soviet régime, he expressed views on the philosophy of natural science that were strikingly similar to those of the mechanists, but without openly commenting on the philosophical controversy.[42] This speech was probably the basis for the abortive Deborinite effort in 1927 to connect the mechanist philosophers with Trotskyism. It was unfortunate for the Deborinites that they did not persist—or were forbidden to persist—in this effort, for they were subsequently charged with surreptitious Trotskyism because of their failure to make an extended critique of Trotsky as a philosopher. It is a fact that the Deborinites were largely silent concerning Trotsky's philosophy, but, viewed in conjunction with their similar failure to critize Bukharin, with the avoidance of such subjects by the mechanists no less than the Deborinites, and with the abstention of Bukharin, Trotsky, and Zinoviev from participation in the philosophical controversy, the fact of Deborinite silence is evidence, not of surreptitious Trotskyism among the

Deborinites, but rather of a common inclination to keep the discussions concerning the philosophy of science separate from the factional quarrels of the Bolshevik Party. Indeed, as shown by the case of the three Deborinites who charged mechanism with being a correlate of Trotskyism, the Deborinites tended to be somewhat closer than the mechanists or the Party chieftains to the mixture of Party politics and the philosophy of science that would be characteristic of the 'thirties.

Another indication of the tendency during NEP to keep the philosophy of science separate from Party politics was Stalin's position in the philosophical controversy. For his part Stalin avoided the controversy no less than Bukharin, Zinoviev, and Trotsky. Stalin's obituary of Stepanov, for example, praised him unreservedly as a Party leader, but said nothing of his controversial philosophical views beyond the vague remark: 'Many thousands of comrades know him as one of the oldest and most popular writers on Marxism'.[43] For their part the philosophical disputants tended to ignore Stalin completely, which was not difficult, since Stalin had even less to say about the philosophy of science than his fellow Party leaders. There was one notable exception to this rule. In the spring of 1924 Stalin attempted to define the foundations of Leninism in a series of lectures at Sverdlov Communist University. His introductory delimitation of the subject set aside 'world view' [*mirovozzrenie*, *Weltanschauung*] as an area in which Lenin contributed nothing new and distinctive 'to the general treasurehouse of Marxism', and the exposition of Leninism that followed was accordingly concerned only with social and political theory.[44] At one point, however, while stressing the importance that Lenin assigned to theory, Stalin reached to the philosophy of science for an illustrative example, and attributed to Lenin and 'no one else' the accomplishment of the 'very serious task of generalizing, in accordance with materialist philosophy, the most important of that which was given by science during the period from Engels to Lenin. . . .'[45] The awkwardness of the Russian, which the present writer has attempted to preserve in translation, as well as the apparent inconsistency with the original limitation of Lenin's creativeness to social theory, and certain other 'lapses' and 'deviousness', were criticized by G. K. Bammel', a prominent Deborinite, in a review of the literature on Leninism that was published in the latter part of 1924.[46] In the same review Bammel' made clear his agreement with Stalin in questions of political and social theory, but he felt free to read lessons in 'methodology', as he called it, not only to Stalin, but also to Bukharin, and the Zinovievist Safarov. Not even Deborin emerged unscathed from Bammel's critical survey.[47]

This curious incident, let it be stressed, occurred in 1924; no Deborinite attempted a similar criticism of Stalin in the years following, as far as the present writer knows.[48] If there was any significance in this early, isolated occurrence, it lay in the philosophical superciliousness that Bammel' evidently combined with political respectfulness in his attitude towards the anti-Trotskyite Party leaders. Subtler evidences of this Deborinite superciliousness towards politicians and others immersed in less exalted occupations than the study of dialectics will be examined below, but the fact of its existence should be noted here, for it was another manifestation of the widespread tendency during NEP to regard Marxist philosophers of science as somewhat autonomous specialists. As far as the public record reveals, Bammel' was not rebuked for presumptuousness;[49] it was not until December, 1929, that he and the other philosophers of science heard from Stalin. They heard—or rather, some of them heard, for many refused to believe that Stalin was addressing them—a complaint of the 'lag of theory behind practice', a complaint of the separation between theoretical work and the Party's practical political tasks.[50] That complaint helped inaugurate Stalin's 'great break' on 'the philosophical front', a time when the 'Bolshevization of philosophy' became the order of the day, and the Deborinite leadership defended the autonomy of philosophers until the Central Committee issued a decree against them. But Stalin's complaint also shed considerable light on the period that it brought to an end. It was a final piece of evidence that until 1929 neither the philosophers of science nor the Party leaders had made significant efforts to connect the discussions of Marxist philosophy of science with the struggle against political deviations within the Bolshevik Party.

The chiefs of Party and state had a 'political' interest in the discussions of the Marxist philosophy of science, but it stemmed from generally accepted policies that were not at issue in the Party's factional strife. These policies will be discussed below, and further attention will be given to the problem of ideological affinities between the philosophers of science and those political leaders who were highly regarded as theorists. For the present chapter was not intended to deny *any* validity to the affinities that have been retrospectively established between the philosophical and political factions of the 'twenties; the purpose has been to show that such affinities do not prove the existence of consciously formed connections between the two fields. A comprehensive intellectual history of Soviet Russia during the 'twenties might well perceive in the contemporaneous arguments of politicians and philosophers of science a pattern of analogous responses to vaguely similar perplexities. Both groups were, after all, struggling to apply a commonly held theoretical heritage to different, but equally refractory areas of Russian life. At the centre of the politicians' arguments was the problem of carrying peasants along in the drive to industrialize and collectivize Russia. At the centre of the philosophers' arguments was the problem of winning scientists to support the Bolshevik ideology that inspired the drive. The mechanist philosopher and the Rightist politician, each in his own area, urged caution, gradualism, coaxing the refractory element, while the Deborinite philosopher and the Trotskyite politician urged boldness, haste, compelling the refractory element. But these affinities are great oversimplifications, apparent only in retrospect. The political and philosophical controversies of the 'twenties may well have proceeded in the same general direction, but they did so along largely separate paths that did not merge until the end of 1929.

4

THE CULTURAL REVOLUTION AND 'BOURGEOIS' SCIENTISTS

IN Soviet Marxist usage 'culture' is 'the aggregate of human achievements in the subjection of nature, in technology, education, social structure'.[1] Since it progresses through stages common to all humanity, the wooden plough, wife-beating, illiteracy, and religion, to take only a few examples, are now signs of cultural backwardness or simply 'non-culturedness' (*nekul'turnost'*). But the wooden plough determines the rest; the chief sign of non-culturedness is the absence of modern technology and science. With this usage in mind it is possible to understand why Lenin warned, as his government beat down its armed opponents, that non-culturedness was the main obstacle to the building of socialism; for the new régime's supporters were overwhelmingly non-cultured, while the minority of Russians who had culture were largely opponents of the new régime.[2] It was against this condition that the Bolsheviks planned their 'Cultural Revolution'. To make cultured people out of revolutionary workmen and peasants, to create a new and numerous generation of professional people and white-collar workers (the Russian word *intelligentsiia* now includes both types) who would be Bolshevik in outlook, that was the goal of the Cultural Revolution.

It would be a mistake to think of this Cultural Revolution as a process beginning in 1917 and continuing in an undeviating line towards its predetermined goal, regardless of the convolutions of Bolshevik policy in the concomitant political and economic revolutions. The Soviet authorities adjusted their cultural policies not only to changing realities within the field of culture, but also to the changing requirements of economics and politics. While a full examination of these relationships is beyond the limits of the present work, some peculiarities of the Cultural Revolution must be considered, for they had a profound influence on the contemporaneous discussions of the philosophy of natural science.

Perhaps the most important peculiarity of the Cultural Revolution from 1917 to 1929 was a fundamental ambivalence in it: concessions to the 'bourgeois' specialist accompanied a determined drive against him. The reasons for this ambivalence are not hard to find. Most professional people and white-collar workers were hostile to Bolshevism (and the Bolsheviks were convinced that the more skilled were the more hostile), but their skills were essential to the new régime. It was even possible to predict a vicious circle. How was a new generation of 'red specialists' to be trained by teachers longing, as Lenin put it, for 'an orderly bourgeois republic'?[3] Or, assuming that proletarians and peasants were the most likely raw material out of which 'red specialists' could be fashioned, how were they to be given higher learning when most were not properly equipped to enter even secondary schools? Pavlov, who was neither a Bolshevik nor a Marxist, was free to consider the Cultural Revolution an 'ineluctable contradiction'.[4] M. N. Pokrovsky, the Bolshevik historian who was probably the leading figure in the transformation of higher education, described the anomalies of the Cultural Revolution more concretely and less pessimistically, when reviewing the first ten years of Soviet education:

> . . . in reality we were faced with two dangers; on the one hand, the danger of remaining in the old rut, since we had a certain fear of too abrupt and decisive demolition, and as a result we could be held prisoner by bourgeois specialists. . . . On the other hand, the danger consisted in the fact that there were some comrades who said: 'All bourgeois education is worth absolutely nothing. It is necessary to throw all this out and to begin anew.'[5]

Pokrovsky believed that a middle course was possible and desirable, and was in the main being followed.

Whether it should be called a middle course or an ambivalent one, it was clearly in evidence from the founding of the Soviet state until 1929, in spite of occasional shifts that portended the end of all concessions to the 'bourgeois' specialist. For example, in 1918 the Council of People's Commissars decreed that all citizens, 16 years and over, had the right to a higher education without regard to sex, nationality, or social origin. No special favours were given to workmen and peasants, other than the order that universities should establish 'workmen's faculties' (*rabfaki* in the Soviet abbreviation) to give workmen and peasants the necessary academic preparation for higher education.[6] But by 1921, when the Civil War was ending, the *rabfaki* were supplying only a small fraction of the students in higher education, and within the *rabfaki* themselves nearly a fourth of the students were of non-proletarian, non-peasant origin. The government thereupon introduced a system of class preference in admissions to institutions of higher education, and purged the *rabfaki*, expelling students who were neither peasant nor proletarian in origin until they declined to 15 per cent of all the students in the *rabfaki*.[7] But drastic measures were exceptional before 1929. In admissions to higher educational institutions class preference was applied, as the law required, but not so vigorously as to effect a rapid transformation. By the academic year 1928–1929 students of proletarian and peasant origin, taken together, were just under 50 per cent of the total student body in non-Communist institutions of higher education. Among graduate students they were still a small minority, on faculties and research staff a tiny one.[8]

Bolshevik efforts to win control of the faculties in higher educational institutions followed an analogous pattern. In 1918, conferences of faculty members refused to adopt Bolshevik proposals that would have given the government effective control of them.[9] In 1921, with the Civil War ended in its favour, the government decreed what it could not accomplish by persuasion; even the right to select graduate assistants was subjected to control by Party and State. The immediate result was actual strikes by the faculties of some leading universities early in 1922.[10] Though the government won its point in principle (the strikes were ended without the 'Statute Concerning Higher Schools' being revoked or revised), some provisions were not enforced until 1929. In the interim, the degree of enforcement, as the extent of Bolshevization in general, varied from one type of institution to another: from the non-Communist to the Communist, with a mixed type in between.

Research institutes, scholarly societies, and university departments in the field of natural science were allowed to remain in the control of non-Communists and non-Marxists—'bourgeois specialists', as they were called in the Soviet press. Until 1929 ideological pressure was exerted upon them only gradually and indirectly, largely through the slowly growing number of Bolshevik-minded graduate students. Through a number of agencies (such as the Union of Scientific Workers, TsEKUBU [the Central Commission for Improving the Condition of Scholars], VARNITSO [the All-Union Association of Workers of Science and Technology for Assistance to the Construction of Socialism], and the societies of materialist natural scientists that were created by the Communist Academy), an effort was made to win the 'bourgeois specialist' in natural science to political sympathy with the régime, and, if possible, to ideological agreement with Marxism.[11] But as late as 1928 natural scientists were assured by Lunacharsky himself, the Commissar of Education, that it was their 'legitimate right' not to be Marxists.[12] Indeed, Kalinin, the President of the Soviet Union's Central Executive Committee, speaking to a doctors' conference in December, 1925, implied a cheerful indifference even to the *political* affinities of the 'bourgeois specialist':

> Communism is being created in the provinces by the man who says: 'I am against Communism, decidedly against Communism.' But by his work that very man is helping the Communist cause, for at present the Communist cause signifies in medicine, for example, raising the people's health. . . . It is precisely in this that the strength of Communism lies: under the Soviet régime Communist work is being performed essentially by everyone who is working honourably, conscientiously, in his own field.[13]

Lunacharsky, to be sure, followed his reassurance to natural scientists with the assertion that

> The victory of the proletariat in this matter [the conversion of scientists to Marxism] is inevitable, for dialectical materialism flows

with a perfect conformity to law [*zakonomernost'*] from the whole structure of contemporary science, considered in its objective and healthy parts.[14]

And Kalinin's assurance that honourable and conscientious work in one's particular field allowed the natural scientist to have what thoughts he chose, could not have been entirely comforting, in view of such a clear announcement as was made by the Twelfth Party Conference of August, 1922:

> . . . the use of repressions cannot be renounced, not only in relation to S-R's and Mensheviks, but also in relation to the intriguing upper strata of the pseudo-non-Party, bourgeois-democratic intelligentsia who, for their counter-revolutionary purposes, abuse the fundamental interests of entire bodies [literally: corporations], and for whom the true interests of science, technology, pedagogy, co-operation, etc., are only a hollow phrase, a political screen.[15]

Nevertheless the evidence indicates pretty clearly that until 1929 repressive measures were not the rule for the government's policy towards natural scientists and their institutions. The voice of a Russian Marat was occasionally heard demanding the closing of the Academy of Sciences, the central research organization largely for natural science, which enjoyed autonomy even to the extent of having the only non-Communist press exempted from prior censorship. But until 1928 the Soviet government financed the Academy without infringing its unique autonomy.[16] Indeed, as the reader will see below, the Academy's Vice-President felt free, in 1923, to lecture the Bolshevik authorities on the necessity of untrammelled academic freedom.

Explicitly Communist institutions of higher learning were a second, and entirely different type. Originally they were centres of teaching and research exclusively in the field of politics and social science. Thus the nominal centre of Marxist research was called the 'Socialist Academy of Social Sciences' when its existence was proclaimed in 1918;[17] the 'Institute of Red Professorship' had no departments of the natural sciences when it was founded in 1921 to train 'red professors' for higher education;[18] and the three-step network of 'Soviet Party Schools', ranging from short-term indoctrination centres for Party agitators to the so-called 'Communist Universities', was charged with the 'training of Party workers' for leading positions at all levels within the government, trade unions, co-operatives, and, of course, the Party itself.[19] From 1921, however, as the Communist Universities began to establish a three-year programme of genuine higher education, courses in mathematics and natural science came to be offered, and some Marxist educators got ideas of training natural scientists and engineers in the Communist Universities as well as 'social and political workers [*deiateli*]', a standard Soviet phrase for public functionaries.[20] In 1923, moreover, the Socialist (soon to be Communist) Academy was directed to go 'beyond the limits of the social sciences' in its research and propaganda, and shortly thereafter the Institute of Red Professorship began courses in the history and philosophy of natural science.[21]

But teaching and research in the natural sciences remained a distinctly subordinate function of these Communist institutions. In 1924, a resolution of the Thirteenth Party Congress reminded the Communist Universities that 'the tasks of specialization', i.e., the training of specialists in technology and natural science, were to be fulfilled by the non-Communist institutions of higher learning.[22] And throughout the 'twenties the Communist Academy had little more than a single laboratory, a place of Pavlovian research, derided by the head of the Marx-Engels Institute, D. B. Riazanov, with a punning epithet (*sobachii institut*) that can be translated only approximately ('bitchy institute' perhaps), for it refers both to the dogs used in the research and to the poor state of the research.[23] A fuller examination of the controversial position of natural science in the Communist institutions of higher learning will come later; here the only point is that their main functions throughout the period under review were the training of public functionaries and the development and propagation of Marxism as a group of social sciences and a general *Weltanschauung*, which did not entail anything like comprehensive and detailed work in the specific disciplines of natural science. Such work was left to the institutions of higher learning where 'bourgeois specialists' predominated.

The third type of institution of higher learning may be described as the mixed or joint type: Communists and non-Communists, Marxists and non-Marxists collaborated—and conflicted—within institutions of this type. In a sense, all the non-Communist universities qualified for this description, if one considers the departments of the humanities and the social sciences, for a 'Commission For the Radical Reform of the Teaching of Social Sciences in the Higher School' began in 1921 a ripping and patching operation that extended throughout the period under review.[24] Courses in philosophy, for example, appear to have been abolished and later re-introduced in altered form, with both Communist and non-Communist teachers, including even a few survivors from the pre-revolutionary era.[25] As late as 1929 there was a report from Bielorussia of 'cases that give rise to the necessity of the liquidation of university courses that are anti-Marxist, hostile to Marxism':

> We still have departments where Marxism is applied by 'automatic Marxists' in very moderate doses, as an 'unavoidable evil,' where it is resorted to as a phraseology that commits one to nothing.[26]

For present purposes the most important instance of the mixed institution was the 'Russian Association of Scientific-Research Institutes of the Social Sciences'. Within RANION, as it was called for short, were about fifteen institutes doing research and training graduate specialists in such fields as history, economics, experimental psychology, and the philosophy of science.[27] The degree of Communist or even Marxist influence varied, as V. M. Friche, the famous Communist art critic who was in charge of RANION, reported to a conference in March, 1928:

> In proportion with the growth of the cadre of graduate students—and here we have about 40 per cent Communists—grow the pressure and thrust from below of the Marxist graduate students on the non-Party, research part [of the staff], hence there is conflict and difficulty in the work. When we have insufficient Marxist leadership in an institute, the graduate students 'lay it on' us, the press lays it on, and so on; but when we put pressure on the non-Marxists, then in the foreign press appears an appropriate article, to the effect that we are strangling science. And thus an extraordinarily difficult situation is obtained.[28]

When D. B. Riazanov, whose contempt for the Communist Academy's venture into natural science has already been noted, twitted Friche on the tardiness of Marxist ascendancy in RANION, Friche was obviously nettled:

It is, of course, very easy for Comrade Riazanov to talk, for, clearly, the Institute of Karl Marx and Engels [headed by Riazanov] consists of Communists and Marxists. (Exclamation from the floor: 'Not only Marxists!') There, of course, the matter is different, but just you try and work where there are not only a few Marxists, but where the majority consists of non-Marxist *spetsy* [colloquial abbreviation for specialists, scholars].[29]

Similar dilemmas were, presumably, the cause of the uncertain status of the Institute of Scientific [*Nauchnoi*] Philosophy. In origin it appears to have been established under the Commissariat of Education's Academic Centre as a sort of special commission, composed largely of non-Party specialists, studying the use of statistics in various sciences.[30] Subsequently it undertook a programme of graduate training and became a member of RANION, when that Association was founded in 1923. But that same year growing Bolshevik concern over the ideological aspects of natural science caused the Institute to be shifted from the control of the Commissariat of Education to that of the Socialist (soon to be Communist) Academy.[31] The Institute seems to have shifted its interests too, from such problems as statistical methods in biology to such as Machism, or Henri Bergson's philosophy of science. In the process it became exclusively Marxist, as Friche boasted in March, 1928, but still it was spoken of as a member of RANION, and not merely as another part of the Communist Academy.[32] This hesitation to cut off the Institute entirely from the mixed or joint type of institution may be interpreted as evidence both of the slow-dying Communist diffidence in the field of higher learning, and of the lingering hope that non-Marxist specialists might be converted to Marxism by working in the Institute.

The Institute of Scientific Philosophy was, indeed, only one example of the Communist authorities' ambivalent feeling that they needed but could not rely on the 'bourgeois specialist'. Even in such fields as economics and history there was a duplication of institutes within RANION, on the one hand, and the Communist Academy, on the other, in spite of the Communists' continual boasts that none but Marxists had anything constructive to offer in these fields.[33] Pokrovsky repeatedly stated, as he participated in the suppression of 'bourgeois' historiography, that he welcomed 'bourgeois' scholarship as a necessary spur to Marxist thought; and he probably was not consciously hypocritical when he did so.[34] Even in 1929, when the Communists were launching a drastically new policy in their relations with the non-Marxist scholar, and Pokrovsky announced that the necessity of relying on non-Marxists and 'fellow-travellers' in higher learning was coming to an end, he added the qualification, 'except, of course, for natural science'.[35]

A similar ambivalence can be discerned in the Soviet government's policy towards publishing during the period to 1929. In a decree of December 12, 1921, the State Press, known as *Gosizdat* or *Giz*, was given life-or-death power over private publishers by its authority to license them and to exercise prior censorship over their products. But this power, soon transferred to a special body called *Glavlit*, was not used to extinguish private publishing until the time of the First Five Year Plan.[36] Furthermore *Giz* itself was far from being a Bolshevik monolith. From 1922 to 1930 its 'Scientific Section' was headed by the non-Party mathematician, V. F. Kagan, who was then only vaguely sympathetic to Marxism.[37] Occasionally *Giz*, or some other governmental press, published such plainly non-Marxist works as the two books by the phenomenologist G. G. Shpet that appeared in 1927.[38] Within the field of the philosophy of natural science, non-Marxist books and articles appeared throughout the 'twenties, although in steadily declining numbers. They might be native Russian works, as, for example, the mathematician A. V. Vasil'ev's popularization of recent physics interpreted according to Eddington;[39] or they might be translations of foreign works. Sometimes the latter suffered unannounced deletions. For example, in 1929 the Academy of Science's journal of popular science, *Priroda* [Nature], quietly excised some of Jeans' patently religious thoughts in a translation of his article, 'The Physics of the Universe'. But even so the reader of *Priroda* could have had little doubt concerning Jeans' fundamentally non-Marxist metaphysics.[40] In short, the Bolsheviks, having placed throttling fingers on the publication of non-Marxist philosophy of science in 1921, choked it only gradually, with the result that

Soviet Marxists faced overt competition in this field throughout the 'twenties.

It would hardly be appropriate to attempt a full description of the currents of non-Marxist philosophy that circulated among Russia's 'bourgeois' specialists. Nevertheless, one needs a more vivid sense of them than a simple assertion of their existence can supply, for it was largely in reaction to them that intensive Marxist discussion of the philosophy of natural science took shape. For present purposes, non-Marxist currents of thought may be grouped in two categories: those that struck Soviet Marxists as basically antithetical to their own philosophy, and those that appealed to them as support for their own philosophy. It must be borne in mind that the distinction between the two types could be made easily only in extreme cases. When, for example, the famous old biologist K. A. Timiriazev startled the scholarly world and 'quite enraptured' Lenin by announcing his fervent support of the Soviet régime, it became an almost unquestioned assumption among Soviet Marxists that Timiriazev's views on the philosophy of science were basically akin to Marxism.[41] (The same *non sequitur* could not be applied to Pavlov before his conversion to Soviet patriotism in the 'thirties; until then he openly scorned the Bolsheviks, though many of them admired his school of psychology.) At the opposite extreme were philosophers of the so-called Russian school, who occasionally ventured into the philosophy of natural science. When Father Florenskii, for example, argued that Einstein's theory of relativity proved the existence of heaven and hell, Soviet Marxists could hardly have considered Florenskii's argument a support for their own philosophy, though some did use it as alleged evidence that Einstein's theory was inherently anti-materialistic.[42] But between the apparently clear-cut cases of affinity and incompatibility with Marxism lay a variety of doubtful cases, provoking the Marxists to a disconcerted, anxious hostility.

Consider, for example, the case of Orest Danilovich Khvol'son, or Chwolson, as he signed his name to publications in West European languages. Already sixty-five when the Bolshevik Revolution came, he had an international reputation not only as a physicist but also as a defender of religious faith.[43] 'The purpose that I have in mind,' he told the students of Petrograd University in the introduction to his course of physics, 'consists in this: . . . I want to forestall an incorrect exaltation of knowledge on your part and a pernicious disparagement of faith.'[44] It is hardly to be wondered at that Lenin, in his *Materialism and Empirio-criticism*, had commented waspishly on Khvol'son, or that Khvol'son was considerably apprehensive in 1918, as he approached the new rulers of Russia to learn the fate of some scientific projects that had enjoyed governmental support before the Revolution.[45] He found that the Bolsheviks, faced by the hostility of the *intelligentsiia* in its majority, were overjoyed at his willingness to work with them; in 1926 they gave him the title of 'Hero of Labour' and decorated him with the 'Labourer's Red Banner'. What may well have touched Khvol'son more deeply was the new government's willingness to spend considerably more on physics, both in research and education, than its predecessors had.

But what were Soviet Marxist philosophers to make of Khvol'son, who had a special genius for popularizations and textbooks? In a popularization of the theory of relativity that he published in 1922, he showed clearly his continuing fideism:

> What right do we have to suppose that everything that exists or takes place in the universe must unfailingly be accessible to our clear perception, must be situated within the horizon of our understanding and comprehension? And is not such a proposition a dangerous relic, an unwitting echo of that arrogance, that mania of greatness that grew out of the idea that man is the centre of the universe, the tsar of nature, to whose all-encompassing reason everything that exists and takes place *must* be accessible? Is it not more seemly for us to take a more *modest* position ('there are more things, Horatio . . .'), granting the possibility that there are limits to our reason, our perceptions, and even our fantasy, which it is not given to us to transgress?[46]

When Soviet Marxists objected to this explicit rejection of their Promethean notion of science, Khvol'son removed the passage from a later edition of his book, and ceased his overt campaign on behalf of fideism.[47] But he did not espouse materialism, and Marxist philosophers continued to find 'idealist errors' in his publications. Nevertheless, one critic conceded in 1927,

Until the materialists produce significant scientific forces, it will be necessary to make peace with such a *Course of Physics* as Professor Khvol'son's, and, while giving credit to all that is scientifically worthy in it, unceasingly to underline and expose what is scientifically harmful.[48]

The only trouble was that Soviet Marxists could not agree on what was 'worthy' and what 'harmful' in the chastened Khvol'son. For example, the critic just cited (Z. A. Tseitlin) thought that Khvol'son's denial of universal ether was a rejection of materialism, but another Soviet Marxist (S. Iu. Semkovskii), who had scoffed at the concept of ether in his own book on the theory of relativity, proudly cited Khvol'son's praise as proof of the book's merit.[49]

Philosophical interpretations of the theory of relativity were by no means the only reason for the mixed awe and anxiety that Soviet Marxists felt towards natural science and its adepts who had a philosophical bent. V. I. Vernadskii, an outstanding geo-chemist with interests in a wide range of subjects including the history and philosophy of natural science, produced a similar effect among the Marxists. His essay, 'On the Scientific *Weltanschauung*', had won him Lenin's praise in *Materialism and Empirio-criticism*, which caused some Marxists to consider him an unwitting dialectical materialist.[50] Others were disturbed by his respect for religion and idealist philosophy, his allegedly vitalist views on the origin of life, and his steadfast unwillingness even to consider Marxism as a philosophy of science.[51] On the political side he had been a *zemstvo* liberal, and had joined the mass resignation from Moscow University in 1911 as a protest against governmental abridgment of the University's autonomy. He spent the Civil War with other anti-Bolshevik scientists in the Crimea, but he did not go into emigration when that last stronghold of the Whites fell. He returned to the Academy of Sciences in Petrograd and resumed his leading role in work of capital importance to the Soviet régime, such as the inventory of the country's natural resources, and the founding of new scientific institutes.[52] He shared the Bolsheviks' desire for a great flowering of science and technology under the sponsorship of 'the state, [which] must give funds, call to life scientific organizations, set tasks for us. But,' he added, we must always remember and know that its intervention in creative scientific work cannot go further than this. Science, like religion, philosophy, or art, is a spiritual field of human creativity, fundamentally more powerful and more profound, more eternal than all the social forms of human life. It suffices unto itself. It is free and does not tolerate any limitations.[53]

Vernadskii was not the only eminent scientist to set such terms on his co-operation with the Soviet régime's promotion of science. When V. A. Steklov, the famous mathematician and Vice-President of the Academy of Sciences, published a brief, popular biography of Galileo in 1923, there could have been little doubt concerning his basic motive: he was reading the Bolsheviks an historical lesson in the philosophy, or perhaps one should say the sociology, of science. The 'party' that controlled the Catholic Church in the seventeenth century, he explained, tried to subject science to their 'party discipline'. 'In social affairs, in politics.' Steklov conceded, 'this party discipline is unavoidable for the achievement of specific action at a given, specific moment, but to science and to scientists it is inapplicable.'[54] Indeed he hammered the last point with considerable force, which is unfortunately weakened in translation by our rejection of the double negative:

. . . The free mind of the exact researcher and thinker can never be subordinated to any previously defined and forever immobile slogans of any party. There has not been, is not, and will not be such a force as would oblige that mind to subordinate itself to this requirement. . . .

Demonic science [a reference to a monk's description of mathematics as a satanic science] must not be trifled with in any respect; it and its servants should be allowed full freedom in all respects, and the whole world should pay close attention to its discoveries of genius, which fly out over the world from century to century, above all the conditions of the given historical moment.[55]

One Soviet Marxist disagreed sharply with this conception of the relationship between science and society, and sneered at Steklov's 'homily for politicians'.[56] Other Soviet Marxists occasionally spoke of 'the dictatorship of Marxism' as already existing in the 'twenties, with the apparent implication that only transient exigencies of the Cultural Revolution explained the freedom of scientists to express non-Marxist views.[57] But still other Marxists showed signs of fearing that Steklov's view might be, to some extent at least, justified. Friche's sensitivity to complaints that the Bolsheviks were 'strangling science' has already been cited, as has Pokrovsky's view that Marxist scholarship needed the competition of non-Marxist scholarship. In a later chapter Trotsky will be found to have expressed a view similar to Steklov's, and further on the reader will find evidence of continued division among Soviet Marxists even in the time of the First Five Year Plan, as the 'dictatorship of Marxism' emerged more and more clearly from the vanishing ambiguities of the 'twenties. If extensive arguments concerning the position of non-Marxist philosophies of science in Soviet Russia are not to be found—and they are not—the reasons are complex, and consideration of them may be postponed. But it should already be clear that even on this critical question of the relationship between science and society, between scientists and politicians, the reaction of Soviet Marxists to the non-Marxist views that were current during NEP was not simple hostility, but complex anxiety.

5

THE CULTURAL REVOLUTION AND MARXIST PHILOSOPHERS

SOVIET Marxists responded to the challenge of non-Marxist interpretations of science not only on a level that may properly be called philosophical discussion, but also on the level of mass propaganda, as part of the drive against religion. Even though the very word, 'philosophy', was generally shunned on this level, the anti-religious campaign exerted a considerable influence on Soviet Marxist philosophy of natural science, and a brief excursion into the history of Soviet atheism is accordingly unavoidable.

People who were not only atheist in thought but joined atheist organizations and participated actively in the 'struggle against religion' were known in Soviet Russia of the 'twenties as *antireligiozniki*, which had best be translated literally as 'anti-religionists', so that the cumbersome neologism may suggest in English as in Russian zealous participation in a new cult. There were astute intellectuals in the movement, but they appear to have been a distinct minority; complaints of the low intellectual level of the average anti-religionist and his propaganda were constant in the anti-religious press throughout the 'twenties.[1] One gets the impression, moreover, that intellectual crudity was often joined with a streak of wildness, an obstreperous eagerness to play the gadfly on religious 'Philistines' (*meshchane*, *obyvateli*). Consider, for example, the anti-religionists' use of the defiantly ambiguous term *bezbozhnik* rather than the precise and relatively decorous *ateist* (atheist). *Bezboz-*

hnik may be rendered literally as 'godless', but this translation does not convey the pejorative sense of the Russian word, which Lenin himself used as invective, indicating shameless unscrupulousness.[2] By calling themselves *bezbozhniki* the anti-religionists were gratuitously 'giving a fig', or thumbing their noses as we should say, to their religious countrymen. One is accordingly not surprised to find *The Anti-religionist*, the theoretical journal of the 'League of Bezbozhniki', warning anti-religious speakers that they did the cause harm by railing at their audiences when questions and comments indicated resistance to the atheist creed.[3]

Another noteworthy feature of the anti-religious campaign was the widespread feeling among anti-religionists that the attack should be limited to political and social aspects of religion, and that anti-religious propaganda among peasants and workmen should not deal with cosmological questions. This 'practicality' was fairly widespread, but it was officially censured.[4] 'The peasant,' wrote V. N. Sarab'ianov, who was a leading figure in philosophical discussions as well as the anti-religious movement,

> also poses questions broadly, methodologically, if you please, and not only narrowly, practically: concerning god [*sic*] and the world, and not only concerning the priest and the peasant's own household. If one succeeds in demonstrating to the peasant the law of the conservation of matter and what follows from this law, in the long run he, even on his own, will see the connection with the question of the world's creation and its end.'[5]

In short, propaganda of the rationalist as well as the anti-clerical elements in atheism seemed desirable in spite of widespread intellectual poverty and emotional wildness among the anti-religionists. It was this ambitious programme that the Central Committee of the Communist Party sanctioned; directives of September, 1921, called for propaganda that would 'set a foundation of natural science under an historical explanation of the question of religon'.[6] A similar decision was embodied in a resolution adopted by the Twelfth Party Congress in April, 1923:

> . . . Most of the literature published thus far cannot satisfy the mass reader. Pamphlets and leaflets must be published that are accessible

to the average worker or peasant, that will, in a form he can understand, give answers to questions concerning the origin of the world and life and the essence of human relationships, that will expose the counter-revolutionary role of religion and the church, especially the Russian Church. . . . It is necessary to give special attention . . . to the attraction of specialists, natural scientists, materialists to the reading of [popular] lectures. . . . All our anti-religious agitation and propaganda cannot really penetrate the people until we actually set in motion work on the formal education of the labouring masses of town and country in the spirit of scholarly, materialist natural science. . . .[7]

In effect the Party was calling on the anti-religionists to propagate a substitute for religion, a comprehensive vision of the universe and man's place in it that would replace the worship of supernatural beings under the guidance of priests by faith in progress towards an earthly paradise under the direction of Communists.[8] In this atheist *Weltanschauung* natural science was called on to play a leading role. In practice, 'scholarly, materialist natural science' was shaped to suit the peculiarities of the anti-religious movement. The crudity and obstreperousness that were widespread in the movement, and the lingering doubts among anti-religionists concerning the efficacy of cosmological propaganda in any form, were probably the main reasons that the atheist *Weltanschauung* of the 'twenties tended to be a low grade of positivism.[9] For the most part philosophy was not even discussed but simply ignored or summarily dismissed as an attempt 'to light up the darkness with the stars people see when they run their heads against a stone wall'.[10] Nevertheless, some anti-religionists, as the reader will see below, did get involved in explicit discussions of the philosophy of science, bringing their positivist bias with them.

On the level of philosophy, properly speaking, Soviet Marxists were rather slow in taking up the challenge of non-Marxist interpretations of natural science. When *Under The Banner of Marxism*, the chief philosophical journal of Soviet Marxism, first appeared in 1922, its concentration was almost entirely on social and political theory. The editorial manifesto and a salutatory letter from Trotsky said nothing about the philosophy of natural science.[11] Only among the book reviews was there a portent: A. K. Timiriazev, a Bolshevik physicist, who had already begun his compaign against the theory of relativity in another journal, commented hostilely on a Russian translation of Einstein.[12] Otherwise, from the flaming editorial at the front—'We are not trying to be researchers . . . we are fighters . . .'—to the woodcut of rebellious proletarians on the back cover, the stress was on Marxist materialism as a militant political and social philosophy.

Rather circumspectly, in a communication that was published in the second issue of the journal, Lenin advised the editors that without attention to the dialectics of natural science 'Materialism cannot be militant materialism. It will be not so much the smiter as the smitten, to use an expression of Shchedrin's'.[13] This communication of Lenin's, 'On the Significance of Militant Materialism', proved to be of critical importance, not simply because it impelled Soviet Marxists towards work in the philosophy of natural science, but also because it impelled them towards two different and ultimately conflicting tendencies. Or perhaps it would be more accurate to say that Lenin's authoritative words hastened the emergence of tendencies that were already germinating among Soviet Marxists. In either case, the article merits close scrutiny.

To begin with, Lenin commended the editors' declared intention of engaging not only Communists but materialists at large to write for *Under the Banner of Marxism*. He also applauded the editors' proclamation of war on religion, and gave them strategic advice.[14] It was in a separate second part of his article that Lenin noted the journal's neglect of the philosophy of natural science. He suggested that perhaps even more important than an alliance with non-Communist materialist philosophers was

> an alliance with those representatives of modern natural science who incline towards materialism and are not afraid to defend and preach it against the modish philosophical wobbling towards idealism and skepticism that is prevalent in so-called 'educated society'.
>
> The article by A. Timiriazev . . . permits us to hope that the journal will succeed in effecting this second alliance too. Greater attention needs to be paid to it. It needs to be remembered that reactionary philosophical schools and quasi schools, trends and pseudo trends, are being born almost continuously, precisely out of the abrupt breakup that contemporary natural science is experiencing.

For this reason, following the problems that are raised by the recent revolution in the field of natural science, and drawing natural scientists to this work in the philosophical journal, are a task without the performance of which militant materialism cannot possibly be either militant or materialism.[15]

Returning to Timiriazev's article as a model, Lenin gave the misleading impression that Timiriazev had expressed hostility not to Einstein's physics but only to the philosophical interpretations that the 'bourgeois intelligentsia' put upon the theory of relativity as upon most of the great innovations in recent natural science.[16] Thus, Lenin's demand for an alliance with natural scientists 'who incline towards materialism' could have been interpreted as having little or no substantive bearing on natural science itself, and requiring little or no specifically dialectical materialism.

But Lenin went on to argue the necessity of 'a solid philosophical ground' both for the natural sciences and for materialism if they were to endure 'the struggle against the onslaught of bourgeois ideas, against the re-establishment of the bourgeois world outlook'. And the 'solid philosophical ground', Lenin specified, must be Hegelian dialectics materialistically re-interpreted:

The group of editors and contributors of the magazine *Under the Banner of Marxism* should, in my opinion, be a kind of 'Society of Materialist Friends of Hegelian Dialectics'. Modern natural scientists will find (if they know how to seek, and if we learn to help them) in the Hegelian dialectics materialistically interpreted a series of answers to the philosophical problems that are being raised by the revolution in natural science making the intelligentsia who admire bourgeois fashion stumble into reaction.[17]

Lenin probably thought of this programme as complementing the one he outlined first. Natural scientists with materialistic inclinations were not only to be enlisted in the struggle against 'reactionary' philosophical interpretations of scientific theories, but they were also to be taught how to be specifically dialectical materialists. In practice, as the reader will see, the two programmes diverged. Promoters of the anti-'bourgeois', materialist alliance with natural scientists tended to become positivist in their world view and reductionist in their methodology, while the 'Friends of Hegelian Dialectics'—and such a society was actually formed—[18] tended towards a metaphysical philosophy that insisted on irreducible 'qualities' or levels of integration. And both groups could claim to be following Lenin's directives.[19]

It seems strange, in view of Lenin's ambitious programme for Soviet Marxist philosophy of natural science, that the Party's formal directives on the subject were so scanty as to be almost unnoticeable. The Party press, to be sure, carried constant articles warning of the anti-materialist indoctrination that the 'red specialists' of the future were receiving from the 'bourgeois' professors of natural science.[20] But when the Twelfth Party Conference in August, 1922, adopted a long resolution on measures to combat 'bourgeois' ideology, the philosophy of natural science was not explicitly mentioned. Only a faint echo of Lenin's article could be found in one paragraph.[21] The Twelfth Party Congress, which met in April, 1923, was a little less vague in a similar resolution concerning the struggle with 'bourgeois' ideology:

> In closest connection with the necessity of organized counteraction to the influence that bourgeois and revisionist-minded professors have above all on the student youth, the task of enlivening the work of scholarly Communist thought must be advanced to a greater degree than at present, by making the Socialist Academy the centre of this work, and by extending the range of the Academy's activity beyond the boundaries of social science.[22]

This brief, vague statement of the Party's requirements in the philosophy of natural science, and a similarly brief but slightly more explicit statement in the Socialist (or Communist) Academy's Statute,[23] were the only formal, public directives that Communist philosophers of science received throughout the 'twenties. This seems strange, until one recalls the Communists' poverty of men and expert knowledge in the field of natural science and its philosophy. Lenin could project the triumph of his philosophy of natural science as readily as he sketched plans for massive electrification or for a government so simplified and democratic that cooks would be able to manage it, but actual directives for immediate action had to be considerably more modest.

This explanation of the difference between Lenin's plan and the Party's directives is borne out by the evidence of delay in 'extending the range of the [Communist] Academy's activity beyond the boundaries of social science', and by the meagerness of the result when the extension did occur. A plenary meeting of the Academy's members, and a conference between the Academy and the Agitation and Propaganda Department of the Communist Party's Central Committee, decided in 1923 to transfer the Institute of Scientific [*nauchnoi*] Philosophy to the control of the Communist Academy; but, as the reader will recall, considerable irresolution was displayed in carrying out the transfer.[24] Moreover, it was not until 1925 that steps were taken to establish within the Academy a Section devoted specifically to *natural* sciences (*estestvoznanie*), and Otto Iulevich Shmidt, the Bolshevik mathematician and explorer who fathered the new Section, had his eye much more on the political ideology of natural scientists than on their science or its philosophy. 'At the time that I came to this work,' he recalled in 1930,

> the situation with respect to Marxist cadres [of natural scientists] was, of course, many times worse than now, although now too there are dreadfully few. I drew some Communists into the Section, but at the same time I considered it possible at that stage to draw in what appeared to be the left front of non-Party people, to try and organize them, and thereby to intensify the division among natural scientists, and to increase the guiding role of the Communist Academy in this essentially political work.[26]

The division among natural scientists that Shmidt tried to intensify was a division between well- and ill-wishers of the Soviet régime. Merely to get a reputable scientist to work within the Communist Academy's Section of the Natural and Exact Sciences was considered a gain almost regardless of what he did there. The result was at least one scandal (Professor Kostitsyn, who was put in charge of the seminar on the structure of matter, emigrated), some hard feeling on the part of Communist graduate students because the eminent specialists tended to ignore Party affiliation in appointing assistants and filling seminars, and in general the anomalous presence in a Communist Academy of 'perfectly decorous scientific establishments', as Shmidt remarked in 1930, when he helped launch a drive to Bolshevize the proliferations of the original Section.[27] To be sure one can perceive a change creeping over the Section already in 1927. An influx of young scientists focused the discussions more and more on problems of the Marxist philosophy of natural science, and Director Shmidt finally became involved in the controversy between the Deborinites and the mechanists. But even to the end of the controversy most of the papers read at meetings sponsored by the Section (or Institute, as it came to be called) of Natural Science dealt with technical subjects that had little bearing on the philosophical controversy, and indeed displayed little that was explicitly Marxist.[28]

Similar results can be observed in other institutions where Soviet Marxists tried to go 'beyond the boundaries of social science'. The 'State Timiriazev Scientific Research Institute for the Study and Propaganda of Natural Science From the Point of View of Dialectical Materialism' got a considerable number of natural scientists to publish little pamphlets of popular science with a more or less materialist slant.[29] But otherwise it belied the promise of its ponderous title until 1925, when it rallied to the defence of the mechanist faction in the philosophical controversy, and became their chief stronghold, publishing what was in effect the organ of the mechanist faction. In the Ukraine the 'Scientific Research Department of Marxism-Leninism' was part of the Ukrainian Academy of Sciences only 'in a dialectical sense', a Marxist once reported to a conference, for the Ukrainian Academy was actually dominated by 'bourgeois' natural scientists.[30] In 1928, when this report was made, the Department of Marxism-Leninism had begun to harangue—or perhaps one should say hector—the rest of the Academy with dialectics;[31] but in 1924, as the reader will see below, the Ukraine's chief Marxist philosopher, S. Iu. Semkovskii, approached 'bourgeois' scientists with the argument that they already were Marxists without realizing it.

This chief philosopher in the Ukraine had been a Trotskyite and 'Machist' before the Revolution; in 1913 Lenin had denounced him as 'a liquidator of Marx's philosophical materialism'.[32] The fact is noteworthy not as a curiosity but as an illustration of a general rule, which can also be attributed to the peculiarities of the Cultural Revolution in the 'twenties. The chief philosophers in Moscow were A. M. Deborin and Liubov' Aksel'rod the Orthodox, both resembling Semkovskii in their history of pre-revolutionary conflict with the Bolsheviks in philosophy and in politics.[33] Only in 1924 did the Bolsheviks feel strong enough to drop the broad appeal implicit in the description of their highest institution of learning as the *Socialist* Academy; they began to call it specifically the Communist Academy.[34] But this restriction, like the contemporaneous decision to assume responsibility for natural science, was considerably more difficult to accomplish than to announce. As late as 1928 twenty-six of the Academy's fifty-nine 'senior scholarly colleagues [*starshie nauchnye sotrudniki*]' were not Party members, and a good many of the remaining thirty-three were very likely former Mensheviks only recently accepted into the Bolshevik Party.[35] In 1924, when Pokrovsky proposed the substitution of 'Communist' for 'Socialist' in the name of the Academy, he explained, 'Our diapason has been somewhat constricted.'[36] But it was not so constricted that any public notice was taken of Riazanov's characteristically non-Leninist comment in agreeing to the change: 'I am not a Bolshevik, I am not a Menshevik, and not a Leninist. I am only a Marxist, and as a Marxist I am a Communist.'[37] Not until 1931 was Riazanov publicly censured for the views that prompted such a remark.

In part, this toleration of heterodoxy was a product of the destruction of the rival revolutionary parties during the Civil War. Many former Mensheviks and Social-Revolutionaries 'recognized the October Revolution', as the current phrase ran, and offered their services to the Bolshevik régime. At first the Bolsheviks welcomed them, but in 1921, with victory assured, a campaign was launched to put Communists on guard against 'Mensheviks', in Lenin's description, 'who have repainted their "façade" but have remained Mensheviks at heart'.[38] The Central Committee, in a less truculent mood, warned that 'Menshevism is a contagious disease; it is not easy for one to be cured of it, even one who sincerely strives for a cure'.[39] This campaign brought a major purge in the Party, but it had relatively little outward effect on the rapidly growing network of Communist institutions of higher learning, where even a former Cadet (Constitutional Democrat) became a leading philosopher.[40]

Some Communist officials had misgivings, but the extreme shortage of undeviating Leninists who were qualified for advanced research and teaching shoved such misgivings into the background. In 1920 or 1921 Sverdlov Communist University asked the Party's Organizational Bureau whether Deborin and Aksel'rod the Orthodox might be allowed to give courses in philosophy. The 'Orgbureau' replied that Deborin might, but not Aksel'rod, presumably because she had been a 'defencist' during the World War, while Deborin had been an 'internationalist' Menshevik. (The 'defencists' had supported Russia's war effort, while the 'internationalists' had urged an immediate peace, but had not agreed with Lenin's programme of 'turning the imperialist war into a civil war'.) But Sverdlov University, pressed for teachers, would not let the matter drop, and the 'Orgbureau' passed the problem on to Lenin, who replied: '*In my opinion, both, unquestionably*. They are *useful*, for they will defend Marxism (if they begin to agitate for Menshevism, we shall catch them; [they] *need to be supervised*).'[41] Such philosophers were accordingly in a position strangely similar to that of the 'bourgeois' specialists; in the eyes of some Bolshevik officials they were a possibly dangerous makeshift to get the Cultural Revolution under way.

It would be wrong to exaggerate this distrustful attitude towards heterodox Marxists, as many have done, reading into the history of the 'twenties the tone of Soviet intellectual life since 1930. Aside from the shortage of qualified personnel there was another reason for the toleration of heterodoxy in the 'twenties: the carry-over from pre-revolutionary times of Lenin's broad understanding of the partyness of philosophy. The very note that told the Organization Bureau to keep an eye on Deborin and Aksel'rod, lest they revive their Menshevik politics, took it for granted that they would defend Marxist philosophy. Apparently Lenin still did not think that the Bolsheviks had a monopoly of orthodox philosophy. Plekhanov died in 1918, an unregenerate Menshevik and opponent of the Soviet régime, but Lenin, while criticizing the Bolshevik Bukharin's understanding of dialectics, advised young Communists to study 'everything Plekhanov wrote on philosophy, for it is the best in all the international literature of Marxism'.[42] Moreover, Lenin's toleration of philosophical heterodoxy within the Bolshevik Party survived the Revolution. Such former *Vperedists* as Lunacharsky, Pokrovsky, Bazarov, and Liadov were given high posts in ideologically sensitive fields without disclaimers of 'Machism' being required. Indeed, Lunacharsky, the Commissar of Education, gave printed evidence of his continuing heterodoxy.[43] And when (in 1918) Bogdanov organized the First All-Russian Congress of Proletarian Culture, Lenin did not repeat his pre-revolutionary denunciations of the concept and the organization as a screen for 'Machism', but politely returned the best wishes that the Congress sent to him.[44]

Yet the paradoxical fact is that in 1920, when Proletcult, as Bogdanov's organization was called for short, gathered for its Second Congress, Lenin invoked the authority of the Party's Central Committee to destroy the organization. Very likely the success of Proletcult was its undoing. It had become a mass movement, claiming to have eighty thousand at work in its studios of proletarian art, and disseminating Bogdanov's 'organizational science', as he called the latest phase of his philosophy, to more than half a million workmen outside the studios.[45] Lenin declared 'merciless hostility' to Proletcult in May, 1919, when he discovered its great influence at a conference of Soviet educators.[46] In effect, the pre-revolutionary conflict over Bogdanov's school on Capri was now repeated on a much grander scale; at stake was the right to teach proletarian philosophy not to a handful of *émigrés* but to a mass audience in Russia. A month before the Second Congress of Proletcult Lenin brought out a second edition of *Materialism and Empirio-criticism*, explaining in a new preface that he did not have time to bring it up to date, but that an appendix by V. I. Nevskii did the job for him. Nevskii's essay, 'Dialectical Materialism and the Philosophy of Dead Reaction,' was a reply to Bogdanov's *Philosophy of Living Experience*, which had just appeared in a second edition.[47]

Not content with pitting book against book, Lenin detailed Lunacharsky to tell the Congress of Proletcult that the organization must become a part of the Commissariat of Education. Lunacharsky, who sympathized with Bogdanov's philosophy, turned this into a request for co-operation between the two bodies, whereupon Lenin took the matter to the Party's Central Committee.[48] Pokrovsky, another former collaborator of Bogdanov's, and Bukharin, a young admirer of Bogdanov's philosophy, were sent to tell the Congress that it must adopt a resolution expressing its desire to become a part of the Commissariat of Education.[49] 'After long, hot debates,' the journal of Proletcult tells us, 'the Communist faction, by force of Party discipline, carried the resolution through the Congress.'[50] As the quotation suggests, the report in the journal of Proletcult was angry at its loss of independence, and the Party's Central Committee replied with a letter in *Pravda*, justifying its intervention on the ground that workmen must be protected against 'Machism', and promising Proletcult 'full autonomy in the field of artistic creation'.[51] (In fact, the organization withered away within the next few years.)

Thus a precedent was set for the Central Committee's intervention in philosophical discussion, but it is significant that the precedent was not invoked as such until 1930. In the meantime Bogdanov continued to publish his 'Machist' views in books and in the journal of the Communist Academy, where he was a full member until his death in 1928. He had been kicked upstairs. He could speak to the intellectuals and participate in the controversy over the philosophy of science, as the reader will see below, but he had been denied a mass audience organized independently of the Communist Party.

In general, the highest Party organs kept a fairly light rein during the 'twenties on the institutions where Marxist philosophers worked. The Communist Academy, which was supposed to be the organizing and directing centre in all fields of Soviet Marxist thought, was formally attached not to the Party's Central Committee but to the Central Executive Committee of the Soviet Union.[52] And in practice the high Party officials who were nominal members of the Academy's governing body, its Presidium, rarely attended its meetings.[53] It fell to the historian M. N. Pokrovsky, another Marxist with a history of deviation from Bolshevism, to report the Party's requirements to the Academy. In October, 1923, for example, he reminded a meeting of the Academy that the Party's Twelfth Congress had, six months ago, resolved that the Academy should extend its commitment 'beyond the boundaries of social science'. But,

Pokrovsky added, 'the Congress' decree binds nothing [specific] on us'. Indeed, he was somewhat embarrassed and apologetic:

> And the fact that at the present time we are charged with this task as a Party function, I think that this is pure gain for us; it will be possible for us, at least when utilizing Party forces in this direction, to operate with the methods of Party discipline. I, like many old Party intellectuals, have reacted querulously and discontentedly to external discipline, but I must say that recently I have begun to notice that some things are not feasible without it, so that in the given instance external discipline will be useful to us.[54]

Towards the end of the period under review the autonomy of the Communist Academy began to shrink. In January, 1927, Stalin, Molotov and Rykov were elected to membership, apparently as a gesture of the Academy's political orthodoxy;[55] and in the summer of 1927 the work of the Academy was subjected to a formal review by the Party's Central Committee. Pokrovsky's attitude, when he reported the results of this review to a meeting of the Academy, was significantly different from his attitude in 1923. He hammered the point that the Academy could not be 'a parliament of opinions', that it must adhere to a single line; and even though it was obvious that he had the Party's factional quarrels uppermost in mind, his view of other matters was affected by the new spirit. The same resolution of the Twelfth Party Congress that he had described in 1923 as 'binding nothing on us', he now characterized as stating 'for the first time absolutely officially' that Marxist thought cannot be confined to the social sciences.[56] Nevertheless, as the reader has already seen, it was not until March, 1928, after Stepanov had resumed active membership, that the Communist Academy expelled Trotsky and other prominent deviationists, 'in accordance with considerations of an ideological character'.[57] And, as far as the philosophy of natural science was concerned, to the end of 1929 some philosophers in the Academy could persuade themselves that they were leading rather than following the Party's Central Committee.[58]

It will be necessary in a later chapter to consider at greater length the related problems of the limited autonomy and of the diversity of opinion among Soviet Marxist philosophers of science, for they were problems, even though the philosophers tended to slur them no less than the analogous problem of the limits of toleration for non-Marxist philosophies. The point here is that the necessities of the Cultural Revolution obliged the Central Committee to entrust the development of a Marxist philosophy of natural science to old intellectuals who, as Pokrovsky indicated, would have reacted querulously to externally imposed discipline. Moreover, as the reader will find below, some of these philosophers experienced an internal conflict between their desire to be 'orthodox' and their irrepressible conviction that objectively existing truth is found by individual minds and is no respecter of orthodoxies. The same thinker who ridiculed his critics as 'zealots of the ancient piety', earned from *them* the epithet 'Talmudist' because of the endless quotations that he culled from Marx and the other 'founders' or 'classics' to prove his points.[59] In the absence of a clearly defined programme and a firm guiding hand—and the Party's high command provided neither—controversy could hardly have been avoided.

PART THREE

THE ANOMALOUS REJECTION OF POSITIVISM

6

MECHANISM AS A TENDENCY

MECHANISTIC materialism in various forms was probably the dominant philosophy among Soviet Marxists until the Deborinites challenged it in the mid-'twenties. In part this was a prodigiously anti-philosophical mechanism, an inflamed reaction to the Civil War and the desperate position the Bolsheviks were in even when the war was won. Staggering from one crisis to another, many Bolsheviks developed an impatience with abstract theory, if not an outright contempt for it, a 'revolutionary pragmatism' or 'Soviet Americanism', as it was sometimes called.[1] The obverse was a hunger for panaceas, a search for simple, all-embracing formulas that would on application quickly dissolve complex difficulties and realize the Communist paradise. Exhibiting both sides, 'Enchmenism' and 'Mininism' are the prime specimens of this anti-philosophical, simplistic mechanism.

Emmanuel Enchmen, a former Social Revolutionary, announced his 'evolutionary theory of historical physiology' during the Civil War itself, by the publication in 1920 of *Eighteen Theses Concerning the Theory of the New Biology*.[2] The theory was a travesty of the neuro-physiological school of psychology.[3] What Pavlov, the cautious scientist, only hinted at occasionally,[4] Enchmen proclaimed as a firmly-established, indeed a final truth:

Our understanding of the process of cognition does not in the least coincide with the meaning that has been given and is given to these words in all doctrines except the theory of the new biology: for us the words knowledge and cognition signify only physiological reactions without any participation of a psyche, i.e., without any participation of non-spatial phenomena.[5]

Mind, spirit, psyche—all such concepts had been invented by the exploiters to befuddle the exploited. The end of these concepts, indeed, the end of concepts in general, was now imminent.

World-view is an invention of exploiters. . . . With the arrival of the epoch of the proletarian dictatorship we are against any 'world-view'. We are for the proletarian, the emerging Communist, single system of organic reflexes. . . . The philosophical world-view called dialectical materialism will be reduced to ashes, for thanks to biological training the necessity of logic will disappear; cognition, thought, will wither away, and only the physiological reflex of the guffaw will recall exploitative dialectical materialism, which befuddles the innocent heads of advanced revolutionary workmen.[6]

It would be profitless, though diverting, to follow Enchmen's detailed account of the utopia that would be realized, if only his fellow revolutionaries would utilize the 'fifteen analysers' in the human nervous system. Ultimately there would be a 'world commune', administered by a *Revnauchsovet* (an acronym for 'Revolutionary Scientific Council') issuing 'physiological passports' to people who would no longer possess the written or even the spoken word but only the 'single system of organic reflexes'. But so long as proletarians did read, Enchmen was 'profoundly convinced that in the not distant future all the works on the philosophy of Marxism that are now considered very authoritative will in the saddest fashion gather an ever thicker layer of dust on the shelves of school libraries', while his own *Eighteen Theses* would go through many editions, 'and many times we shall witness workmen studying this pamphlet, turning its pages with their calloused hands"[7]

Enchmen's expectations were cruelly disappointed. After the first and only edition (1920) of the pamphlet just mentioned, he published in 1923 'Issue One' of a projected 'Encyclopedia of the New Biology'.[8] And that was the end. He had, to be sure, acquired a following among the revolutionary youth who were turning from war work and soldiering to higher education.[9] In some Communist universities his followers had formed clubs of '*te-en-bisty*' (T-N-B'ers', for 'Theory of New Biology').[10] But among the people who taught in the Communist universities, edited Communist publications, and headed Communist research and propaganda organizations, Enchmen had, so far as the present research has revealed, not a single supporter. In fact, established authority was mobilized against him. Bukharin was among the leading Party theorists who published polemics against him, and an editorial in *Pravda* took a slap at Enchmenism, giving the impression in a passing reference that only those opposed to the Party's general line could be Enchmenist. [11]

Fortune was only slightly kinder to S. Minin, the revolutionary who used Comte as Enchmen used Pavlov. Minin broke into print not two but three times, and subsequently the Deborinites attemped to discredit their mechanist opponents by calling him the founder of the mechanist bloc; later still this polemical exaggeration was elevated to the status of dogma, and careless scholarship outside the Soviet Union continues to propagate the error.[12] It would be as accurate to count Father Coughlin among the founders of the New Conservatism in America.

There is reason to believe that Minin was arguing his views to sympathetic students at Sverdlov Communist University during 1920 and 1921, while the Civil War was still on.[13] But it was not until the middle of 1922 that he broke through the hostility of Communist editors, first in a provincial journal,[14] then in *Under the Banner of Marxism* itself,[15] the newly established journal of Soviet Marxist philosophy. In a note prefixed to Minin's manifesto, the editors of this journal warned that they were printing it—abridged, they said, owing to a lack of space—only as a symptom of anti-philosophical currents of opinion, and immediately after it they placed a rebuttal more than twice the length of Minin's article.[16] Towards the end of 1922 they allowed Minin space once again for a reply to his many critics.[17] And then silence. For a time he may have nourished hopes of vindication. The leading anti-religionist of Kharkov, and an anonymous author in Tsaritsyn (now Stalingrad), published articles in defence of his views.[18] But by the end of 1923 defences of Minin had disappeared even from provincial journals.[19]

In substance, Minin's theory was not nearly as revolutionary as Enchmen's, but his style made up for that lack. Brief, peremptory paragraphs asserted his simple argument. Religion was a tool of landed ruling classes, who needed no science; philosophy had been a tool of the bourgeoisie, who needed science for production but needed also to befuddle the masses. The proletariat alone needed neither religion nor philosophy, but only science.

> PHILOSOPHY IS A PROP OF THE BOURGEOISIE. Not idealist, not metaphysical philosophy only, but precisely philosophy in general, philosophy as such. . . . In a word the proletariat retains and must retain science, only science, but no kind of philosophy. SCIENCE TO THE BRIDGE—PHILOSOPHY OVERBOARD!
> S. Minin, 26/IV, 1922, 0225 [i.e., 2.25 a.m.][20]

To the charge that he was merely dressing Comte's scheme of intellectual history in revolutionary phrases, Minin replied, with Plekhanov as his authority, that Comte had stolen the theory from the socialist Saint-Simon.[21] To the argument that Marx and the other 'founders of Marxism' had repeatedly indicated their belief in the necessity of philosophy, Minin replied that they had been children of their times, and that, even so, their writings contained hints of the newly-revealed truth that science needs no philosophy. 'Away with the "dirty linen" of philosophy!' Minin exclaimed, paraphrasing Lenin's famous remark concerning Social Democracy. 'We need science, only science, simply science.'[22]

The influence of Enchmen and Minin was almost entirely confined to the Communist student youth. With the single exception of the Kharkov anti-religionist, established Communist theorists, including those who would become leading members of the mechanist faction, displayed hostility or indifference to these ultra-revolutionary positivists. But evidences of other, less crude varieties of mechanism as a tendency abound in the very polemics against Minin and Enchmen. Bukharin, for example, countered Enchmen's reduction of social science and psychology to biology by asking whether biology could not be reduced to chemistry and physics.[23] A critic writing in *Pravda* found Enchmenism and Mininism to be essentially idealist, and set up against them 'the mechanico-materialist view'.[24] And the Rector of Sverdlov Communist University responded to the vogue of Enchmenism and Mininism among his charges by calling for the abolition of courses in philosophy. At this point something quite different from mechanism emerged; another Communist educator contended that the study of *Hegel* was the only antidote for Mininism and Enchmenism, and a brief dispute arose bewteen him and the Rector of Sverdlov, which helped to precipitate the extended controversy between the mechanist and Deborinite factions.[25]

Two other Soviet Marxists who are frequently associated with the clash of the Deborinite and mechanist factions, even though they took no part in it, are much better known than Minin and Enchmen. They are Trotsky and Bukharin, who contributed to the dominance of mechanism in the first half of the 'twenties, but did not come to its defence when the Deborinites attacked it.

Trotsky's mechanism may cause surprise to those who facilely assume that his celebrated 'voluntarism' in political theory must have precluded a mechanistic philosophy. Actually, political 'voluntarism' and philosophical mechanism can be joined in various ways, as the quarrel between Bukharin and Sarab'ianov showed.[26] Trotsky, who was less interested in abstract theory than Bukharin, made no extended effort to justify his political theory with philosophical arguments,[27] but his mechanistic outlook was clearly revealed on the one occasion when he became seriously involved in the philosophy of science.[28]

In 1925 he tried to persuade a congress of chemists that they had no cause to feel strange towards Marxism, for they were Marxists themselves. He told them that Mendeleev—and by implication the contemporary chemist also—was, in chemistry, without realizing it, 'nothing else but a dialectical materialist'.[29] On that basis he offered a programme of mutual non-interference between Communists and chemists:

> From the field of chemistry there is no *direct and immediate* exit to social perspectives. . . . An objective method of social cognition is necessary. Marxism is that method.

> When any Marxist tried to convert Marx's theory into a universal skeleton key and flitted through other fields of knowledge, Vladimir Il'ich [Lenin] would rebuke him with the expressive little phrase, 'Communist conceit'. This would signify in particular: Communism does not replace chemistry. But the converse theorem is also true. The attempt to step across Marxism, on the pretext that chemistry (or natural science in general) must solve all problems, is a peculiar *chemical* conceit, which is theoretically no less erroneous and practically no more likeable, than *Communist* conceit.[30]

Trotsky drew a picture of individual scientists setting their own research goals without regard to the demands of social utility, yet, without willing it, producing enormous social utility.[31] Adam Smith's famous 'invisible hand', which created social harmony out of the seeming anarchy of individualism, was of course replaced in Trotsky's vision by the immanent 'laws of social development'.

Thus Trotsky's views embodied a paradox that was almost universal among the mechanist philosophers. The proclamation of dialectical materialism as a universal philosophy, applicable to natural as well as social science, was actually, for *natural* science, little more than a benediction. The 'revisionist' belief that Marxism is not a universal philosophy but only a social theory was ceremonially rejected and implicitly accepted; Marxism in natural science meant what natural scientists were already doing. But Trotsky expressed more of the mechanistic philosophy than that. He sketched out a *Weltanschauung* that would justify his simultaneous assurances to Communist social theorists and non-Communist natural scientists that their separate and apparently unconnected endeavours were actually complementary.

Material particles were, in Trotsky's view, the ultimate reality, to which science would 'in the last analysis', 'on reaching its final goal', reduce all phenomena, whether chemical, biological, social, or intellectual.[32] This was in effect a justification of the 'chemical conceit' that Trotsky rejected. He therefore limited his mechanism. Reductionist assumptions, Trotsky argued, were necessary for a 'complete world view that connects all phenomena into a single system', but these assumptions must be qualified by such phrases as 'in the last analysis', so that the various sciences may continue to apply their own special methods to the special types of phenomena that they study.[33] The *insistence* on reductionism protected Marxist social theory from spiritualization: human affairs were still to be reduced to class interests, and beyond that, to forces of production. The *limitation* of reductionism protected Marxist social theory from a Pavlov or a Bekhterev, who sought the causes of 'wars, revolutions and their horrors' in the physiology of the central nervous system,[34] or in sunspots.[35]

One other aspect of Trotsky's thought is worthy of note, and that is his assessment of Lenin's place in the development of Marxism. This was in large part a political question, growing out of the clash of parties before the Revolution and sustained by the factional struggles of the 'twenties. The political leaders of the 'twenties defined their attitudes towards Menshevism in the past and towards the problem of 'building socialism' in the present partly by their estimates of Lenin's innovations in Marxist theory. It is accordingly not surprising that this question was barely mentioned by the philosophers of natural science in the period before 1930; the Party's internal politics entered the philosophy of natural science only late in 1929. At that time their entry was effected in part by the charge that the Deborinites had been surreptitiously abetting Trotsky's alleged disparagement of Lenin as a source of Marxist philosophy.

The tangled problem of Trotsky's attitude towards Lenin as a philosopher may be omitted here, but two points should be clearly understood. If Trotsky did consider Lenin considerably less important as a philosopher than the other 'founders of Marxism'—and the evidence indicates that he did[36]—the thought was by no means 'Trotskyite'. When Bukharin, for example, commemorated Lenin's death in January, 1924, by a long address on the theme, 'Lenin as a Marxist', he had nothing to say about Lenin as a philosopher.[37] And Stalin's famous lectures, 'The Foundations of Leninism', delivered in April, 1924, began with an explicit denial that Lenin had contributed anything 'distinctive and new' to the Marxist *Weltanschauung*.[30] His inconsistent assertion, further on in the same lectures, that Lenin had 'generalized, in accordance with materialist philosophy', the major developments of natural science since Engels' time, was only the faint beginning of reverence for

Lenin as the most important philosopher after Marx and Engels, an aspect of the cult of Lenin that was not clearly announced until 1930. It is therefore wrong to see Trotsky's influence whenever a philosopher of science ranked Plekhanov ahead of Lenin, or even neglected to mention Lenin, in the list of 'founders'.

In the second place, Trotsky's contribution to the ranking of Plekhanov or Kautsky or Lafargue ahead of Lenin as sources of Marxist philosophy was probably more helpful to the mechanist than to the Deborinite cause, for Lenin of the *Philosophical Notebooks*, a posthumous work that appeared during the controversy, was pronouncedly 'Hegelian' by comparison with Plekhanov and the other Marxists just mentioned. In short, the retrospective charge that the Deborinites' 'underestimation of Lenin' was evidence of Trotskyite influence is quite wrongheaded. In this matter as in the others Trotsky's influence on Soviet Marxist philosophy of science was probably more mechanistic than otherwise.

Bukharin probably contributed more than Trotsky to the formation of the intellectual atmosphere within which the philosophers of science worked, for he wrote at length on abstract problems of Marxist philosophy, and in 1923 was actually entangled in a polemic with two future leaders of the philosophical factions (the mechanist Sarab'ianov and the Deborinite Gonikman). But in the period 1924–1929, as the reader has already seen, an artificial silence prevailed between Bukharin and the philosophers of science. One may nevertheless suppose that during this period the philosophers experienced a continuous influence from Bukharin's widely-read theoretical writings.

The essential nature of that influence can be perceived by considering one of the issues on which Bukharin explicitly disagreed with Bogdanov, who was Bukharin's mentor in philosophy.[39] Both accepted the view that the mode of production determines the organization and development of society, but Bukharin feared that Bogdanov had 'psychologized' the human side of the mode of production, that is, the relationships of production. Bukharin corrected the mistake as follows:

In my view, the relationships of production are the labour *co-ordination of people* (*considered as 'living machines', in space and time.* The system of these relationships is as 'psychical' as a system of planets together with their sun. *Determinateness of place* at each chronological point—that is what makes the system a system.[40]

The implicit vision of a single universal science, expressed perhaps in a set of formulas from which all particular phenomena could be derived, did not offend Bogdanov; it had indeed been his before it was Bukharin's. They could not agree on where to seek this 'scientific monism', as Bogdanov called it. Bukharin seems to have visualized the ultimate unity of science in a sort of energetics, which, in the particular case of human society, would calculate the absorption of energy from non-human nature by the 'living machines', whose organization in society would be derived from their spatial distribution in the productive process, itself determined by the non-living machines at which they laboured.[41]

The reductionism of this philosophy is not its only noteworthy feature. Like many other mechanists, Bukharin believed that dialectical materialism meant little more for natural science than a conscious formulation of what natural scientists were already doing. It is curious that he approached this conclusion by an argument very similar to one often used to discredit dialectical materialism. The distinction between mechanical and dialectical materialism is a false one, the argument runs, since the characteristic features of dialectics—so far as these features are rational and not a vestige of Hegelian mysticism—can be found in mechanics as well as any other science. Hence, the opponent of Marxism concludes, the celebrated dialectical materialism is exploded.[42] Hence, Bukharin concluded, dialectical materialism is essentially mechanical materialism, a philosophy confirmed by the triumphant progress of natural science, one that natural scientists are already following spontaneously.[43] The first argument ends with a sneer at Marxism, the second in a posture of reverence, and the version a Russian natural scientist might choose to accept would therefore be a matter of considerable political importance to the Bolsheviks. But it seems clear that the scientist could accept either version without altering his methodology—unless, to be sure, he felt that mechanistic methods had become inadequate for science's further development.

It would probably be incorrect to assume that Bukharin elaborated his philosophy with an eye on natural scientists and their political attitudes. The present writer gets the impression that Bukharin was primarily a social theorist who took seriously, and tried to obviate, many of the criticisms of Marxism.[44] The startling derivation of relationships of production from the spatial distribution of the producers was, in effect, a reply to the argument that historical materialists equivocate when they include social relationships within the allegedly *material* 'mode of production' that determines human development. The search for a purely mechanical world view in the laws of thermodynamics—and that is the essence of the notorious principle of *equilibrium*, which Bukharin took from Bogdanov, who had taken it in turn from Gibbs and Le Châtelier[45]—was an effort to forestall several famous criticisms of Marxism. The critics argued that dialectics is not a truly universal methodology; what could be more truly universal than the laws of thermodynamics? The critics said that Marxism is based on a religious faith in infinite progress, that its allegedly scientific prediction of social development is actually a new eschatology; Bukharin affirmed the reversibility of universal processes (human included), pointing to mechanics as his model. The critics called dialectical terminology a gibberish, used by its adepts to mean whatever they might choose; Bukharin offered 'the theory of equilibrium' as 'a formulation [of the laws of material moving systems] *cleansed of idealist elements*'.

Marx and Engels emancipated dialectics from its mystical husk in *action*, i.e., by *applying* the dialectical method materialistically in the investigation of various fields of nature and society. The problem now concerns the theoretico-systematical exposition of that method and the placing of a foundation, also theoretico-systematical, under it. Just that is furnished by the *theory of equilibrium*.[46]

Disturbed by the ambiguities in Marxism, Bukharin tried, in effect, to eliminate the Hegelian, the dialectical parts.[47] But, like Trotsky, he did this in order to achieve a satisfactory vision of the universe rather than a set of practicable methods for scientists. Theoretically, social science and biology could be reduced to mechanics; but in one place Bukharin argued that it would be truly monstrous to suppose that, let us say, the law of the conservation of energy makes the law of labour value, or the doctrine of the base and the superstructure, or the laws of monetary circulation superfluous.... The latest generalizations of science... do not liquidate the particular laws, they merely establish the connection among them, they express the principled monism of science and the unity of its method. But this does not in any measure abolish any division of scientific labour, i.e., the special scientific disciplines.[48]

In any assessment of Bukharin's share in the dissemination of mechanistic views, one must bear in mind that the Party youth greatly admired him, even if the established theorists tended to be hostile or condescending before 1924, and respectfully indifferent from 1924 to 1929.[49] But one may not leap from that fact to the conclusion that Bukharin drew many theoretically minded young Communists towards support of the faction that defended mechanist philosophy against the Deborinite attack. The one philosophical disputant who can be proved to have begun his intellectual career as a protégé of Bukharin was Podvolotskii, and he became a leading *Deborinite*. The fact cannot be avoided that Bukharin did not associate himself with the mechanist faction and was not assigned to that faction by its members or its adversaries until the end of 1929, when the mechanist faction had already been condemned. Thus, his influence on the controversy of the Deborinites and the mechanists becomes an imponderable, depending on something that the educational psychologists call *transfer*. Enthusiastic about Bukharin as theoretically minded young Bolsheviks may have been, were they reflective enough to realize that his textbook on *sociology* presented much the same views on the *philosophy of science* as the mechanist faction was defending against the Deborinites? When it became possible for the Deborinites to criticize Bukharin as a mechanist, in the latter part of 1929, a Deborinite claimed that students had 'besieged' their teachers with questions reflecting their doubts 'concerning the purity of the Marxist chasuble on Comrade Bukharin'.[50] If this is true, Bolshevik students did see the connection between Bukharin's views and those of the mechanist faction, even though no one pointed it out while the controversy over mechanism was in progress. But, of course, one cannot tell whether or to what extent the story is true.

In general it is extremely difficult to determine where significant transfer was made, where mechanism as a tendency in fields of thought outside the philosophy of natural science was an important source and continuing support of mechanism as a faction in the philosophy of natural science. The problem is really twofold. Soviet Marxists who quarrelled over the philosophy of natural science during the 'twenties were subject not only to mechanistic influences flowing from their fellow-Marxists in the various fields of social theory; they were also subject to mechanistic influences coming from 'the most inaccessible citadel of the ruling classes', as, in 1931, natural science was still described.[51] Reading over the literature of the philosophical controversy, one cannot avoid the paradoxical conclusion that 'bourgeois' natural scientists, even though they were considered an alien group, exercised on Soviet Marxist philosophy of natural science a more considerable influence than fellow-Marxists in the philosophy of social science did. One of the most illuminating details in Bukharin's writings is the footnote that he appended, early in 1924, to a long polemic against Pavlov, who offended the Bolsheviks during the 'twenties by his political hostility to them. 'The author of these lines,' Bukharin wrote, 'who expounds dialectical materialism from the point of view of *equilibrium*, is particularly pleased to note the following statements of Professor I. Pavlov.' The statements that followed were Pavlov's profession of faith in the mechanist theory of equilibrium, essentially the same as the one that Bukharin had borrowed from Bogdanov.[52] Thus Bukharin, smarting from the attacks that fellow-Marxist social theorists had recently made on his philosophy, boasted of the support he found in a 'bourgeois' scientist whom he was criticizing as an opponent of Marxism.

In general, one is impressed by the predominance of mechanistic over non-mechanistic influences coming into Soviet Marxist philosophy of natural science from biology and physics during the 'twenties.[53] To be sure, the writings of Soviet Marxists were a curved mirror in which the currents of philosophical opinion among biologists and physicists were reflected with considerable distortion. But they were for all that a mirror, and on looking into them it is possible to say that there was a great deal of mechanistic thinking among biologists and physicists of the 'twenties.

One notes, for example, the fairly uniform assessment that the Marxist disputants made of the philosophical outlook prevalent among natural scientists. Deborinites and mechanists who mentioned the subject—and they did so frequently—were almost entirely agreed that most natural scientists who gave a thought to the philosophy of science were mechanist or positivist in one form or another.[54] On the part of the Deborinites this was a damaging admission, for they were in effect confessing that theirs was more a philosophy *for* than *of* natural science. It therefore seems probable that Professor A. F. Samoilov was indeed expressing '*communis doctorum opinio*'—as a Deborinite admitted[55]—when that eminent physiologist explained why he sympathized with the mechanist philosophers.

> Natural scientists do not form some kind of caste [or] union, whose members are bound by some definite philosophical world view. They are bound together essentially only by a method of research; whoever understands the peculiarities of the contemporary experimental method of natural science and knows how to use it in his chosen specialty we call a natural scientist. But here it is fated that the method of contemporary natural science drives the natural scientist spontaneously to the side of mechanistic materialism. Once he has learned and grown accustomed to turn for the explanation of the complex to simpler relationships, once he has learned to reduce the more complicated phenomena to simpler ones, he spontaneously comes to the thought that all phenomena [*iavleniia*] can be reduced to a rather simple phenomenon [*fenomen*], to the motion of a material point in space and time. The method and the world view of the natural scientist are organically inter-connected, and without concerting together on this matter, natural scientists, in the mass, in a preponderant majority, profess the mechanical view concerning all phenomena of dead and living nature.[56]

Even if we question the quantitative accuracy of Professor Samoilov's judgment, there can be no doubt that a good many of Russia's natural scientists were as he described most of them, and these must be counted among the mechanists by tendency —overwhelmingly non-Marxist in this case. Among them, to take one of the few Marxist examples, was M. M. Zavadovskii, a biologist who concluded a popular survey of his science's history with the declaration that the trend from primitive fantasy to science was a movement from external manifestations to physico-chemical and mathematical analysis; from mysticism and the prejudices of naïve vitalism to the rigorous analysis of the consistent mechanistic interpretation. And everywhere from quality to quantity. We can say boldly that contemporary biology has developed under the sign of the numeral and the experiment.[57]

This credo was stated in 1924 with no thought of starting or entering a controversy, and, so far as the present writer knows, no Soviet Marxist took exception to Zavadovskii's credo. But in 1923 and 1924 other writers who expressed similarly mechanistic views were being challenged in a number of small separate controversies, merging into a clash of two factions. M. M. Zavadovskii and many other mechanists in tendency did not join in that clash, but, Marxist and non-Marxist alike, they must have been a considerable influence on it.

7

THE FIRST CHALLENGES TO MECHANISM, 1922-1924

Mechanist thought was challenged, before 1925, in a series of isolated disputes concerning diverse issues, disputes that only gradually developed into a clearly defined struggle of two factions with the philosophy of natural science as their main concern. These preliminary clashes were so numerous, haphazard, and apparently disconnected, that the one which precipitated the formation of philosophical factions was generally credited by contemporaries with causing the formation of factions, and the role of other earlier disputes, which had expired without apparent issue, was overlooked. In retrospect one can see a pattern.

One can see Soviet Marxists proceeding towards controversy over the philosophy of natural science as they became enmeshed in the problems of training 'red specialists', and the process was a complicated one. A. K. Timiriazev, for example, a physicist who turned Bolshevik in 1921 to become one of the chief figures in the reconstruction of education and subsequently one of the chiefs of the mechanist faction, aroused no discord in 1923 when he addressed a gathering of the editors of *Under the Banner of Marxism* on a question 'not infrequently discussed in the [Party's] Moscow Committee': whether the teaching of natural science in the Communist Universities could be entrusted to non-Marxists.[1] Timiriazev believed that it could, for natural scientists, he said, were unwitting dialectical materialists in their specialties. He argued the point at length to an audience that probably included some of his future opponents in the philosophical controversy, but there is no record of contention either at the meeting or when Timiriazev's speech was published. Evidently the people who would later denounce Timiriazev's views as positivist were, in 1923, largely indifferent to the philosophy of natural science.

A different effect was produced when M. N. Liadov, a publicist who traced his Bolshevism back to 1903,[2] examined the problem of training 'red specialists'. He had been made Rector of Sverdlov Communist University in the spring of 1923, and, after observing the students for several months, he became disturbed by their intellectual tendencies. He found them susceptible to Enchmenism, to the bourgeois ideologies of their professors of natural science, and to political deviations from the Party line. The main cause, he wrote in *Pravda*, was that the students' training in Marxist theory was too abstract, isolated in separate courses called 'dialectical materialism', 'historical materialism', and 'the history of philosophy'. To Liadov this curriculum was one symptom of a general disease in the Communist educational system that he called 'academism', attributing it to the ascendancy in the system of non-Communist Marxists. He warned that they would transform the revolutionary student youth into

> Menshevising intellectuals who of course will never understand and cannot understand our Bolshevik tactical and organizational line, but will always willingly follow any opposition to that line. . . . I am entirely agreed with Comrade Lenin that without 'solid philosophical training' we 'shall not win the struggle against the onslaught of bourgeois ideas'. But precisely for that reason I suggest that philosophy, as a separate discipline, should be decisively stricken from our programmes.[3]

Liadov tried to get around the apparent inconsistency of his recommendation by utilizing Marx's dictum that truth is always concrete:

> A concretely thinking proletarian student without any philosophical training is invulnerable to bourgeois ideology. He will never be permeated with it. But that same student, having taken an abstract course of Marxist philosophy, will easily become a victim of any arch-bourgeois philosophical school.[4]

It will be recalled that the most important teachers of philosophy, not merely at Sverdlov University but in the Communist educational system as a whole, were Deborin and Aksel'rod the Orthodox, ex-Mensheviks, whose appointment to Sverdlov in 1921 had been the subject of anxious exchanges among Communist officials, including Lenin himself.[5] Now it seemed not only that Lenin's recommendation in favour of their appointment might be undone, but that the entire programme of philosophical education might be destroyed. Deborin and Aksel'rod, neither of whom was a Bolshevik at this time, held their peace. But S. L. Gonikman, a student of Deborin's who had an almost unblemished Party record, and was already at the age of twenty-six, a teacher of philosophy, defended his profession in *Pravda.*[6] He found the *non-Marxist professors of natural science* to be the main source of the corruption of proletarian students, and he not only defended courses in philosophy as the only antidote, but argued strongly for Hegel as the main concern of such courses. Indeed, in a contemporaneous paper that grew out of his work in Deborin's siminar, Gonikman called Hegel 'the only springboard from which the leap into Marxism is possible'.[7]

The reader who has noted Gonikman's disagreement not only with Liadov but also with A. K. Timiriazev is more perceptive than Liadov, who replied to Gonikman in *Pravda* but neglected to call Timiriazev to witness, and then dropped the question of philosophy altogether.[8] Perhaps Liadov desisted because N. A. Karev entered the field against Gonikman. Karev was himself a student of Deborin's and a fledgling professor of philosophy, who argued that Gonikman exaggerated the importance of Hegel to Marxist philosophy, that, indeed, Gonikman was echoing the error of George Lukacs and other West European 'Hegelianizing' Marxists.[9] Karev did not, to be sure, defend Liadov's recommendation that courses in philosophy be closed, and within a relatively short time, in the course of a different dispute with still another philosopher, he discovered that he was in fundamental agreement with Gonikman on the prime importance of Hegel to Marxist philosophy.[10]

These brief disputes, between Liadov and Gonikman, and then between Gonikman and Karev, were thus minor incidents in the drift of Soviet Marxists towards the formation of philosophical factions. They are significant not only because they show that the process was a drift, and a rather complex one with subsidiary eddies, but also because they reveal one of the forces pushing Deborin and his students towards their challenge to mechanism. Proposals to terminate university courses in philosophy, and complaints about the esoteric uselessness of philosophical writings,[11] could hardly be ignored by the teachers of such courses and the authors of such writings.

Concern over the spread of Enchmenism and analogous doctrines also aroused some disputes in 1923 and 1924 between a couple of psychologists and three philosophers. The psychologists, deriving from Pavlov a psychology that was somewhat similar to Enchmenism, denied the reality of traditional concepts not only in psychology, but also, perhaps without intending it, in philosophy.[12] That two associates of Deborin took up the defence of philosophy is noteworthy only for the fact that this was one of the very few occasions on which Deborinite philosophers intervened in the discussions of Marxist psychologists.[13] But the third man to defend philosophy was not an associate of Deborin; he was Sarab'ianov, a prominent anti-religionist and subsequently a leader of the mechanist faction. He was willing to accept a radically reductionist psychology, but demurred at its extension to *Weltanschauung* and epistemology:

> Epistemology is not biology, and conversely.
>
> That epistemology cannot but rest on the achievements of the natural sciences is indisputable, but it is also indisputable that the ends sought by epistemology are other than those in biology. That is why the conclusions of Academician Pavlov, which are entirely correct in our view, cannot be interpreted in the extended fashion of [the Enchmenists].[14]

These words, written before Sarab'ianov had to defend himself against Deborinite charges of denying the right of philosophy to exist as an independent discipline, show that the clash of the Deborinite and mechanist factions would not be a simple clash of friends and enemies of philosophy. The psychologists who were the enemies of philosophy in this early dispute took no part in the subsequent controversy of mechanists and Deborinites, and the friends of philosophy turned up on opposite sides of that later and larger controversy.[15]

Another series of minor disputes that brought Soviet Marxists to the verge of their major controversy was set in motion by V. V. Adoratskii, a friend of Lenin's whose tastes in philosophy, however, were rather different. While convalescing from a serious illness during the Civil War, he followed Lenin's suggestion to compile a one-volume selection of the correspondence of Marx and Engels. 'Only scholarly fools like you and me,' Lenin said, 'ever read the four fat volumes published by Bebel and Bernstein.'[16] The *Selected Correspondence* that resulted is probably familiar to students of Marxism all over the world in one of its many editions or translations, but the first editions of the early 'twenties, which contained a long preface by Adoratskii, were not universally received as a standard work.[17]

In his preface Adoratskii tried to prove Lenin's remark that materialist dialectics is the focus of the entire correspondence.[18] But instead of seeking an understanding of dialectics in Hegel, as Lenin had done,[19] Adoratskii concentrated on the positivistic utterances of Marx and Engels. Without realizing it, he came close to the 'revisionist' view that Marxism is a social theory rather than a universal philosophy. To be sure he proclaimed his adherence to the 'orthodox' view that Marxism rests on materialistic dialectics, which is universal in its application, but he declared that this universal methodology is merely a handful of commonsense rules of thought. Applied in one area it gave rise to the natural sciences; consciously applied in another area by Marx, Engels and their followers, it gave rise to the social and political theory of the Bolsheviks, to which Adoratskii devoted the bulk of his preface. Referring to Marx's desire to present the 'rational kernel' of dialectics in a pamphlet,[20] Adoratskii denied that more than a pamphlet was necessary:

> In these matters [i.e., in dialectics] everything essential has been said. No new philosophy of Marxism need be created. We need the dialectical method, the consistent application of the materialist point of view, and the positive sciences, which deal with concrete material, study objective reality, and do not weave a web of ideology.[21]

In another place he declared simply that 'philosophy, as an ideology, must be overcome [*preodolena*]'.[22]

In asserting that dialectics was no 'ideology', which he

defined as any partial and therefore unrealistic scheme of thought,[23] Adoratskii helped initiate a special controversy among Soviet Marxists that dragged on for years in various forms. One writer, for example, roused considerable furore by turning to Freudian psychology as a necessary supplement to Marxism in explaining the origins and nature of ideologies.[24] Among other philosophers the debate centred on the definition of the concept itself: Was ideology necessarily a distortion of reality? Was human thought *either* science, disinterested and therefore objective, *or* ideology, interested and therefore unreal; or was it in some sense both? Adoratskii was not active in the controversy he helped initiate, beyond publishing a single reply to a critic who had called him a 'Mininist'.[25]

A teacher of historical materialism at Saratov State University, I. P. Razumovskii, came to Adoratskii's defence with a pejorative explanation of ideology that included philosophy in it, and left no place for either in the classless society of the future. Only the positive sciences would be necessary to the dispassionate minds of that time.[26] When a student of Deborin's pointed out the similarity of Razumovskii's views with the 'simple Mininist scheme, "religion, philosophy, science",' Razumovskii complained that he had been misinterpreted and began shifting his position—until he turned into a Deborinite.[27] The reason for his anomalous shift, and for Adoratskii's grudging endorsement of the Deborinite faction in 1929, will be examined below. But these opponents of ideology had an anomalous connection with the start, as well as the close, of the Deborinite compaign against mechanism. They helped involve the Deborinites in the philosophy of natural science, even though neither they nor the Deborinites had any special interest in the subject. When A. K. Timiriazev argued at length that the natural sciences were classless, disinterested branches of knowledge on which bourgeois philosophers wilfully fastened their ideological nonsense, Deborin and his students tacitly agreed. Adoratskii and Razumovskii merely suggested this view while arguing that ideology, philosophy included, must be eliminated, and the Deborinites took up arms. Of course, the Deborinites' main object was the defence of philosophy, but they could not help raising the question whether the natural sciences *were* pure of ideology.

The Deborinites were not eager to get involved with this question. Indeed, in 1922–1924 they did not always object when *philosophy* was rejected as 'ideology'. The determinant of Deborinite silence or protest seems to have been whose or which philosophy was the target of positivistic arguments. Thus, in 1922, when the future *enfant terrible* of the mechanist faction used thoroughly positivistic arguments to discredit the religious philosophy of the academic philosophers then being exiled, no Deborinite protested.[28] And as late as 1924 the editorial board of the Communist Academy's journal, presumably with Deborin's approval, took the unusual step of placing its formal endorsement on a series of articles in which the terms 'philosophy' and 'ideology' were regularly used as pejorative synonyms.[29] But these articles were directed against the philosophy of the Hungarian Communist, George Lukacs, with whom the Deborinites apparently felt no bonds of intellectual sympathy or professional interest.

As teachers, the Deborinites limited their responsibilities to general expositions of philosophy; as scholars, trying to carry out Lenin's behest to elaborate the theory of dialectics, they concentrated on the history of philosophy, especially on the philosophical antecedents of Marxism. In both phases of their activity they were increasingly drawn, apparently with some reluctance, towards controversy over the philosophy of natural science. The case of G. K. Bammel' and V. N. Ivanovskii is instructive with respect to the general expositions of philosophy. Bammel' was a junior associate of Deborin's in the Marxist research institutes and publications of Moscow; Ivanovskii, at the Bielorussian State University in Minsk, was one of the few academic philosophers of pre-revolutionary standing still teaching under the Soviet régime. In 1923 he published a *Methodological Introduction to Science [Nauka] and Philosophy*, and Bammel', who reviewed it, showed a mixture of respectful criticism and unqualified admiration.[30] The criticism centred on Ivanovskii's unwillingness to argue an ontological basis for the methodology of science, on his indifference both to materialism and to idealism. When Ivanovskii published a reply, making explicit his indifference to ontology, arguing the classless nature of scientific truth, and plainly stating his view that Marx's theories were applicable to social phenomena but not to the philosophy of natural science, Bammel' lost some of his deference, and criticized Ivanovskii rather sharply as a positivist.[31]

But Bammel' was still far from declaring war on such positivism. In 1924 he sponsored the translation of a popular exposition of Marxist philosophy by Karl Korsch, a German Communist who specifically denied the applicability of dialectics to natural science.[32] It may be that Bammel' warned the reader against this positivism in his Preface, which has been inaccessible to the present author; certainly the students of Deborin who reviewed ths book did.[33] But none of these future Deborinites felt that Korsch's lapse with respect to the philosophy of natural science destroyed the value of the book. Moreover, nobody objected to the fact that the Communist Academy's journal carried an entirely favourable review of Korsch by Adelbert Fogarasi, a Hungarian Communist who also felt that dialectics was inapplicable to natural science.[34]

The arguments of the future Deborinites concerning the history of philosophy were, intellectually, the richest of all their disputes, for the history of philosophy was their special field. Here it is appropriate to recall only some aspects of those debates, and not necessarily the most interesting ones from a philosopher's point of view. Deborin, an editor of the Communist Academy's journal, was apparently indifferent to the positivist elements in the polemic against George Lukacs and in Fogarasi's review of Korsch. But when the journal of the Communist Academy published an article denying the 'idealist legend' that Kant had made a significant contribution to the advancement of knowledge, an editorial note promised a reply from Deborin, which shortly appeared.[35] It was difficult for Deborin to stay clear of the philosophy of natural science in this case, for the offending article used the history of philosophy to expound a radically positivistic philosophy of natural science. But Deborin managed the difficult feat by turning over to I. E. Orlov the task of criticizing the article's philosophy of natural science.[36] One finds Deborin repeatedly turning such problems over to Orlov or A. K. Timiriazev—until both Orlov and Timiriazev became leaders of the mechanist faction, thereby obliging Deborin to do his own thinking about the philosophy of natural science.

The division of philosophical labour among Soviet Marxists during this early period, and the growing strain on that division, may be seen in the tangle of Orlov and two other Marxists concerning Descartes and physics. Orlov, who was one of the few Soviet Marxists specializing in the philosophy of natural science in the early 'twenties, published a popular historical survey of the concept of matter. He held Descartes responsible for the 'purely descriptive' or 'formal' method in physics, a method that had been Orlov's *bête noire* ever since his pre-revolutionary debut as a philosopher of science.[37] It happened that Z. A. Tseitlin, a young Marxist physicist who would be a leader of the mechanist faction, was just then coming to the conclusion that Descartes' method was essentially the progenitor of all that was *worthy* in natural science, and in social science as well. Moreover, Tseitlin was convinced that Einstein's general theory of relativity was acceptable to dialectical materialism, and he had already debated the issue with Orlov, who felt that both the special and the general theories were contrary to sound physics and philosophy.[38] Now, in 1924, these future leaders of the mechanist faction extended their debate to Descartes and the history of philosophy. Thereupon a student of Deborin's specializing in the early modern history of materialism entered the dispute. And he did more than challenge Tseitlin's interpretation of Descartes. He disagreed with Tseitlin's view on the relationship of matter and space.[39] But it is indicative of the continuing Deborinite reluctance to assume full responsibility for the philosophy of natural science that he refrained from comment on Tseitlin's views concerning the theory of relativity. Indeed, as late as 1927 he was still neutral on *that* issue.[40]

As the example just given suggests, the disputes of 1922, 1923, and 1924 that were concerned specifically with the philosophy of natural science tended to occur not between Deborinites, on one side, and mechanists, on the other, but among people who later joined the mechanist faction or took no part in the major philosophical controversy of the 'twenties. The Deborinites did not participate in discussions of the philosophy of natural science during these early years. Indeed, Deborinite indifference or neutrality towards two special issues, genetics and the theory of relativity, lasted until the late 'twenties, even though discussion of these issues began in 1922. Marxist debate over genetics and relativity influenced the formation and clash of the philosophical factions only by bringing individual Soviet Marxists to an awareness of their own philosophical views. Some were brought in this manner to the mechanist faction, and a very few to the Deborinites, but the bulk of the participants in these special disputes joined neither philosophical faction.[41]

Thus, the Deborinite challenge to the mechanistic philosophy of natural science grew out of disputes over such issues as the definition of 'ideology', or Descartes' place in the history of Marxist philosophy, or the place of philosophy in a Communist university, rather than Einstein's theory or Mendelian genetics. Further reflections on this anomaly may be left to a later chapter, where, in conjunction with other peculiarities of the controversy, it may help to reveal the essence of the contending positions. But the reader may gain a keener sense of this anomaly by considering the startling foreshadowing of the Deborinite and mechanist positions in an exchange printed in January, 1924, before the rival factions had formed. A. Gol'tsman, a publicist who described Einstein's theory as 'the apotheosis of dialectical materialism', replied as follows to Timiriazev's assertion that the theory lacked experimental proof:

> And in any case it is not the Marxist, or the materialist, who are condemned to demand of philosophy a limitation to 'facts' and experiments alone. Hence runs the direct, well-beaten path to positivism, the narrow-minded 'philosophy' of natural scientists with a bourgeois orientation.
>
> What does materialism demand of physics? Not very much. In order for a physical theory to satisfy dialectical materialism, *it is necessary and sufficient that it reduce all occurrences in nature to a process of material substances.*[42]

Timiriazev's rejoinder not only reasserted the requirements of experimental proof, but transcended the question of Einstein's theory, as Gol'tsman had, to assert a general philosophy of natural science:

> Now I think that it is clear to everybody, and especially to us and you, that the root of our disagreements does not lie in certain details. You think that materialist philosophy should live and develop

apart from natural science, allowing natural scientists to do their job and only from time to time slightly reining them in. While I think that Engels was right when he said that 'as soon as each separate science is required to get clarity as to its position in the great totality of things and of our knowledge of things, a special science dealing with this totality is superfluous. What still independently survives of all former philosophy is the science of thought and its laws—formal logic and dialectics. Everything else is merged in the positive science of Nature and history'.[43]

Two or three years after January, 1924, Soviet Marxists would have unhesitatingly indentified Gol'tsman's statement as Deborinite and Timiriazev's as mechanist. At the time, the future Deborinites paid no attention to either statement. Subsequently, Gol'tsman did not join the Deborinites, nor was he honoured as one of their pioneers. Indeed, when the Deborinites finally took a position on the theory of relativity, they ignored not only Gol'tsman, who was their predecessor in the characterization of the theory as dialectical materialism in physics, but also the German Communist, August Thalheimer, who preceded them in crediting Hegel with a forecast of the theory.[44] The final irony came in 1929, when Semkovskii, who had earned the enmity of the Deborinites in mid-1924 by coupling support of Einstein's theory in physics with a denunciation of Hegelianism in philosophy, went over to the Deborinite faction and proudly announced that *he* had initiated both support for Einstein's theory as dialectical materialism in physics, and the search for its philosophical explanation in Hegel.[45]

Yet it would probably be a mistake to regard the foreshadowing of the rival philosophies of natural science by Gol'tsman and Timiriazev as a freak, and the dispute among Soviet Marxists over the theory of relativity as having no connection with the origins of the philosophical controversy. After all, Timiriazev did become a leader of the mechanist faction, and there is little room for doubt that he was brought to a realization of his position in the philosophical controversy partly by his discussion of Einstein's theory. Oddly enough, the same can probably be said of the physicist Z. A. Tseitlin, who was defending the theory of relativity against Timiriazev's attack at the same time that both of them were becoming leaders of the mechanist faction in philosophy. For Timiriazev did not reply with a rival philosophy when Tseitlin hailed a successful repetition of Michelson and Morley's experiment as 'a great victory of the mechanical picture of the world and, accordingly, of dialectical materialism, which assumes that all the phenomena of nature are the motion of matter . . . i.e., are subject to the laws of mechanics'.[46] Timiriazev, tacitly accepting Tseitlin's philosophy, rejected his physics by pointing to Dayton Miller's alleged refutation of Michelson and Morley's experiment.[47] On the other hand, the students of Deborin, embroiled with Tseitlin's philosophy of natural science by an argument over Descartes, bumped against a very similar philosophy of natural science when they quarrelled with Timiriazev over a philosophical handbook for anti-religionists, which will be described in the next chapter. Obliged in this manner to fashion their own philosophy of natural science, the Deborinites raised a cry for Hegel and against mechanism, and Timiriazev and Tseitlin sank their differences over Einstein's physics to defend the mechanist philosophy they held in common.

8

THE FORMATION OF FACTIONS, 1924-1926

DEMANDS for the curtailment of philosophy as an independent discipline, and problems in the historiography and popularization of Marxist philosophy provoked Deborin and his students to unconnected polemics in the first half of the 'twenties. In 1924 and 1925 two new elements were introduced into the continuing stream of such minor disputes. Deborin's two most distinguished colleagues among the professional philosophers expressed a muffled sympathy with the demands for the curtailment of philosophy, and Deborin's students challenged mechanism in a book by I. I. Stepanov, a Party chieftain and leading anti-religionist. The unexpected result was the crystallization of philosophical factions that were divided by issues centring on the philosophy of natural science.

There can be little doubt that L. I. Aksel'rod the Orthodox and S. Iu. Semkovskii were, along with Deborin, the most distinguished professional philosophers in the Soviet Marxist community during the 'twenties. (Bogdanov, of course, had a more famous history, but his status during the 'twenties was low, even if he did participate in the work of the Communist Academy.) Semkovskii, the chief Marxist philosopher in the Ukraine, had exhibited positivist tendencies before the Revolution, when Deborin was a disciple of Plekhanov's orthodox line in philosophy. But it was probably more than this inherited difference that led Semkovskii in 1924 to insinuate that Deborin was a scholastic. For Semkovskii, the leading figure in the campaign to win the natural scientists of the Ukraine, was intent on proving to them that their method and ontology as natural scientists were already Marxist in all but name. At the Kharkov Academy of Sciences in February, 1924, he read a paper arguing that the theory of relativity was the realization of dialectical materialism in physics. Presumably as a simultaneous reassurance to the natural scientists and as an admonition to Deborin in Moscow, he disowned metaphysics in the following words: 'For Marxists, dialectics is not a dead stereotype, *even if it is scholastically "deepened" from Hegel himself*, but a living method, rooted in the achievements of science at each stage of its development.'[1] Deborin and his students were of course the target of the phrase in italics, for they were earnestly following Lenin's advice to elaborate dialectics from Hegel. And they did not overlook the admonition, however subtle and parenthetical it was.[2] But they ignored it outwardly, and in October, 1924, Semkovskii repeated his thought more forcefully and more prominently, including it in a Postscript to a collection of essays:

> In recent years we in the Soviet Union have had to experience in the field of Marxist philosophy a flowering of research that is externally sumptuous. But this entire flowering threatens to remain a sterile blossom in view of the *scholastic deviation* that draws it backwards, away from the actual, living problems of dialectical materialism, which are raised by the development of the natural and social sciences, back to the fruitless scholasticism of school philosophy.[3]

Semkovskii still did not mention names, but it was plain that he meant Deborin and his followers. About the same time a similar complaint came from Aksel'rod the Orthodox. In an Introduction to some things of Plekhanov's that were published in 1924, she wrote:

> At times the old abstract scholasticism, clothed in Marxist terminology, emerges before us with extreme clarity. The concrete, real 'base' is set off to one side, and all thought is drowned in barren and meaningless metaphysical abstractions. . . .[4]

The implied criticism of Deborin and his students was even more remote than Semkovskii's remark of February, but the Deborinites took immediate offence, and began a campaign against Aksel'rod before they took public notice of Semkovskii's criticism. Aksel'rod and Deborin were, it will be recalled, jointly responsible for education in philosophy at the Institute of Red Professorship, the graduate school preparing Communist professors, and at Sverdlov Communist University, the model for a growing network of Communist universities. Perhaps Aksel'rod was beginning to have second thoughts about the concentration on Hegel at the Institute. Whatever the reason, Deborin and his followers must have been angry, for they began a rather underhanded campaign against her. A Deborinite review of a booklet by Askel'rod insinuated that she had abandoned the orthodox philosophy she once learned from Plekhanov, but, ostensibly as a kindness to her, refrained from offering evidence.[5]

In the fall of 1924, Ian Sten, probably a student of Deborin's,[6] went further, in a review of a Polish Marxist's popularization of philosophy. Semkovskii had written the Introduction for the Russian translation, and Aksel'rod had added footnotes, both taking exception to the Polish Marxist's positivistic tendencies.[7] But Sten criticized the positivism in the book without mentioning Semkovskii's and Aksel'rod's disavowal of it, and concluded his review with the opinion that a new edition with a 'genuine Introduction and genuine notes' could be a useful addition to the literature.[8] Even in 1925 the clashes of Aksel'rod and the Deborinites continued to be somewhat veiled by indirection, as in the examples just given.[9] It was not until the spring of 1926, at the debates within the Communist Academy's Institute of Scientific Philosophy, that the break between Aksel'rod and Deborin came fully into the open.[10]

The incidents that have just been detailed may seem like petty back-biting more than the beginning of a philosophical controversy. Certainly they do not have as much claim to philosophical significance as some of the minor disputes that were, during 1924 and 1925, continuing to erupt. When Sandor Varjas, for example, became embroiled with two or three Deborinites over his *History of Modern Philosophy*, the issues were much more clearly stated, fully developed, and philosophically significant, than they were in the furtive duelling of Deborin's students with Aksel'rod and Semkovskii. And the dispute over Varjas' book did contribute to the formation of the opposing factions. Bogdanov and Varjas learned that they were closer to each other's views than they had imagined when Varjas first projected his ideas to the Communist Academy in the summer of 1924, and Bogdanov had criticized them. At any rate, they found themselves closer to each other than they were to the philosophy of Deborin and his students, whom Varjas finally accused if Hegelianism.[11] And Karev, who had been criticizing a fellow-student of Deborin's for Hegelianism, found his true feelings in this minor dispute with Varjas. 'Yes!' he shouted at Varjas, '. . . we are "Hegelians"! Everything great in modern history has been in one way or another connected with Hegel's name.'[12]

By comparison with such a bold confrontation, the bickering of the Deborinites with Aksel'rod and Semkovskii seems petty indeed. Yet the bickering was probably more important in precipitating the formation of factions. In later references to the beginning of the controversy the Deborinites usually neglected to mention the particular disputes that were described in the preceding chapter, or even the dispute over Varjas' book, which merged directly into the major controversy. But these later Deborinite accounts nearly always referred to the complaints of Aksel'rod and Semkovskii concerning scholasticism, representing them as the culmination of a 'steadily growing stream' of similar complaints.[13]

Deborin and his students conceived their lifework to be the elaboration of dialectics from Hegel and his predecessors. It is therefore not hard to understand what they must have felt when Deborin's two most eminent colleagues muttered their sympathy with the widespread positivistic aversion to Hegelian dialectics. However indistinct or trivial the mutterings may seem to one viewing the affair only as it was recorded in print, they must have struck Deborin and his students as incipient apostasy in high places. Otherwise one can hardly explain either the stir aroused by mere *obiter dicta* or the vigour with which the Deborinites began to press an organized campaign against mechanism. A dispute over a handbook for anti-religionists was, to be sure, the trigger that set off the Deborinites' campaign, but the spring that it released had probably been given its final compression by the obscure quarrel with Semkovskii and Aksel'rod.

The dispute over the handbook is susceptible of more than one interpretation. Face the facts all at once. In September, 1924, Ian Ernestovich Sten, a fairly prominent Party publicist, probably a student of Deborin's at the Institute of Red Professorship, and one of the two leading Deborinites who were Trotskyists in politics, published in *Bol'shevik*, the theoretical journal of the Party's Central Committee, a critique of a handbook for anti-religionists by I. I. Stepanov-Skvortsov, a famous old Bolshevik publicist, a leading theorist of the anti-religious movement, editor of *Izvestiia*, and one of the chief members of the Central Committee's anti-Trotskyist majority.[14] To face these facts all at once is to realize the multiple possibilities, and therefore the futility, of speculative interpretations. Perhaps Sten's attack was the consciously planned beginning of a Deborinite campaign against mechanistic thought in the anti-religious movement. Whether or not one supposes this to have been the case, there may or may not have been consultations with high authorities—the Agitation-Propaganda Department of the Central Committee, for example.[15] There is also the remote possibility that Sten acted as a Trotskyist, that his chief motive was to discredit Stepanov as a Stalinist theorist.[16] And there is the further possibility that no elaborate manœuvring was involved in Sten's polemic against Stepanov, which may have happened, as so many other little disputes were happening, simply because one Marxist disagreed with another. Such speculations may be a diverting exercise, but they are also fruitless, as long as one has no access to archival materials, and no relevant information can be found in the printed record.

The printed record does show very clearly that the necessities of the anti-religious struggle had evoked Stepanov's *Historical Materialism and Contemporary Natural Science*. As the title implied, his *Weltanschauung* for atheists had two parts: historical materialism (the Marxist doctrine of human society), and 'philosophical materialism, or, to employ a clearer and more direct expression, contemporary natural science. . . .'[17] Stepanov, as the reader will see below in greater detail, explicitly denied the need for philosophy as a separate discipline. Sten advised him to study Hegel, as Lenin had recommended, and defined 'the chief positive task of philosophy' as 'the elaboration of the methodology of science or, as Hegel puts it, the elaboration of "the science of logic".'[18] In substance this quarrel was hardly different from some others that have been reviewed above. What was different was the temper of the disputants. Stepanov was sufficiently incensed to continue the quarrel past one or two replies and rejoinders. And Sten, along with the other Deborinites, not only defended philosophy but launched a campaign against its detractors. Indeed, Deborin himself, who had previously been aloof from the disputes of Soviet Marxists,[19] proclaimed the campaign.

By January, 1925, when Deborin published his proclamation, he was in effect, though not yet in title, the managing editor of *Under the Banner of Marxism*. The editors of *Bol'shevik* had closed their columns to the controversy between Sten and Stepanov, expressly transferring it to Deborin's journal.[20] Even if Deborin had not felt threatened and aggressive because of the affair with Semkovskii and Aksel'rod, he could hardly have played the neutral referee. After all, Sten was defending a conception of philosophy that he had probably learned from Deborin, and Stepanov had quoted from Deborin's best-known book a passage that could be interpreted *against* Sten.[21] Deborin, however, went beyond a note of clarification, or a defence of Sten. Indeed, he did not even mention Sten or Stepanov. Using the first published instalment of Lenin's philosophical notes as a wall, he posted a general manifesto.

'Lenin's notes,' he announced in his editorial preface to them,

> are very timely in their appearance. There is no point in concealing the evil: not all is right in our ranks. Some comrades are inclined towards a simplification and vulgarization of Marxism, dialectics included. Others indulge in vulgar and mechanistic materialism. A third group, inclined towards a peculiar positivism, are prepared to declare dialectics 'scholasticism'.[22]

In a fourth group, no doubt for reasons of symmetry, Deborin placed those who had deserted to Hegelianism and other forms of idealism. He may well have intended the Hungarian Marxist, George Lukacs, for this category, for Lukacs was regularly used as a whipping-boy by the Deborinites when they were charged with Hegelianism. But Deborin mentioned no names in the fourth group, just as he did not name Aksel'rod and Semkovskii in the third. His vagueness was the result not of timidity but the reverse, a desire to challenge all possible contenders:

Obviously a fierce ideational struggle will erupt against all these deviations, and against any vulgarization, if a turn does not set in towards genuine dialectical materialism. Moreover, we have reason to think and hope that such a turn will come, perhaps as a result of struggle, but inevitably all the same.[23]

The doctrinal core of this manifesto, which will be examined below, stressed what Deborin's students had been stressing earlier: the autonomy of philosophy, and Hegel as its chief source. Deborin, indeed, asserted the *supremacy* of philosophy over the other, particular disciplines.

One is accordingly not surprised to find Stepanov turning for support to the Institute for the Study and Propaganda of Natural Science from the Point of View of Dialectical Materialism, or the Timiriazev Institute, as it was known for short, in honour of K. A. Timiriazev, the famous biologist. As its title implied, this institution put the major stress on the natural sciences rather than philosophy, and was staffed in the main by natural scientists of vaguely Marxist sympathies.[24] In February, 1925, after a discussion that produced a single defender of the Deborinite position, a formal resolution of support for Stepanov's book and its views was adopted with only two dissenting votes:

The open session of the Council of the . . . Institute welcomes the appearance of I. I. Stepanov's book. It is the opinion of the meeting that this book, though it contains a number of easily corrigible errors concerning particular questions of natural science, elucidates the foundation of mechanistic natural science quite correctly, and truly projects the connection of such science with the dialectical materialist world view.[25]

Thus the nucleus of the mechanist faction was formed, giving rise to the idea that the controversy was between Marxist philosophers and Marxist natural scientists, a notion that was fairly widespread at the beginning of the controversy, and has certain elements of truth. The Timiriazev Institute did become the centre of the mechanist faction; it published, beginning in 1926, what amounted to the faction's organ, a series of collections (*sborniki* or *recueils*) under the title *Dialectics in Nature.*

But an examination of the opposing factions and their philosophies in the following chapters will show that Marxist natural scientists were by no means the only important element in the mechanist faction, and further, that the Deborinites managed to acquire some natural scientists of their own.

The early view of the controversy as a clash of philosophers and natural scientists was based on two significant facts: organized defiance of Deborin's manifesto came first from a group of natural scientists, and the main issue dividing the embryonic factions was the relationship of Marxist philosophy and natural science. In a sense both these facts are surprising, for, as the reader may recall, neither the natural scientists nor the issue of natural science had played much part in the minor disputes that led up to this factional conflict. Even when the conflict had begun, the Deborinites continued, for a time, to avoid specific issues in the philosophy of natural science. As early as July, 1924, the Deborinite editors of *Under the Banner of Marxism* censured an article on epistemology by I. E. Orlov for its alleged failure to criticize 'the shortcomings of Mill's empiricism'.[26] But as late as the summer of 1926, when Orlov was a leader of the mechanist faction they disowned an attack on his theory of contingency, even though the young Bolshevik mathematician who wrote the attack quoted Deborin on the problem's philosophical aspects. And Orlov's reply to the young Bolshevik was printed in the fall of 1926 without editorial comment.[27]

But however reluctant Deborin may have been to go where the issues and some of his students were leading, he could hardly have avoided ever deeper involvement in the philosophy of natural science. Stepanov and many others understood Marxist philosophy to be a social theory (historical materialism) plus 'the most general results of contemporary natural science'. If Deborin wanted to prove the need of something more, a separate discipline trying to describe 'the universal connection of everything with everything', as he once defined the task of dialectics,[28] then he had to prove this need in the context of natural science. It was generally taken for granted that Marx's Soviet followers were consciously elaborating social science on the basis of 'the universal connection'. Soviet Marxists also took it for granted that natural scientists were unwittingly working in harmony with 'the universal connection'. The problem was to convince the natural scientists of the harmony, to persuade them that they were wrong to be indifferent or even hostile to Marxism. Thus the Deborinite defence of an autonomous discipline of philosophy came to centre on natural science.

The participants in the controversy perceived the essential nature of their conflict only gradually. The Section of Natural Science, which the Bolshevik mathematician O. Iu. Shmidt established within the Communist Academy in 1925, did not become an arena of conflict until late in the controversy. In the beginning, immediately following Deborin's manifesto and the resolution of the Timiriazev Institute in January and February, 1925, the controversy did not penetrate even into the Institute of Scientific Philosophy, the strange hybrid of the Communist Academy and RANION that was described in Chapter Four. But when the controversy did erupt within this Institute, in the spring of 1926, it provoked a series of weekly public debates, each lasting about four hours, in the course of which the opposing factions assumed their mature form.

As far as one can reconstruct the debates from subsequent allusions and printed versions of isolated speeches, the affair began quite unexpectedly in the discussion of a paper that one of Deborin's students had read.[29] The paper was a critique of Henri Bergson. Bogdanov objected to criticism based on what he called 'empty ratiocination' (his pejorative, *rassuzhdatel'stvo*, is actually untranslatable), and called instead for criticism based on the data of 'positive science', whereupon Bergson was forgotten.[30] In two months of debates the speakers ranged from the nature of universals to the place of Kant, Fichte, and Spinoza in the evolution of dialectical materialism, and to the question—only faintly philosophical—whether Soviet Marxists were actually divided into two groups. A Deborinite had just published in *Pravda* an unfavourable review of a mechanist publication, and the mechanists were convinced that he had violated the ethics of comradely discussion.[31] Deborin insisted that the mechanists were 'revisionists' and not entitled to comradely consideration.[32] Aksel'rod recalled the two years of criticism she had borne in silence, because she was reluctant to get into a controversy. But now

Deborin has declared war by characterizing me in his speech as a revisionist and positivist. . . .

You have turned one and all into 'revisionists': Semkovskii, Stepanov, and others, and finally you have begun on my humble self. You have smitten all, you have excommunicated all. There remains Deborin alone, and his students.[33]

The major philosophical question that revealed itself as the debates wore on was whether or not dialectics should be elaborated largely from Hegel, apart from the concrete material of the natural sciences. In giving a negative reply to this question, such different thinkers as Aksel'rod, A. K. Timiriazev, Bogdanov, and Varjas found themselves in substantial agreement, probably to their own surprise. The Deborinites were convinced from the beginning of the debates that Hegel was the chief source of dialectics, and that all who did not agree, however different their views might be in other respects, were 'mechanists', opponents of 'the penetration of Marxism into new scientific fields . . . in union with open and concealed opponents of Marxism. . . .'[34] The mechanists impatiently tried to brush off the label of 'mechanist', as indeed their opponents did the label 'Deborinite', for each faction claimed that it was simply Marxist.[35] The mechanists denied that they were opposed to the penetration of Marxism into natural science. They were opposed, they said, to the 'neo-Hegelianism' or 'scholastic realism' that the Deborinites were trying to bind upon natural science.[36]

About two months after the debates within the Institute of Scientific Philosophy ended, V. V. Sorin, a fairly prominent Bolshevik administrator and publicist,[37] wrote an account for *Pravda*, in which he not only credited the debates with crystallizing the factions in philosophy, but described 'Deborin's school' as 'the orthodox Marxists', and called for Party intervention on their behalf:

It would be a departure from our Party's tradition, if the struggle of ideas now proceeding in the field of philosophy were regarded as having no significance for the Party and as being unworthy of its attention. Ideational hegemony for orthodox Marxism and Leninism must be secured here too. And one cannot consider one's self—or be —a supporter of orthodox Leninism without sharing *all of Lenin's fundamental philosophical positions without exception, both in general and regarding Hegel in particular.*[38]

By his allusion, in another place, to the Central Committee's Resolution of July 1, 1925, concerning the literary controversy, Sorin seemed to be calling for a most authoritative form of Party intervention, which did not actually appear in the field of philosophy until January, 1931.[39]

Until that time, one can only guess what 'the Party's' policy was towards the controversy over the philosophy of natural science, if, indeed, 'the Party' had a policy in the matter. One thing is clear: already in 1926, when the controversy was just beginning, a trend of Party thought was observable in the Deborinite direction. Whether or to what extent this trend was initiated and directed from some high authority—perhaps the Agitation and Propaganda Department of the Central Committee—is a question that cannot be answered without access to archival materials. But the evidence of the trend is indisputable, in the Deborinite editorial policy of a growing number of periodicals and the policy of various Marxist institutions and societies of higher learning, as well as the growing number of individual Marxist intellectuals who joined what Deborin called 'the social movement around dialectics'.[40] A detailed analysis of this trend will be undertaken in a further chapter, in connection with the victory of the Deborinites, but it can be noted here even in one accession of strength that the *mechanists* enjoyed in 1926, when the anti-religious movement founded a theoretical journal.

One would expect the anti-religious movement to have contributed at least the degree of formal, organized support to the mechanists that the Timiriazev Institute did. Not only the sources and component elements of the anti-religious movement, but the fact that two of its most important theorists were also leading members of the mechanist faction, while none were Deborinites, lead one to this expectation.[41] The 'League of the Militant Godless' was being organized at the very time that its leading theorist, Stepanov, was being attacked as a revisionist for his anti-religious handbook.[42] And early in 1926, as the philosophical factions were taking final shape, the League launched its theoretical journal, *The Anti-Religionist*, with indications of support for the mechanists, as one might expect. But they were veiled and indirect indications of support. The editorial manifesto that proclaimed the journal's existence stressed '*the elaboration of a rigorous, scientific world view*' as its chief aim, and one can note in such phrasing the same aversion to the term 'philosophy', or even 'dialectical materialism' that was observable in Stepanov's book.[43] The manifesto appealed for aid to the Society of Militant Materialists, and especially to materialist natural scientists, pointedly omitting an appeal to Marxist philosophers, who were regarded as uniformly Deborinite in the early phase of the controversy. *The Anti-Religionist's* first book review was explicit in its support for the mechanist faction only by comparison with the truly subtle hinting of the editorial. The review was an encomium of a second edition of Stepanov's disputed book. Noting that the first edition occasioned 'lively debates in philosophical circles', and completely ignoring the issues of the debates, as though they were without significance to the anti-religionist, the reviewer declared that Stepanov's book 'correctly elucidates the foundations of the Marxist world view'.[44] That was the full extent of the journal's support for the mechanist faction.

For a long time *The Anti-Religionist* was mechanist *in tendency*: not only the frequent popularizations of natural science but even the periodic lists of anti-religious literature displayed this tendency, for these bibliographies had no category marked philosophy, and the works of Deborinites were not mentioned, as a rule. The editors, Iaroslavskii and Lukachevskii, did not excise from their journal such remarks as occurred, for example, in a report on the public debates between priests and anti-religionists: '. . . sometimes the theme takes a philosophical deviation, the speakers toy with foreign words, quarrel about terms. . . .'[45] But there was no open support for the mechanist *faction*. Sarab'ianov, who was a prolific anti-religious writer as well as a leading opponent of the Deborinites, published his philosophical polemics in other journals than *The Anti-Religionist*.

Speculation concerning the causes of this restraint—particularly remarkable for such an obstreperous group as the anti-religionists—may be left to a later chapter, where a number of similar anomalies will be considered jointly. The fact itself is noted here as a sign of the mechanist faction's peculiar failure, at the very outset of the controversy, to realize fully its potential resources of strength. A similar failure can be noted in the case of natural scientists with Marxist sympathies. From the general agreement in both factions that the majority of Marxist natural scientists sympathized with the mechanist faction,[46] we may assume such sympathy to have been a fact. The failure of significant numbers of Marxist natural scientists to speak up strongly and persistently in support of the mechanist faction was accordingly no less anomalous than the restraint shown by the anti-religious movement. The silence of Bukharin, Trotsky, and other social theorists of a mechanist tendency was thus part of a general pattern. Mechanism in various forms was probably the dominant philosophy among Soviet Marxists, but relatively few came to its defence when the Deborinites challenged it.

9

THE MECHANIST FACTION: PROPAGANDISTS AND PHILOSOPHERS

To call the mechanists a faction may suggest an organized coherence of thought and membership that they sadly lacked. Most of the Soviet Marxists who associated themselves with the faction did so only occasionally and briefly, and those who may be considered the faction's leaders, either because of their personal eminence or because of their persistence in the faction's cause, were far from professing a uniform or clearly defined creed. Accordingly, a series of intellectual biographies is the only approach to the faction's basic standpoint.

'Ivan Ivanovich,' we are told in an obituary of I. I. Stepanov, 'was never of a "scholarly" nature, he was a fighter. . . .'[1] The comment was undoubtedly justified, but it should be borne in mind that Stepanov's weapons were ideas. Gorky knew him as a 'Marxist scholar. He eschewed all books except *Das Kapital*—he made a boast of that. . . . A short man with a greyish complexion, his light blue eyes smiled, however, the triumphant smile of the lucky man who has reached a truth inaccessible to others'.[2] Before and after the Revolution he was one of the Party's leading 'propagandists', as the disseminators of basic doctrine were called, in distinction from 'agitators', who carried the Party's line on immediate issues to 'the broad masses'.[3] Already forty-seven at the time of the Bolshevik Revolution, Stepanov was best known for his popularizations of Marxist economics, some of which had been written with Bogdanov, and for the translation of Marx's *Capital* (all three volumes) that he had done with Bazarov. The association with Bogdanov and Bazarov must not be interpreted to imply agreement with their 'Machist' philosophy, for Stepanov had helped Lenin bring out *Materialism and Empirio-criticism.*[4] (His subsequent wavering towards support of Bogdanov's *Vperedists* was probably prompted by their political appeal as arch-revolutionaries.[5])

In 1922, when Russia's production of iron and steel had fallen to the level that Peter the Great had reached two hundred years earlier, and a disorganized economy helped a drought to produce a terrible famine, Stepanov contributed to Bolshevik self-confidence with a book, *The Electrification of the R.S.F.S.R. in Connection with the Transitional Phase of the World Economy.* It was perhaps the first of what has since become an important genre of propagandistic literature in the Soviet Union: the 'plan for the transformation of nature and society', a mixture of popular science, Communist politics, and science fiction of the optimistic variety. Lenin himself wrote an enthusiastic preface, and Stepanov came to be regarded as an authority on the Party's programme in the field of natural science and technology.[6] The book's attitude towards the philosophy of natural science was pretty clearly mechanist and positivist, but such views were taken for granted more often than not in the first half of the 'twenties.[7] Stepanov was therefore surprised and indignant when an organized campaign against such mechanism and positivism was precipitated by his publication of a handbook for anti-religionists in 1924.

The demands for such a manual, Stepanov later explained,

> made themselves felt in dozens of notes that poured in after every speech on 'religious themes' in Moscow, in the Urals, the Ukraine; they besieged us almost at every meeting of the anti-religionists; they erupted, finally, in the questions that every anti-religious agitator is obliged to answer, and in the face of which he often feels his helplessness. At each step it was revealed that the Marxist cannot avoid contemporary natural science, that he must give at least a preliminary answer to several basic problems of natural science.[8]

The central problem was to refute the 'argument from design', to show that life and the human psyche could be explained without recourse to a supernatural intelligence or creator.

Stepanov assumed that an ontology was necessary for this purpose, and he found it in the popular science of his day. The entire universe, in his view, was made up of microscopic particles that differed in a few physical qualities such as charge and mass but were otherwise identical. Ultimately all phenomena, biological and intellectual as well as physical, would be reduced to the combination and separation of electrons and protons. Although the completion of such a unified science lay still in the remote future, Stepanov believed that major steps towards it had already been taken, and indeed that the basic law had been discovered: the law of the conservation and transformation of matter and energy. Some day, he believed, all the theories of the physical, biological, and social sciences would be shown to be particular instances of this basic law.[9]

This ontology has been inferred and pieced together from a number of remarks that Stepanov dropped on various occasions. If it cannot be found clearly and fully developed in any one of his writings, the reason is not that he was ashamed of it. When his opponents challenged him to comment on Engels' statement that undifferentiated matter, 'matter as such', is an abstraction only, a mental construct, while only particular kinds of matter exist in reality, Stepanov replied boldly that twentieth-century physics had proved Engels wrong. 'Matter as such,' he said, had been found experimentally; it was electrons and protons, the building blocks of the universe.[10] Or, when reproached for not mentioning dialectics in his manual for anti-religionists, he replied quite forthrightly:

> The dialectical understanding [of nature] is too general a name. For the present time the dialectical understanding of nature takes concrete form precisely as the mechanical understanding, i.e., as the reduction of all nature's processes exclusively to the action and transformation of those forms of energy that are studied by physics and chemistry.[11]

But if he was unashamedly mechanistic or reductionistic, he was also pretty thoroughly positivistic, and this is presumably the reason that his ontology is nowhere clearly and fully developed:

> Historical materialism *continues* the work that is done in one part by philosophical materialism, or, to use a clearer and more direct ex-

pression, by contemporary *natural science*; for Marxists there exists no field of some kind of 'philosophizing', separate and aloof from science: materialist philosophy is for Marxists the latest and most general conclusions of contemporary science.[12]

He was convinced that the Deborinites were helping the cause of religion by denying the adequacy of physical and chemical explanations of life, and he feared that the 'Hegelian scholasticism' of the Deborinites was alienating rather than winning the natural scientists:

If we dialectical materialists want, not in words, not in empty chatter, but in fact to lead the natural scientists, will it not be correct to show [them] . . . that *Engels almost half a century ago foresaw that science can expect great conquests precisely along the path that they are travelling*?[13]

In other words, Stepanov believed that Soviet natural scientists should be won to Marxism by being shown that they were already dialectical materialists without realizing it.

Stepanov's one-time collaborator in the field of economic theory, Alexander Aleksandrovich Bogdanov (*né* Malinovskii in 1873), was an ardent believer in collectivism who insisted to the end on going his individual way. Repeatedly offered the choice of submitting to Lenin's 'rigid discipline' or standing aside, he repeatedly chose to stand aside.[14] The last choice, as it turned out, was made in 1920, when the Central Committee ordered Bogdanov's *Proletcult* to submit to the Commissar of Education —his brother-in-law and fellow 'Machist', Lunacharsky, who had chosen to submit to Lenin's discipline while retaining his intellectual independence. 'My *Proletcult* work was cut short,' Bogdanov tells us in his laconic autobiography; 'I devoted myself entirely to scientific work.'[15] As a medical doctor, he headed an Institute for Research in Blood Transfusion; as a member of the Communist Academy he participated in that body's discussions of philosophy and published in its journal, until he killed himself in 1928 by performing an unsuccessful experiment in blood transfusion on his own body.[16] But he had no part in the teaching or popularization of Marxist philosophy, where, indeed, his name and theories were endlessly condemned.[17] His occasional words on behalf of the mechanists were accordingly an embarrassment rather than an aid to their cause.

When a Deborinite speaker, in the course of one discussion within the Academy, said that the mechanists would cite a paper of Bogdanov's in their behalf, Bogdanov called out: 'You won't find people so stupid as to cite me; it's disadvantageous. . . . There is unity [between my views and the mechanists'], but they won't cite me.'[18]

In reality 'unity' was too strong a word to describe the relationship between Bogdanov's views and those of most mechanists. For one thing, most of the mechanists had a thirst for orthodoxy, a tendency to 'prove' their arguments by citations from the 'classics', which was repulsive to Bogdanov, who had learned in a militaristic high school 'to fear and hate dominators and to reject authorities'.[19] 'One needs to study,' he told the Communist Academy in 1927,

> not old chains of reasoning, even though the best of them; one needs to go to nature, to life, to prove every chain of reasoning in nature, in life; there is the criterion, there is truth, and all the rest, all expenditures of energy on what has outlived itself and cannot be revived—that is grave-robbing, that is vampirism of what is moribund but does not want to die.[20]

When he was censured for his 'revisionist' attitude towards 'orthodox Marxism', he grew sarcastic:

> Precisely my little old godmother, who first excommunicated me, was orthodox; and now she too has proved to be a heretic. . . . In Russia, when you speak of dialectics, you do so most often in order to deny in fact the possibility of development.[21]

Not even mechanists like Stepanov, Sarab'ianov, or Aksel'rod, who were bold enough to express explicit disagreement with the 'classics' on specific issues, were willing to take this completely mocking attitude towards orthodoxy.

The basis for Bogdanov's 'unity' with the mechanists must be sought in certain aspects of his philosophy of science, which was most extensively elaborated. Indeed, of all the mechanists and Deborinites, Bogdanov is probably the most deserving of detailed study as a philosopher; he is probably the one who is least adequately described in a brief summary. Accordingly, the following synopsis of his views is offered tentatively, without any claim to being a definitive account.

Bogdanov felt that Marx and Engels had many perspicacious ideas that could be developed further if their antiquated philosophy, borrowed from Hegel, were superseded by a philosophy more in harmony with the methods and spirit of modern natural science. Dialectics he considered neither a universal theory of knowledge nor a universal ontology, but a needlessly confused version of the commonplace idea that opposing forces can sometimes produce motion and change. Materialism was to Bogdanov a near truth that had been made an erroneous dogma by Lenin and his followers. In Bogdanov's view the 'external', the 'objectively real' were what the human mind must contend with, what the human mind perceives as a regularity (*zakonomernost'*, *Gesetzmässigkeit*) different from its own. Or, approaching the problem in a different way, he defined the 'objectively real' as the regularities (*zakonomernosti*, *Gesetzmässigkeiten*) constructed by the human mind: the socially verified, collective experience of humanity.

The basic trend of human thought, he believed, was towards a 'scientific monism'. First, the fantastic monisms called religions, whose main function was to subject nature to a crude level of production and to subject exploited producers to the will of the organizers of production; then, the speculative monisms of metaphysical philosophies, which attempted to overleap the inadequacies of nascent science—these were the past stages of human thought. The time had come, he believed, to construct a monism that could be tested empirically. He selected the thermodynamic theories of Gibbs and Le Châtelier as furnishing the most likely chief hypothesis for such a monism, and tried to surmise how the main theories of other sciences—Marxian economics included—might sometime be converted into instances of the 'universal law of equilibrium'.[22]

It was not the relativism as much as the positivism in this philosophy that brought Bogdanov into 'unity' with the other mechanists. The reader has already seen how, in the spring of 1926, he helped precipitate the debates within the Institute of Scientific Philosophy by objecting to a Deborinite criticism of Henri Bergson. The criticism was based, Bogdanov said, not on the data of positive science but on 'empty ratiocination' (*rassuzhdatel'stvo*). The Deborinites dared the other mechanists at the debates to agree with Bogdanov's 'rejection of philosophy', and L. I. Aksel'rod, one of the two mechanists whose remarks in the debates were subsequently printed, did so, brushing aside the invidious allegation that Bogdanov and she were 'rejecting philosophy'.[23] A. K. Timiriazev, the other mechanist whose remarks were printed, ignored the Deborinite challenge. Timiriazev was entirely opposed to relativism in philosophy and to the theory of relativity in physics, and had previously clashed with Bogdanov on both subjects, but he did not disown Bogdanov on this occasion for he shared Bogdanov's repugnance towards 'empty ratiocination'.[24] Timiriazev and most other mechanists usually expressed this repugnance by the demand that dialectics should not be elaborated apart from concrete data, that it should flow out of a detailed study of the positive sciences, while Bogdanov, lacking reverence for the terminology of 'orthodox Marxism', did not ordinarily speak of dialectics. He demanded simply that the generalizations and hypotheses of philosophy be subject to empirical verification. But the difference in the formulations does not conceal the basic affinity or 'unity' of views. Moreover, Aksel'rod and Sarab'ianov, the two mechanists who publicly acknowledged their affinity with Bogdanov, shared some of his relativism as well as his positivism.

It is surprising that Liubov' Isaakovna Aksel'rod (aged fifty-eight in 1926) expressed agreement with Bogdanov, for she had been the first Marxist to criticize his revisionist philosophy before the Revolution. She was known as Aksel'rod the Orthodox, or simply as Orthodox, because of her devotion to 'our classics and masters', as she called Marx, Engels, and Plekhanov.[25] Indeed, a full biographical study of Aksel'rod would probably show that devotion to Plekhanov was the central principle of her intellectual life. Her politics faithfully reflected Plekhanov's somewhat erratic Menshevism throughout the pre-revolutionary period of her life; with him she quit politics after the Bolshevik Revolution, and she probably retained his attitude towards the Soviet régime even after he died in 1918.[26] At any rate she did not join the Bolshevik Party, and she publicly refused to repudiate her pre-revolutionary Menshevik views when she was pressed to do so.[27]

The reader has already seen that Lenin nevertheless urged

her appointment to an important place in the teaching of philosophy at Sverdlov Communist University and the Institute of Red Professorship, and that she had a falling out with A. M. Deborin, her colleague at these institutions and at the Communist Academy. In the ensuing dispute she tried to prove that her philosophy was orthodox, that it was derived from Plekhanov, 'whose piety towards Marx and Engels is well known'.[28] But her orthodoxy was more Protestant than Catholic; what her individual reason found in the revered texts and in the new revelations of science, not the collective wisdom of instituted authority, defined the orthodox position for her. In 1929, when a Conference of Marxist-Leninist Institutions condemned mechanism, and even after 1931, when the Party's Central Committee officially endorsed the condemnation, she did not renounce her views.[29] She has already been quoted sneering at her opponents for 'excommunicating everybody . . . [but] Deborin and his students'.[30] On one occasion she expressed herself more forcibly:

> If one is to regard as 'revisionism' differences [from the classics] in the evaluation of one or another *particular* phenomenon, then it is perfectly clear that nothing remains for the followers of Marxism but to repeat, and to repeat endlessly and with painful tedium, what was said by the masters. In other words the living, eternally creative theory of dialectical materialism would have to be converted into a deadening religious dogmatism.[31]

Orthodox Marxism, she explained, was adherence to 'the method and spirit, not the letter of the doctrine'.[32]

Aksel'rod's understanding of the method and spirit of Marxism found positive expression mostly in the fields of aesthetics, social theory, and the history of philosophy; her largest work, a doctoral dissertation that had appeared in 1901, was a study of Tolstoy.[33] But she was better known as a polemicist. Lenin had esteemed her attacks on various forms of 'idealism' and revisionism, which were reprinted several times in the Soviet period—until the habit of judging books by their authors' politics grew even stronger than reverence for Lenin's judgment.[34] Her polemics with the Deborinites, who were originally no more interested in the philosophy of natural science than she, began with arguments about Spinoza and Hegel.[35] But in defining their respective attitudes towards Hegel, Aksel'rod and her opponents defined their divergent views on dialectics; and by their debates over Spinoza, they clarified their positions on materialism.

At the centre of her quarrel with the Deborinites Aksel'rod found 'the problem of dialectics, or rather, not the problem itself, but the different views on the relationship of dialectical materialism to Hegel's dialectics'.[36] She urged, in effect, that Hegelian dialectics be forgotten:

Deborin says, for example: purposefulness is *transcended* [*tselesoobraznost' snimaetsia, Zweckmässigkeit ist aufgehoben*] by causality. This sounds chic and entirely Hegelian. But such terminology binds one to nothing and suggests no questions. It is another matter if we say: purposefulness is a variety of causality. In this case a whole series of concrete questions immediately rises, to wit: what kind of variety, what are its sources, etc.?[37]

Her chief contention, constantly repeated in various forms, was that

dialectics must receive *the content of contemporary science*. The simple transmission of some well-known Hegelian propositions is devoid of any serious significance. . . . Dialectics acquires its true significance only when it is intrinsically and indissolubly connected with concrete content.[38]

Though she denied the Deborinite charge that she wanted the discipline (*nauka*) of philosophy to be limited to ethics, aesthetics, and the history of philosophy, and otherwise to be dissolved in the positive sciences, her underlying attitude clearly seems to have tended in that direction. As a methodology of science, she said, '*dialectics taken in its entirely abstract form leads inevitably to empty and meaningless scholasticism*'.[39] And when she was pressed for her views on dialectics as the most general synthesis of the findings of science, she exclaimed:

I dare to reply to this question: the synthesis is in the aeroplane, in the radio receiver, and in general in all the great practical results of contemporary natural science. I dare to reply in this way because I do not forget Marx's great thought that one may explain the world one way or another, but the most important thing is to change it.[40]

Aksel'rod's dispute with the Deborinites concerning Spinoza in effect resumed her pre-revolutionary disagreement with what she called 'naïve realism'.[41] Her chief opponent was now Deborin, who had in the pre-revolutionary period agreed with her criticism of Lenin as a 'naïve realist', and Aksel'rod now described the main issue as an historical problem:

My grievous sin, from my opponents' point of view, consists of this, that I regard Spinoza's substance as the course of conformity to law [*zakonomernost'*, *Gesetzmässigkeit*]. From my opponents' point of view Spinoza's god or substance is nature, identical with matter. In other words, substance is matter.[42]

But she, no less than the Deborinites, accepted Plekhanov's dictum that 'Marxism is a variety of Spinozism', and their disagreement over Spinoza's concept of substance was accordingly a disagreement over the source and nature of Marxist materialism. To Aksel'rod the 'soul of materialism' was its recognition in nature of a universal regularity or conformity to law (*zakonomernost'*, *Gesetzmässigkeit*), and its consequent rejection of teleology and religion. She would not agree that the hallmark of materialism was Lenin's 'theory of reflection'; she rejected his theory that human knowledge '*reflects*' the '*essence*' of external matter. She believed that matter exists independently of, and prior to, the human mind, but she felt that our sensations and knowledge '*correspond*' to certain '*properties*' of external matter. She repeated the old distinction between the primary and secondary qualities of matter, the former alone being subject to measurement and hence to scientific study:

Imagine that my neighbour's brain is exposed and that I, as a researcher, am observing the activity of the nerve centres just at the moment when he is experiencing some sensation. What then? Under this circumstance, ideal for cognition, I should nevertheless come to know merely the *motion* that causes my neighbour's sensation, but I should not obtain the slightest conception of his internal condition, i.e., of his experience [*perezhivanie*]; in other words, this would remain closed to me by a dense, impenetrable cloud.[43]

In a similar vein Aksel'rod objected to the view that universal concepts reflect independently existing universals; her position on this ancient problem was that universal concepts correspond to objectively existing relationships or connections among individuals.[44] In general, then, she was a Marxist who construed materialism to be largely a synonym for empiricism and irreligious determinism.

A similar understanding of materialism can be found in the writings of Semen Iulevich Semkovskii (*né* Bronshtein in 1882) who had been a prominent Menshevik and Trotskyist before the Bolshevik Revolution. Even though Semkovskii's politics and his 'Machist' philosophy had involved him before 1917 in fairly sharp clashes with Lenin, the necessities of the 'cultural revolution' made him the chief Marxist philosopher in the Soviet Ukraine. His efforts to win the natural scientists in the Ukrainian Academy of Sciences, by telling them that they already were Marxists without realizing it, have already been mentioned; for the incidental rejection of Hegelian dialectics with which he bolstered his appeals to scientists helped precipitate the formation of rival factions among Soviet Marxists. The main burden of his argument to the Ukrainian scientists was that Einstein's theory of relativity was the concrete form of dialectical materialism in the physical sciences; that Darwinism was the realization of the same philosophy in the biological sciences; and that Marxist social theory was its crowning triumph in the scientific study of man.[45]

The identification of dialectical materialism and the theory of relativity brought Semkovskii into conflict with a leading mechanist, A. K. Timiriazev,[46] at the same time that the rejection of Hegelian dialectics involved him in controversy with the Deborinites. He called for a 'third force' in Soviet Marxist philosophy, which would avoid both 'vulgar materialism' and 'Hegelian scholasticism'.[47] But he could not escape identification with the mechanist faction. For he attacked the Deborinites on the central issue (the elaboration of dialectics from Hegel, apart from the positive sciences), and defended the theory of relativity when it was no issue at all in the clash of the two factions. The Deborinites were largely indifferent to Einstein's physics, and few mechanists approved of Timiriazev's hostility to it. Even in 1928, when the Deborinites were moving towards Semkovskii's position on relativity, he repeated the chief mechanist criticism of them: they were '"pure" philosophers, separated from the material of the sciences, [who] philosophize fruitlessly. . . .'[48]

In 1929, at the Conference that formally condemned the

mechanist faction, Semkovskii spoke as a *Deborinite.*[49] He spoke as if the views that the Conference was endorsing had been his all along. Had he not been arguing that the theory of relativity was the realization of dialectical materialism in physics? As for the general issue of the controversy, the relationship between dialectical materialism and natural science, he recalled an ingeniously ambiguous metaphor:

In this question I usually employ the following figure, which, in my opinion, most correctly expresses the relationship between dialectical philosophy and the individual sciences. This relationship can be compared with the relationship of *the Party and the class* in the Leninist view. 'The Party,' I have written . . ., 'does not go behind, does not passively "summarize", but goes in front, actively leads. But for this, Lenin taught, the Party itself must be the concentrated expression of the class.' And such is also the dialectics of the interaction between philosophy and science.[50]

The ambiguity of the metaphor was its chief virtue to Semkovskii; it spanned his previous views as a mechanist and his present endorsement of the Deborinite position. Had he not been trying to establish Marxist leadership in natural science by showing that Marxist philosophy was the 'concentrated expression' of science? The Deborinites, to be sure, sought this 'concentrated expression' mainly in Hegelian dialectics, which Semkovskii had rejected in 1924 as 'scholasticism'. In April, 1929, he did not repeat that epithet, but he did warn that Hegelian dialectics was idealist and therefore could not be transferred to Marxist philosophy without basic alterations.[51] It seems clear, therefore, that his shift of philosophical factions was accompanied by little change of philosophical views.[52]

While Semkovskii was searching for a philosophy that would win natural scientists to Marxism in Kharkov and Kiev, Aleksandr Ignat'evich Var'iash was doing much the same in Moscow. Sandor Varjas, to use his proper Hungarian name, was a professor of philosophy who had been an official of the union of teachers and civil servants during Bela Kun's régime. Admiral Horthy's régime had imprisoned him, but an exchange of prisoners between Russia and Hungary brought him in 1922, aged thirty-seven, to Moscow, where he was placed in a munber of teaching and administrative positions, all involving efforts to win intellectuals to Marxism.[53]

For the most part there was little to distinguish his philosophy of science from the mechanist versions that have been surveyed so far. The unity of science was his great dream no less than the others'. 'I stand,' he said, 'on Marx's point of view that "we know one single science. . . ."'[54] Maxwell had taken a giant step towards this single science by reducing electricity and magnetism to a single set of laws; most recently Einstein's theory of relativity had moved the physical sciences closer yet. Ultimately scientists would achieve 'the unification of physics and chemistry, then of chemistry and biology, then of biology and psychology'.[55] The curious omission of the expected final item (social science) together with Varjas' explicit rejection on two occasions of efforts to reduce social science to biology,[56] was an anomaly that was fairly common among the mechanists. But the anomaly was especially marked in Varjas' case, for he was given to sweeping assertions of reductionism:

> There is not one basic law or regularity [*zakonomernost'*, *Gesetzmässigkeit*] that is not physico-chemical. . . . The so-called 'mechanists' . . . do not accept more fundamental, irreducible qualities than are absolutely necessary for objective explanation. We accept the fundamental qualities of the electrical charge in the two forms of positive and negative electricity; we accept the various possible distributions of electrons and other material particles; we accept also a whole series of properties (motion and extension as attributes of matter), but naturally we do not accept as many properties as man's naked eye is capable of noting.[57]

Moreover, when Varjas discussed social theory, he was again prone to rigidly reductionist sentiments, as in the following thematic statement from his *History of Modern Philosophy*:

> . . . *between ideas and the social process of production there exists only one correlation, namely that ideas are entirely and unequivocally and causally determined by the process of production.*[58]

Yet he denied the suggestion that social phenomena might be reduced further than 'the process of production', to biological or even physical laws.

In other matters too Varjas had views that were typical of the mechanist faction. In the spirit of empiricism he derided the 'Hegelian panlogism' of the Deborinites, and insisted that dialectical materialism was not to be extracted from Hegel and applied to the sciences. He said that dialectical materialism was to be found in the work of an Einstein or a Weyl, who might imagine themselves to be following the philosophies of Mach or Husserl but were actually using the method of dialectical matərialism:

For *dialectical materialism is not a norm of some categorical imperative, which one must follow*, according to which it is necessary to proceed in order to obtain good results; it is, as far as possible, an exact description of the actual path along which humanity is actually proceeding. Dialectics is not a logical postulate, not a norm, but the actual path.[59]

Yet this foe of 'Hegelian panlogism' was charged by the Deborinites with hypostasizing logic and mathematics, with seeking materialist dialectics in Bertrand Russell's *Principles of Mathematics*.[60] And there was some justice in the charge. Varjas did have a high regard for Russell's contribution, which, as Varjas saw it, transcended formal logic and was actually dialectical logic.[61] In general he believed that contemporary logicians and mathematicians, whether they were 'bourgeois' or not in their social philosophy, were making important contributions to materialist dialectics, which he regarded as a universal methodology that summed up the most general characteristics of the universe. In other words, Varjas had a trend of thought similar to what has already been noted in Lafargue and will be found again in Tseitlin: a tendency to fuse rationalism with mechanistic materialism. If reason is a mirror of material reality, he argued, then one can see material reality in it. In the most abstract products of reason (mathematics and logic) one can perceive the most general characteristics of material reality, and thereby arrive at a methodology for science as a whole.[62]

Unusual combinations of ideas were also characteristic of Vladimir Nikolaevich Sarab'ianov, who was a very prolific propagandist in a variety of fields, from anti-religion to economics. As late as 1926 the Deborinite Luppol found reason to say that Sarab'ianov, who was then forty, was 'a rather good dialectician, and in a whole series of the theoretical disputes of these days he has taken a correct dialectical [i.e., Deborinite] position'.[63] Perhaps Luppol had in mind Sarab'ianov's criticism of Bukharin, which has already been noted.[64] It is more likely however that Luppol was pleased by Sarab'ianov's explicit defence of philosophy as an autonomous discipline, in marked contrast with the tendency of some mechanists to dissolve philosophy into the natural and social sciences.[65] Moreover, Sarab'ianov had no objection to the use of Hegelian terminology. To be sure, he held a number of views that the Deborinites opposed as mechanistic. He defined 'accident' simply as an event whose cause is unknown, while they favoured the Hegelian notion that accident is a form of necessity;[66] he felt that mental phenomena would ultimately be reduced to chemical and physical processes;[67] and he had a tendency to interpret the abstruse Hegelian concepts that he did accept in such a way that they became indistinguishable from ordinary concepts—'self-movement', for example, was to Sarab'ianov little more than a synonym for development or growth.[68] But more than anything else, it was his views on the concept of quality that involved him in dispute with the Deborinites. For both Sarab'ianov and the Deborinites placed heavy emphasis on the concept of quality, but each understood it differently.

Sarab'ianov assigned great importance to the concept of quality as a result of his difficulties in the propaganda of historical materialism (Marxist social theory) and anti-religion. He blamed these difficulties on the vagueness that he found in Marxism. Social base and social superstructure, ran the formula of historical materialism, influence each other reciprocally, but the base is the determining influence 'in the long run', or 'in the last analysis'. Marxist anti-religion taught that living and thinking beings have developed out of non-living and non-thinking matter in a manner that can be explained by physical and chemical processes, without recourse to the supernatural; but the Marxist 'classics' also denied that such explanation by reduction 'will exhaust the essence' of life or thought.[69] Sarab'ianov argued that analysis of both formulas turns on the concept of quality. Superstructure and base, he reasoned, can influence each other quantitatively, but *qualitative* change can come only from the base. Higher levels of the organization of matter rise out of lower ones, but each level has a special quality that is not entirely reducible to the levels below it. The problem was to define quality unequivocally, and he believed that he had found such a definition: 'Quality is the relationship of the aggregate of properties of a given thing, a given phenomenon, to the aggregate of properties of another thing, another phenomenon.'[70] The empiricist shunning of ontology was as obvious in this definition of quality as it was in Aksel'rod's; both avoided speaking of the essential attributes of things. To be sure, Sarab'ianov repeated Aksel'rod's exception for 'motion, bulk, [and] weight', which he considered inherent in matter and not merely 'the result of the interaction between the external world and our own'.[71] But on the whole, Sarab'ianov stressed the agnostic view of quality as a relationship; he endorsed Plekhanov's characterization of sensations and knowledge as 'hieroglyphs', and he even repeated Hobbes' argument that the sound we hear when a bell rings is a 'phantom'.[72]

Into this understanding of quality Sarab'ianov introduced what he considered a scientific version of dialectics; the relationship that constitutes a quality he called a relationship of 'opposition' or 'contradiction'. His purpose was to make quality a measurable concept by describing it as the 'predominance' of one aggregate over the other. When quantitative, measurable changes in either aggregate reversed the 'predominance' of one over the other, Sarab'ianov argued, a 'leap' occurred, a new quality came into existence. To his opponents it seemed at times that he was trying to ridicule dialectics, as in the following illustration of the 'leap' from one quality to another:

> I am hungry, I eat a loaf of bread—less hungry; I eat another—still less hungry; I eat a biscuit—I am full. Quantity has passed into quality. In a certain relationship I have become a new quality. But at this moment (moment of the leap), when I have become full, I still want to drink; in this relationship I shall become another quality (don't want to drink) only some time later. And so on.
>
> Was there a leap? Yes and no. Dialectics.[73]

There can, however, be little doubt that Sarab'ianov had serious purposes in mind. In the process of cognition, he argued, men select for study only a few of the infinite qualities of things. The basis for the selection is the purpose, the practical goal of the person making the selection. This train of thought led him to recall and stress the remark that Lenin had dropped while criticizing Struve, to the effect that 'materialism includes in itself partyness [*partiinost'*], requiring that one take the point of view of a definite group or class'.[74] He interpreted this dictum, which was ignored by most Soviet Marxists in the 'twenties, in a highly subjectivist manner. Both Lenin (the ideologist of the proletariat) and Struve (the ideologist of Russia's nascent industrial bourgeoisie) were right, each from the point of view of the class he chose to speak for. One chooses one's allegiance and interests, and truth is altered accordingly.[75] Sarab'ianov called this epistemology 'subjectivism on the basis of objectivism', and argued that it alone could justify Bolshevik social theory and political action.[76]

Of more interest to the present study is his claim that the anti-religious campaign and the effort to win natural scientists were seriously hampered by fundamental obscurities in Marxism:

> In any university audience to a lecture on dialectical materialism, when you give examples of the transition [of a thing] into its opposite they will reply to you: 'What kind of an opposite is that!' Meantime the teacher of natural science strengthens this 'scepticism' by a series of very learned proofs that the plant *is not* the opposite of the seed, that nature does not make leaps, and so on.[77]

Sarab'ianov admitted that, in his efforts to overcome such arguments by defining Marxist concepts more precisely, he was departing somewhat from Engels' or Lenin's understanding of dialectical materialism and approaching the views of Bogdanov. But, like Aksel'rod or Bogdanov himself, though less forcibly, he argued that criticism of the 'classics' was required not only by new times and new knowledge but also by devotion to the true spirit of the 'classics'.[78]

It should by now be apparent that the leading mechanist philosophers and propagandists adhered not to a uniform creed but to diverse formulations of a common attitude. The problem of defining this common attitude may be postponed until the views of the natural scientists who were leading mechanists have been examined. But already it should be clear that it is a considerable over-simplification to characterize the mechanists *en masse* simply as opponents of dialectical materialism, as advocates of the view that science is its own philosophy. It is noteworthy that the one mechanist whose views can fairly be described in that fashion was associated only briefly with the faction. In February, 1925, when the Timiriazev Institute discussed and formally approved Stepanov's handbook, Ivan Adamovich Borichevskii (aged thirty-three) added his vigorous assent, but objected to a fellow-mechanist's use of such 'indisputably "philosophical" terms' as quality:

'Qualities' are nothing but certain changes of 'quantities'. And for scientific materialism it is entirely sufficient to recognize that these changes arise not only by means of slow, 'continuous' development, but also by explosions, by leaps. All the rest is the purest 'philosophy', which is entirely unnecessary, whether for positive science or for scientific materialism.[79]

'Scientific materialism,' he explained, is a *Weltanschauung* that is 'intra-scientific, that wants to be one of the limiting generalizations of positive science as a whole, and strives to emancipate science from any philosophical phantasms'.[80] Borichevskii even objected to the 'philosophical legend, invented by Engels and repeated by Lenin', that Büchner and Moleschott were 'vulgar' materialists.[81]

The Deborinites repeatedly cited Borichevskii's statements as prime examples of the mechanists' views, but actually such extreme positivism was unusual for the mechanist faction. Borichevskii himself, who had been very active as a writer on philosophical topics up to 1926, published nothing more until 1930, when the clash between the Deborinite and mechanist factions was superseded by a controversy between the Deborinites and a new group.[82] One suspects that in the period from 1926 to 1930 he was *persona non grata* to the mechanists as well as the Deborinites.[83] Even Stepanov, who was frankly opposed to the cultivation of philosophy as a separate discipline, took philosophy more seriously than Borichevskii, treating its traditional concepts with something more than an irritated gesture of rejection. But the natural scientists in the mechanist faction must be canvassed before its least common denominator can be discovered.

10

THE MECHANIST FACTION: NATURAL SCIENTISTS

THERE were two distinct phases in the association of biological scientists with the mechanist faction. In the first, several biologists who were opposed to vitalism supported the mechanist faction because it seemed clearly committed to physical and chemical explanations of living phenomena, while the Deborinite position on this problem seemed dubious. Gradually, however, these mechanist biologists dropped out of the philosophical controversy; at the close in 1929 only one of them was still actively supporting the mechanist faction. The second phase set in rather abruptly towards the end of the controversy, when the genetical theory called 'Lamarckism' was attached to the mechanist philosophy, and a few previously aloof biologists joined the faction, arguing the inheritability of acquired characteristics rather than the possibility of reducing life to the chemistry of proteins. The two phases thus involved an almost complete turnover both of individuals and of issues.

Among the Marxists who may be considered charter members of the mechanist faction because of their part in the Timiriazev Institute's formal approval of Stepanov's little book, a considerable number must have been biologists. Of the eight speakers who took Stepanov's side in the discussion preceding the vote, four were biologists, including the Assistant Director of the Institute.[1] He was the botanist, George Gustavovich Bosse, aged thirty-eight in 1925, when he helped found the mechanist faction, and apparently a tireless teacher and popularizer both of biology and of Marxist philosophy.[2] In both areas a thoroughgoing reductionism was his basic principle:

Theoretically and in the last analysis social phenomena are also open not only to qualitative (sociological) analysis, but also to quantitative (physico-chemico-biological) analysis. It is another matter whether, apart from the method of historical materialism, that genial method for the analysis of social phenomena, we shall sometime succeed in working out a physico-chemico-biological method that will be applicable to social phenomena. As yet we have no basis for this hope. *We have not made a single step in this direction.* . . . But past unsuccessful approaches *do not prove* the impossibility *in principle* of building a quantitative, dialectical materialist, mechanical foundation under sociology.[3]

The Deborinites made much of this frankly expressed dream of the absorption of historical, no less than dialectical, materialism by a quantitative, 'physico-chemico-biological' science. Marxist social theory was not truly scientific, they interpreted Bosse as implying, as long as such a reduction was not effected. Even when Bosse was sent to South America to study rubber-bearing plants, the Deborinites continued to cite him as a typical mechanist, until three leading mechanists felt it necessary to publish a joint disclaimer of responsibility for Bosse's philosophical views.[4] They dismissed him as 'only a popularizer'; though in fact Bosse was ceasing to be even that, except in the rather narrow field of rubber-bearing plants. He became an eminent specialist, but he entirely dropped philosophical writing and popularizations of general biology.[5]

The other biologists who spoke for the mechanists at the start of the controversy and then fell silent were not publicly repudiated by the leaders of their own faction, and one can only surmise their reasons for quitting the fight.[6] It may be that the increasing abstractness and complexity of the issues as the controversy unfolded made them feel beyond their depth. Their chief interest, the conflict of vitalism and mechanism as biological theories, was very soon transcended by such issues as the relationship of quantity and quality, or the nature of dialectical contradiction.[7] Moreover the Deborinites succeeded, very early in the controversy, in fixing upon the mechanist faction the brand of 'revisionism', of opposing the penetration of Marxism into natural science. It became increasingly difficult for a biologist who considered himself a Marxist to defend the mechanist faction, especially after Deborin's master-stroke, towards the middle of 1926, in publishing the defence of the mechanists by Professor A. F. Samoilov. For Samoilov, a famous 'bourgeois' physiologist, explained that he agreed with the mechanists because he himself was *not* a Marxist.[8]

Samoilov's article, 'The Dialectics of Nature and Natural Science', was originally a paper read in the Scholars' House of Kazan, one of many such clubs that the Soviet government had established to court the goodwill of 'bourgeois' scholars. Samoilov described the philosophical controversy among Soviet Marxists in a tone of amused condescension, and his audience probably felt as he did: the controversy was pointless, for it was largely a clash of citations from Engels, who was often vague and inconsistent.[9] Nevertheless Samoilov made a serious effort to evaluate Engels' philosophy of natural science, as expressed in the recently published *Dialectics of Nature*. He claimed that in some important matters Engels had brilliantly anticipated the development of natural science since his time.[10] But Samoilov reasoned that even in such matters Engels' dialectical materialism *proved* nothing and was accordingly superfluous; only empirical science could prove or disprove the theories criticized by Engels. In other matters he found Engels' philosophy to be actually harmful. The tendency 'to bind abstract formulas on nature', which the mechanists attributed to the Deborinites, Samoilov attributed to Engels. His main objection was to Engels' assertion, in criticism of reductionism, that there are irreducible qualities in nature beyond those of homogeneous particles in motion. At one point, to be sure, he agreed with Engels that the reduction of thought to molecular and chemical processes in the brain would not 'exhaust the essence' of thought; Samoilov required only that the term 'essence' be replaced by 'content'.[11] But he would not agree with Engels that physical and chemical methods were inadequate in biological research, and he concluded by challenging 'those Marxists who are inspired by faith in the power of the dialectical method'—by implication the mechanists were clearly excluded—to prove their method by achieving results in natural science that would be difficult or impossible to achieve with another method. If they could, Samoilov said, then the dialectical method would be automatically adopted by all scientists, 'without fruitless, insulting polemics'.[12]

It must have wrung the hearts of the leading mechanists to see their main contentions cogently stated and endorsed by an eminent physiologist as an argument against the penetration of dialectical materialism into natural science. The Deborinite picture of the mechanists as apologists for the 'bourgeois' natural scientists seemed to have received independent confirmation.[13] Perhaps this was the main reason that the Marxist biologists who had originally supported the mechanist faction ceased to do so about 1926, for the most part without giving up their reductionist views in biology.[14]

The case of Fedor Filaretovich Duchinskii tends to confirm this picture of Marxist biologists pulled to the mechanist faction by their philosophy of science but alienated from it by political ideology. In 1926, as Duchinskii, a forty-two-year-old veterinary, participated in the debate among Soviet Marxist biologists over theories of heredity, he did not mention the philosophical controversy, but he revealed his essential sympathies nonetheless.[15] He used the terms 'mechanist' and 'materialist' interchangeably, and he quoted with unqualified approval Lamarck's statement: '"It is not true that in nature there are special laws for organisms, opposite to those governing the changes of dead bodies."'[16] Stepanov quite understandably claimed Duchinskii as a mechanist, but a Deborinite countered with a characterization of Duchinskii as a 'natural-scientific' (*estestvenno-nauchnyi*) materialist on the road to genuine dialectical materialism, i.e., to the Deborinite philosophy.[17] Duchinskii, protesting that he was already a dialectical materialist and had been since 1903, when he first embraced 'orthodox dialectical materialism', complained that neither Stepanov nor the Deborinite understood his biology or his philosophy.[18] One can believe the protestation of philosophical differences with the Deborinite, but the repulse to Stepanov was probably due to the stigma of 'revisionism' that had been attached to the mechanist faction.[19]

Though Duchinskii's views on reduction, which were clearly similar to those of the mechanist faction, could not induce him to join it, his 'Lamarckist' theory of heredity, which had little logical connection with the philosophical controversy, brought him belatedly to attack the Deborinites.[20] And this was not an isolated or unique irony. The genetical controversy among Soviet Marxists had begun about 1925 quite independently of the philosophical controversy, largely as a result of the Communist Academy's success in drawing considerable numbers of Moscow's physicians and biologists to lectures and discussions. In these discussions no reference was made to the contemporaneous philosophical controversy, nor was there much fundamental similarity in issues. 'Lamarckists' and 'Morganists' felt no need to decide whether the Marxist philosophy of science should be an independent discipline, or whether the chief source for its elaboration should be Hegel or empirical science. Nor did they argue whether or where to place a limit on reduction as a method; both groups took the method for granted and argued about its application.[21] Only in 1928 some Deborinites began to claim that their philosophy sanctioned the 'Morganist' position in genetics, while 'Lamarckism' was allegedly a correlate of mechanistic materialism. Three leaders of the mechanist faction, who had previously been indifferent to the genetics controversy, thereupon endorsed 'Lamarckism', and Duchinskii, at the very end of the philosophical controversy, attacked the Deborinites.[22] The most paradoxical aspect of these manœuvres was the general failure to demonstrate a logical connection between the biological and philosophical positions that were being paired.[23]

If the mechanist leaders who rather abruptly went to the aid of 'Lamarckism' did so in order to win the support of 'Lamarckist' biologists, they had little cause to feel successful. Counting Duchinskii only three or four 'Lamarckists' were won to a brief, tenuous association with the mechanist faction. Indeed, they did not actually express approval of the mechanist faction; they confined themselves to a defence of 'Lamarckism', and a criticism of the individual Deborinites who were attacking it.[24] In 1930 Duchinskii deserted the mechanists altogether. He tried to prove that the Deborinite philosophy supported the 'Lamarckist' position in biology.[25] But these were exceptional cases. In general the 'Lamarckists' did not find it necessary to compromise their claim to orthodoxy by coming to the defence of the condemned mechanist faction. It was possible to be prudently silent in philosophy while defending 'Lamarckism' in biology.

The one Marxist biologist who showed no skittishness or vacillation in the philosophical controversy, whether at the beginning or the end, was Sergei Stepanovich Perov. In February, 1925, this thirty-six-year-old biochemist helped to found the mechanist faction by his part in the Timiriazev Institute's endorsement of Stepanov's book. He also took the mechanist part in subsequent debates, and in April, 1929, he still defended the mechanist faction at the Conference that condemned it as 'revisionist.'[26] Indeed, Perov was one of the mechanist leaders who refused to renounce his views even in the 'thirties, and one can detect a continuation of his war with the Deborinites in his participation, during the 'thirties and 'forties, in Lysenko's campaign against 'Morganism'.[27]

This is not to say that genetical theory was Perov's main concern, whether as a scientist or a philosopher of science; until 1929 he appears to have ignored the problem of heredity and variability altogether.[28] His main concern, as befitted a biochemist, was with chemical and physical analyses of living phenomena, and he nursed a grudge not merely against the Deborinites for their philosophical denial of the adequacy of such analyses, but even against fellow-biochemists who, Perov argued, were unnecessarily multiplying the types of proteins and thereby giving support to the vitalist argument that ordinary chemistry cannot explain life.[29] Like the other leading mechanists, he regarded dialectics as a 'constant guiding principle . . . and not the wretched dregs of Hegel. . . .'[30] Like them, he searched in the empirical sciences for the universal principle that would furnish an ontological and methodological basis for the unity of all sciences, and he claimed to find it, somewhat as Bogdanov did, in the work of the American chemist, Willard Gibbs.[31] Even at the Conference of April, 1929, when Perov for the first time stood up for 'Lamarckism', he defended it in terms of political ideology ('Weissmanism', he said, was the basis of racism), and he was no more successful than the other speakers who touched on this theme in demonstrating a connection with the main philosophical issues of the controversy.[32] To Perov those main issues were still the reduction of life to the chemistry of a single protein, and the harmfulness of Hegelian metaphysics. Only considerably later, while participating in Lysenko's struggle against 'Morganism', did he argue a plausible connection between 'Lamarckism' (rechristened 'Michurinism') and the chemical reductionism that was his original inspiration.[33] But that lies beyond the boundary of the present study.

The Communist Academy's efforts to draw natural scientists to Marxism had considerably less success among physicists, chemists, and mathematicians, than among biological scientists. Why this should have been the case is a puzzle, but the fact itself, of considerable importance to the present subject, is readily established. The Circle or Society of Materialist Physicists and Mathematicians was formed only at the very end of 1927, four years after comparable organizations for physicians and biologists had been founded.[34] At the end of 1928, the Society of Materialist Physicians could claim forty-one 'actual' [*deistvitel'nye*] or full members, in spite of the requirement of publication for that rank; while the Circle of Materialist Physicists and Mathematicians refrained from publishing figures on membership, and admitted that even the few scientists who had been persuaded to speak under the Circle's auspices had read technical papers or popularizations, with little or nothing to say about materialist philosophy or Marxism.[35] Participation in the discussions of Marxism and natural science that the Communist Academy sponsored during the 'twenties tells a similar story. Topics like 'Marxism and Darwinism' brought out considerable numbers of people, and largely non-Party scientists with Marxist sympathies, while topics like 'Marxism and the Theory of Relativity' drew very few participants, among whom scientists were comparatively rare.[36] Perhaps it was this scarcity of Marxist physical scientists that was responsible for their playing a much more important role in the philosophical controversy than did the more numerous Marxist biologists. Perhaps it was the markedly philosophical nature of theoretical physics and mathematics, or perhaps quirks of personality, but whatever the causes may have been, the paradoxical fact is that the physical sciences contributed several leaders to both factions in the philosophical controversy.

The most eminent of these physical scientists, until the very end of the controversy, was Arkadii Klimentovich Timiriazev, who, more than Stepanov, deserves to be considered the chief leader of the mechanist faction. Already forty in 1921, when he joined the Bolshevik Party, Timiriazev had little record of previous political or philosophical activity.[37] He had probably absorbed some Leftist inclinations from his famous father, the biologist Kliment Arkad'evich Timiriazev, who boasted that *his* father had been a sympathetic witness of the Decembrist rising.[38] In 1911, when the Tsar's Minister of Education abridged the autonomy of Moscow State University, both Timiriazevs were among the staff members who resigned in protest.[39] But this was as nothing compared to the combination of political, educational, and ideological activity that engulfed A. K. Timiriazev after he saluted the Bolshevik Revolution and joined the Communist Party. Professor of physics at Moscow State University, head of the Department (*Kafedra*) of Natural Science at Sverdlov Communist University, a leading member of the State Council of Scholarship (*GUS*, the organ of the Commissariat of Education that imposed a new abridgement of autonomy on Russian universities in 1921–1922), an important and active member of the Communist Academy, the editor of *Under the Banner of Marxism* in charge of its offerings in the field of natural science, lecturer, translator, and writer—here, one might say, was an incarnation of 'the release of energy' that was the main result of the Russian Revolution, according to Maynard and Pares.[40]

In 1923 he published *The Philosophy of Science*, an anthology of excerpts not from the writings of philosophers (these Timiriazev brushed aside as useless to the student of natural science, for whom he had prepared the anthology), but from the works of famous natural scientists, who were, he explained, unwitting dialectical materialists.[41] The resulting impression, that Timiriazev's dialectical materialism was a sort of positivism, was strengthened by his favourable reference to Comte,[42] and by his assurance to Marxist students of natural science that they would find in the theoretical writings of the great natural scientists very little 'tar' of class ideology and much 'honey' of materialist philosophy.[43] ('Tar in honey' is the Russian equivalent of our 'fly in the ointment'.) He expressed similar views in contemporaneous speeches to the State Council of Scholarship (*GUS*), the Moscow Committee of the Communist

Party, and the editors of *Under the Banner of Marxism.*[44] He reassured those Marxists who feared to entrust the teaching of natural science in Communist universities to non-Marxist scientists, and also those who felt that 'Marxists have no reason to poke their noses' into natural science.[45] Natural scientists, he explained, were already dialectical materialists, and it would be easy to demonstrate this to them, for the principles of dialectical materialism were bound on the scientist by nature. In Timiriazev's interpretation, these dialectical principles were simply that all things are subject to motion or change, that all things are interdependent, and finally, that 'contradiction' is the essence of the endless change of interdependent things. The examples that Timiriazev used to illustrate the meaning of dialectical 'contradiction' (D'Alembert's principle, and the principle of Le Châtelier-Braun) show that he understood this crucial term to mean the opposition of forces moving in different directions, or the disturbance and re-establishment of mechanical equilibrium. He doubted that the notorious triad (thesis, antithesis and synthesis) was a meaningful concept for natural science, and he expressed the same skepticism concerning the principle of quantitative changes becoming qualitative changes.

It was the last issue that he emphasized in February, 1925, when he helped found the mechanist faction. The Deborinites, he claimed, by their assertion of irreducible qualities at each level of integration, were trying to limit or deny the scientific method of reducing complex phenomena to simple ones.[46] Moreover, Timiriazev felt that the method of reduction could not be justified without a corresponding ontology, in which universal ether and two kinds of electricity were regarded as the ultimate reality.[47] Since the Deborinites cast doubt on such ontologies as well as on the unlimited applicability of the method of reduction, Timiriazev's defence of both filled his polemics in the philosophical controversy. He tried to connect his crusades in physics and philosophy (he was at war with the theory of relativity) by arguing that the Deborinite position in philosophy encouraged 'formal constructs in physics, of the Einsteinian variety'.[48] But paradoxically, by 1927, when the Deborinites finally abandoned their initial evasion and endorsed the theory of relativity, he himself had moved close to the Deborinite position on the philosophical problem of reduction, though he was still opposed to the term 'quality':

The 'reduction' of physical and chemical phenomena to mechanics in contemporary theoretical natural science does not in the least signify the *identification* of these phenomena with *mechanics*, as was the case among the French materialists of the 18th century. This reduction signifies the utilization of the equations of mechanics, and even more than that: the expression of the laws of these 'super-mechanical' phenomena in the form of the equations of mechanics, *but with the condition of utilizing also such laws as are not derived from these equations.*[49]

Yet, however closely Timiriazev's view of reduction approached the Deborinite position, his positivist understanding of philosophy as a whole sustained his hostility to the Deborinites, as he explained in 1928:

Now we come to the very root of our disagreements. Those who are called mechanists propose that the study of the concrete facts and phenomena of nature and society should be brought to such a level that the dialectics of these processes would emerge from the processes themselves. Our opponents believe that in the field of natural science one needs to formulate once and for all general propositions like the following: 'There is no positive electricity without negative; there is no dispersion of energy without its concentration; there is no action without a reaction, etc.' . . . 'The task consists in encompassing from the dialectical point of view natural science as a whole.'[50]

'These vague and diffuse formulations', as Timiriazev called them, were actual quotations from Deborin, whose version of dialectics Timiriazev treated sometimes with ridicule, as above, and sometimes with apprehension.

As one might expect, Timiriazev was accused of 'tailism' (*khvostizm*), a Leninist pejorative signifying, in this case, adaptation to the currents of opinion among 'bourgeois' scientists, renunciation of Marxist or Party leadership in the field of natural science.[51] But paradoxically he was also accused of 'nihilism' or more appropriately for the nineteen-twenties, of 'On-Guardism' (*napostovstvo*), a pejorative taken from contemporaneous literary disputes, signifying a tendency to reject 'bourgeois' culture indiscriminately and to advocate unreasoning force in promoting 'proletarian' culture.[52] The explanation of this paradox does not lie in Timiriazev's theoretical justification of Communist intervention in the field of natural science. Few members of either philosophical faction would have disagreed in principle with his argument that

along with valuable material one encounters all kinds of rubbish in contemporary science. . . . In this field as in politics we must learn to distinguish the correct line from all possible 'deviations'. When, in the field of natural science, we learn to distinguish genuine theory, genuine science, from their counterfeits, with the same success that we have now in distinguishing Marxism and Leninism from each and every deviation, and moreover when we do this more quickly and better than the specialists in natural science who do not study dialectics, then a significant part of the task now standing before the Marxist methodologist in the field of natural science will be solved.[53]

Soviet Marxists might approve of these principles and disagree concerning specific cases of alleged 'rubbish in contemporary science'. Or they might try, as the Deborinites did, to be vague in their methodological counsel to natural scientists, to avoid or postpone entanglement in specific issues within the scientific disciplines. Timiriazev himself had two minds. On a general level, as a Marxist philosopher, he invited the charge of 'tailism' by preaching a sort of *laissez faire*; dialectical materialism was simply the philosophy that natural scientists were already using in their work. But Timiriazev was also an old-fashioned physicist who had been committed to war on the 'formal method' of the theory of relativity before he learned to call it 'bourgeois', or to cite Engels and Lenin as authorities on the philosophy of science.[54] When the Deborinites in 1927 finally overcame their diffidence in relation to Einstein's theory, it had won nearly universal acceptance among physicists, and Timiriazev, the self-proclaimed defender of contemporary science against an alien metaphysics, was quite naturally pictured by the Deborinites as a 'nihilist' or 'On-Guardist', threatening the progress of science with his outworn dogmas.

A similar charge could hardly have been levelled at Timiriazev's fellow-mechanist, Zakhar Aronovich Tseitlin, for this young physicist (born in 1892) became involved in the philosophical controversy in large part as a result of his *defence* of the new physics.[55] In 1924 and 1925, while defending the theory of relativity against Timiriazev, and the 'formal method' as a whole against I. E. Orlov, Tseitlin became convinced of an essential similarity between Marxism and the philosophy of Descartes. He thereby incurred the displeasure of the Deborinites, who were, in 1925, still largely uninterested in problems of physics but were very much alive to issues in the history of philosophy and in general characterizations of Marxism.[56] By 1927, when the Deborinites finally endorsed the new physics, Tseitlin and Timiriazev found that their different evaluations of Einstein's theory were grounded on mutual devotion to the Newtonian or mechanist heritage in physics. Both were determined to defend mechanism against Deborinite attack. Indeed, after 1927 Tseitlin muffled his defence of the theory of relativity, or recast it, using language that might not offend Timiriazev.[57]

Tseitlin's reading of the mechanist tradition in physics and philosophy involved him in quarrels with fellow-mechanists over more matters than the theory of relativity, before the increasingly critical conflict with the Deborinites caused him and the other mechanists to forget their differences with each other. For example, Tseitlin described as equivalent Engels' argument that space is a form of matter's being, for we never experience space without matter, and Einstein's argument that '"space possesses physical qualities"', for otherwise space would be unthinkable.[58] The former argument appeals to sensuous experience to prove that space is an attribute of the substance matter; the latter argument, if taken literally, appeals to reason or logic to prove that matter is an attribute of the substance space. In an effort to prove that these apparently different views are actually equivalent, Tseitlin argued that materialism, empiricism, and rationalism could be reconciled, as they allegedly were in the philosophy of Descartes. Lafargue's empiricist and materialist defence of Descartes' belief in innate ideas was repeated. Generations of human experience had allegedly fixed in the evolving human brain an ever more accurate reflection of the laws of the universe; the child of civilized people learns arithmetic easily because of the inherited distribution of molecules in his brain.[59]

In several articles and a remarkable book, *Science and Hypothesis*, Tseitlin tried to show that the great physicists from

Newton to Einstein had been applying and extending the method of Descartes, and, moreover, that Marxism was simply the extension of the Cartesian method to social phenomena.[60] Indeed, Tseitlin went so far as to seek a one-to-one correspondence between certain concepts and laws of physics and those of Marxist social science:

The physical atom (or ether) of political economy is money (gold). . . . This discovery of the real meaning of money justly places Marx on a level with Democritus. But Marx went further than Democritus: like Newton, he gave a mathematical theory of economic atomism (ether). [Tseitlin regarded atoms as vortices in the universal ether.]

Newton's atomism, as we have seen, made it possible for him to define mass as the product of a body's volume and its density, i.e., the number of atomic units in a unit of volume. Precisely in the same way Marx's hypothesis makes it possible to define the mass of capital rigorously as the product of capital's volume, i.e., the general monetary expression of capital, and its density, i.e., the number of units of common abstract labour in a given monetary unit of capital's volume.[61]

It is hardly surprising that such strange comparisons provoked criticism from mechanists as well as Deborinites.[62]

As Tseitlin became deeply involved in controversy with the Deborinites, his views on specific issues of natural science, the principal source of dissension between him and fellow-mechanists, receded into the background, and more abstract problems in the philosophy of science came to the fore. If Marxist social theory is truly scientific, he argued against the Deborinites, a continuity of method must be demonstrated between it and the natural sciences. There could not be a lower method for the natural sciences, based on 'metaphysical' or mechanistic materialism, and a higher fully dialectical method for social thought and philosophy, as the Deborinites seemed to be saying with their doctrine of irreducible qualities at ascending levels of integration. The unitary scientific method, Tseitlin believed, was quantitative analysis. Non-mathematical analysis in terms of qualities was tolerable only when quantitative methods could not be employed because of the current limitations of the human mind. Theoretically, in principle, it was not impossible for 'a super-genius of a mathematician' to 'express in quantitative mathematical form the complex of material motion that we call Tolstoy's *War and Peace*'.[63] Like most other mechanists, Tseitlin did not advocate actual efforts to reduce all knowledge to 'quantitative mathematical form'. Only madmen, he stressed, would want *War and Peace* in mathematical formulas.[64] He considered the denial of irreducible qualities vitally important in the defence of empirical natural science against the Deborinite metaphysics, not in preparation for a reductionist attack on social science.

Indeed, it was Tseitlin who stated the central mechanist grievance against the Deborinites most vividly at the climactic Conference of April, 1929:

> Natural scientists will not accept Deborin's formalism. It is inapplicable to exact scientific research. . . . Here formalism leads absolutely nowhere, and those materialist natural scientists who want to carry on a genuine, dialectical materialist investigation of nature, and not a game of jackstraws, not a game of 'dialectics', inevitably clash with the philosophers who are proceeding along the wrong path. All this is explained by the fact that our philosophers do not study the natural sciences, do not study concrete natural science: they learn Hegel and other philosophers by rote from Kuno Fischer. I should advise them as follows: Lenin studied Hegel when he was 45, and he had the head of a genius: Comrade Karev [a prominent Deborinite] I should forbid as yet to study Hegel. When he is 90, let him study Hegel, but first let him pass an examination in all the concrete disciplines of natural and social science![65]

Tseitlin's sarcasm did not prevent the Conference from condemning the mechanist faction as anti-Marxist. Indeed, he played into the hands of the Deborinites, who pictured the mechanists as aiding the resistance of 'bourgeois' specialists to the penetration of Marxism into natural science. Deborin singled out Tseitlin's speech as 'extraordinarily characteristic of the narrow-minded natural scientists, the *spetsy*' (Soviet slang for experts), and from the floor someone called out the further specification: 'Crawling empiricists.'[66] It was of course a considerable distortion to lump Tseitlin with 'crawling empiricists' (a Soviet cliché corresponding to our own 'mindless empiricists'), for he all but identified materialist empiricism with Cartesian rationalism.

The basic attitudes shared by the mechanist faction as a

whole can be perceived clearly in the writings of the chemist, Ivan Efimovich Orlov. The reason is not that he was a cautiously reticent conformist who took care to be always in step with the rest of his faction, but that his points of basic agreement stood out by contrast with his many disagreements. From the appearance of his first philosophical articles in 1916 (he was thirty at the time) until 1929, when his publications abruptly ceased, I. E. Orlov showed himself to be a rather unusual thinker with much to say about many things.[67] Ethics, logic, the philosophy of mathematics, esthetics, the theory of relativity, popularizations of natural science and its history—all these topics received consideration in Orlov's two books and twenty-odd articles, nearly all of them published in the five years preceding 1929, when, for reasons unknown to the present writer, he fell silent. (He was probably not condemned as an 'enemy of the people', for some of his writings are still listed in Soviet bibliographies.)[68]

The views he expressed in these fields were often at variance with those of other mechanists. As early as 1923, when the future Deborinites were still ignoring natural science, Orlov published 'The Dialectics of the Experiment', in which he called for the 'elaboration of materialist dialectics *by itself*, as a method of research'.[69] The italics have been added to point up the contrast with the usual mechanist view that the philosophy of science, dialectics, must not be elaborated out of the context of the specific sciences. To be sure, a close reading of Orlov's article shows that he was by no means a premature Deborinite; he was merely repeating, this time with Marxist phraseology, the critique of traditional logic, including Mill's 'rules of induction', that he had published already in 1916.[70] The basis of the critique was not Hegelian metaphysics but a kind of empiricism that led Orlov in other publications to reject the so-called 'formal' or 'symbolic' method in physics, quite in the spirit of A. K. Timiriazev's objections to the theory of relativity.[71] Opposition to Einstein's theory was by no means a hallmark of the mechanist faction; indeed, Timiriazev and Orlov were almost alone in this matter. Nor was Orlov's vigorous and unequivocal assertion that dialectical materialism should 'intervene directly in the thick of the struggle of physical theories'[72] characteristic of the mechanists—or of the Deborinites, for that matter. The principle that Marxism must penetrate the natural sciences was generally accepted, but it was generally understood and applied in a vague and diffident way. Orlov was sneering at nearly all Soviet Marxists of the 'twenties, without distinction of philosophical faction, when he asked scornfully whether dialectical materialism

> can lay down its line [in natural science] . . . merely by following after the change of theories, remaining outside special disputes, saluting the theory that has evidently triumphed, and explaining it in the spirit of the materialist world view.[73]

Orlov was also rather unusual in his views on accident and necessity, though here too he was in agreement with A. K. Timiriazev. Using arguments from nineteenth-century physics, but referring also to the traditions of philosophical atomism going back to Epicurus and Democritus, Orlov reasoned that necessity, sequences of causes and ineluctable effects, were actually summaries of myriad chance occurrences. He scoffed at what he called the Philistine (*meshchanskii*) notion of universal determinism, declaring that it had been

> pitilessly destroyed by science. The laws of nature are natural and elemental [*stikhiiny*] in the full sense of the word: they are only the general result of innumerable disorderly motions of particles.[74]

On this basis Orlov asserted that Hegel was right in regarding accident or contingency as an objective category, inseparable indeed from the equally objective category of necessity.[75] Nevertheless, it was possible for a fledgling Deborinite mathematician to attack Orlov's view as mechanistic, for Orlov expressly approved 'a mechanical picture . . . [that] reduces all phenomena to matter and motion'.[76] But in regard to contingency the fledgling mathematician, for all his citations from Deborin and Engels, preferred the tame old definition of accident as the intersection of two or more necessary processes that originate independently of each other; and the Deborinite editors of *Under the Banner of Marxism*, apparently in a quandary, disclaimed responsibility for his article.[77] Obviously then I. E. Orlov was no ordinary mechanist. At the very time he must have been preparing his first polemic on behalf of the mechanist faction, he was probably writing an article on cybernetics, as it would now be called, in which he derided the notion that thought can be performed by calculating machines.[78] It is therefore highly instructive to examine Orlov's expressed reasons for siding with the mechanists and against the Deborinites, for one can find in these reasons the least common denominator of the mechanist faction's philosophy.

Orlov said that the necessity of 'mechanical models' in the physical sciences made him a member of the mechanist faction.[79] This was part of his special animus against the 'formal' method in physics, an animus that he had been expressing since 1916, when he did not yet speak as a Marxist.[80] Actually this favourite issue of Orlov's was irrelevant to his dispute with the Deborinites, for they did not object to the use of 'mechanical models' in the physical sciences. When Orlov defined the term 'mechanical model' carefully, this alleged reason for adhering to the mechanist faction became transformed into his second, and truly fundamental reason. 'Mechanical models,' he explained, 'are heterogeneities of matter that yield to spatial differentiation [*poddaiushchiesia prostranstvennomu razlicheniiu neodnorodnosti materii*].'[81] Putting the matter another way, he insisted that quantitative analysis was the *sine qua non* of natural science; that qualitative differences and changes in matter are always connected with, and expressible in terms of measurable transpositions of homogeneous elements. Or he could put the matter more simply yet, as he did in his final polemic against the Deborinites, isolating the thought in a separate paragraph for special emphasis:

> Let us note the most important specific peculiarities in the methodology of the natural sciences. Such peculiarities are the absolute necessity of reducing the complex to the simple, and the important significance of studying mechanical motion.[82]

Probably all the mechanists would have agreed with this statement on reduction, though some would have found it too modest. It may seem difficult to understand why they regarded this as the basic position that they were defending against Deborinite attack. For the Deborinites granted the 'absolute necessity' of reduction in the natural sciences, and 'the important significance of studying mechanical motion'.[83] The Deborinites limited the applicability of reduction, but so did

Orlov. Indeed, in his final polemic he made a special point of the necessity of different methods in different disciplines, and expressly denied that his cherished 'mechanical models' were applicable in the social sciences. One wonders accordingly why he said that 'the position on which we shall accept battle is the connection between a change of quality and the transposition of matter in space'.[84] Apparently the Deborinites' repeated assurances that they were not attacking this position, that they were perfectly content to see natural scientists searching for quantitative explanations of qualitative differences in nature, did not allay the hostility of Orlov and the other mechanists. What vitiated these reassurances was the concomitant Deborinite insistence that natural scientists must also search for something else, or affirm the existence of something else on the authority of Hegel and his Deborinite interpreters. In other words, reduction was a disputed method not in natural science itself but in the construction of a *Weltanschauung* for natural scientists.

Orlov was too sophisticated to say, as one or another of his fellow-mechanists did, that nothing existed but quantitative relationships, or that there were only atoms and the void, or only an ethereal ocean in endless turmoil, and that the goal of science was the reduction of all phenomena, social and mental included, to the mechanics of the *Urstoff*. His final polemic was modestly entitled, 'Concerning Dialectical Tactics in Natural Science'. It was a homily on the necessity of circumspection in transferring concepts from Marxism, which was primarily a social doctrine, to the field of natural science.[85] In effect he was warning the Deborinites to be cautious in telling natural scientists what to believe about natural science even on the abstract level of *Weltanschauung*. The Deborinite talk of irreducible qualities, Orlov felt, was more likely to alienate than to attract natural scientists, whom he pictured as inherently hostile to metaphysics. If one may view the matter cynically, as in all probability Orlov did not, he was calling for a monistic Marxist *Weltanschauung* that would stress qualitative analysis and non-mathematical laws in the social sciences, but would expediently forget such concepts when seeking converts among natural scientists.

In general, it may be said that a pliant positivism was the

chief distinguishing characteristic of the otherwise diverse mechanist faction. Their varieties of positivism were all part of a common effort to syncretize Marxism and the views of natural scientists. One of the last joint declarations of the faction, published in 1928, revealed this common characteristic with unusual clarity:

> Only *a living dialectics that takes into account the singularity of each field of phenomena, that is extracted from the factual interdependence of phenomena and is not an* a priori *construction introduced into them from without, can yield positive results*, i.e., results that bear testing by facts, by experiment, results that the natural scientist accordingly cannot and will not mark off from himself with the argument that they do not concern his discipline . . . The study of [the problems of natural science] requires great perseverance, *aptitude, and profound preparation in a specialty*. It is most appropriate of all, if the Marxist physicist makes positive researches concerning dialectics in physics, the Marxist biologist in biology. One need not imagine that *anybody can write on any subject on the basis of a study of Hegel's logic.*[86]

In effect, then, each discipline was to have its philosophy defined by its own Marxist specialists in terms of its own content, with the results gathered under the rubric of dialectical materialism. The knotty problem of locating final authority in disputed issues, which would have arisen if this programme had been adopted by the majority of Soviet Marxists, and even more if it had been established as state policy by the Bolshevik leaders, was not examined.

The chief effort of nearly every mechanist was to justify as 'orthodoxy' some variety of syncretic positivism, which was considered indispensable if natural science and its adepts were to be fused with Communist ideology and its advocates. Avowed positivism, in the sense of explicit rejection of any and every ontology, was clearly 'revisionist' and accordingly impossible, except for a few peripheral figures such as Bogdanov or Borichevskii. An unacknowledged positivism that made an ontology by lumping recent physical and biological theories with Bolshevik social thought seemed quite proper for 'orthodox' Marxists; as long as the term 'positivism' was not used, and adherence to a universal, dialectical materialist philosophy was explicitly declared. That is why reduction seemed the chief issue (and, incidentally, why the faction was called 'mechanist', i.e., reductionist, rather than 'positivist').[87] Marxist philosophy of science, dialectical materialism, was acknowledged in principle to be a universal ontology and methodology, only to be reduced in further definition to theories and methods of the particular sciences. The justification offered for this anomaly—or dialectical contradiction, as some mechanists chose to call it—was that the basic method of science is reduction, the explanation of complex phenomena in terms of the interaction of their constituent elements. The mechanist faction was not positivist avowedly but only in effect, and the effect was achieved by its stress on reduction.

11

DEBORIN AND HIS STUDENTS

THE Deborinite faction differed from the mechanist not only in its understanding of Marxism but also in its composition. Instead of a rather haphazard assortment of theorists with divergent views, the Deborinite faction consisted of a homogeneous nucleus of militants, who began the controversy and pushed it to a successful conclusion, and an outer section of complaisant supporters accumulated in the later stages of the controversy. The mechanist description of the nucleus as 'Deborin and his students' was literally true, regardless of the malice that prompted it. Nearly all the leaders of the Deborinite faction were products of Deborin's seminar at the Institute of Red Professorship.[1] What is more, as they left his seminar and became teachers and writers in the field of philosophy, they continued to regard Deborin as disciples do a master. Indeed, there was a growing tendency among them to look to Deborin as a latter-day Engels or Lenin in the field of philosophy. 'It is A. Deborin,' declared I. K. Luppol in 1927,

> who revealed an understanding of the philosophical tasks of the epoch. The tendency both of his pedagogical work and also of his literary activity has consisted in emphasizing the significance of dialectical materialism as the methodology of science, in elaborating materialist dialectics. This led . . . to the analysis and materialist reworking of Hegel. One may say that, after Lenin's instructions, it was by the works of A. Deborin . . . that a new, third period in the philosophical thought of the U.S.S.R. was begun. An enormous

interest in Hegel the *dialectician* has been aroused, and at the same time, if one may express one's self so, a regular struggle for Hegel has begun.[2]

In 1928, another of Deborin's students put this attitude more simply:

Unfortunately, neither Marx nor Engels nor Lenin, in spite of their desire to do so, left us a systematized theory of materialist dialectics. The continuer of their work in this field, Comrade A. M. Deborin, has . . . begun such a work.[3]

Thus, the philosophy of the Deborinite faction was to a large extent a single individual's interpretation of the Marxist heritage, broadcast by his disciples.

The individual, Abram Moiseevich Deborin (*né* Ioffe, 1881), was in origin a revolutionary metal-worker who had sided with Lenin in 1903, when the Russian Social Democrats split into the Bolshevik and Menshevik factions. In 1907, under the influence of Plekhanov and formal study in a Swiss university, he had turned Menshevik, and had criticized Machism as the philosophical correlate of Bolshevik voluntarism in politics.[4] After the Bolshevik Revolution he quit the Menshevik Party and offered his services as a Marxist philosopher to the Soviet régime. In 1921 Lenin approved Deborin's appointment to a teaching position in Sverdlov Communist University, but warned that Deborin must be watched, lest he smuggle Menshevism into his teaching.[5] Despite this inauspicious beginning, Deborin's career as a Soviet Marxist was for nearly a decade an unchecked rise in authority and prestige. Even before 1925, when the philosophical factions took shape, he was in effectual control both of the Section of Philosophy at the Institute of Red Professorship, the centre of graduate training in Marxist philosophy, and of the chief journal of Soviet Marxist philosophy, *Under the Banner of Marxism*. In short, he was already the most influential Soviet Marxist in the field of academic or 'pure' philosophy, though he was not yet a Party member. (He joined in 1928.)

It is somewhat misleading to describe Deborin as an academic or 'pure' philosopher, even in speaking of those years when he was not yet publishing articles on alleged connections between philosophical issues and the current Party line. (His first such article appeared in November, 1929.[6]) For he did write polemics of a markedly political nature, in the broad sense of the word, against the theories of Spengler, Freud, or Christian socialists; and in all his writings he did stress alleged connections between philosophical issues and class conflict.[7] Nevertheless, the bulk of his non-polemical writings dealt with such topics as materialism in early modern times, dialectics in Kant and Fichte, and the comparison of dialectics in Hegel and in Marx.[8] Nor was this assortment of topics merely an expression of personal taste. At the Institute of Red Professorship Deborin put future 'red professors' of philosophy through the following curriculum: the first year was given to the French materialists and Kant (Spinoza was added in 1923); the second year entirely to Hegel; and the last year to historical materialism, conceived as dialectical materialism (the fusion of materialism with dialectics) applied to social phenomena.[9] The philosophy of natural science got incidental treatment in the second year, as an illustration of Hegel's universal dialectic;[10] for, until the controversy with the mechanists, Deborin was not especially interested in the philosophy of natural science.[11] He clearly revealed his academic, historical, tendentious conception of philosophy when he defined the basic idea underlying the curriculum he had established: 'by means of a critical study of the history of philosophy and dialectics, to make clear the necessity of dialectical materialism as the inevitable outcome of all modern philosophy.'[12] His conception of philosophy was thus political, in the sense that it aimed to serve the Communist movement, but it was also academic, in the sense that it regarded rather abstruse studies in the history of philosophy as its main service.

In this connection one must note Deborin's remarkable ability to bend with the wind and remain rooted in one spot. He had already established his programme of education, research, and publication in philosophy, when Lenin called on Soviet Marxist philosophers for services that Deborin's programme hardly envisaged.[13] This was especially true of aid to the anti-religious movement and efforts to win natural scientists to Marxism; Lenin placed great stress on both, while Deborin was little interested in either. Deborin responded by making repeated obeisance to Lenin's instructions, but he continued his work in philosophy much as before. He and his students translated early modern atheists, which fitted in perfectly with their previously established programme, but beyond that they contributed almost nothing to the anti-religious campaign.[14] When the controversy with the mechanists dragged them into the campaign for natural scientists, Deborin found his philosophy of natural science, as Lenin had suggested, in a materialist interpretation of Hegelian dialectics. And in 1929, when Marxist philosophers were called upon to help the Party's immediate struggles, Deborin complied with this behest too: he described the elaboration of dialectics from Hegel as the philosophical correlate of the Party's drive against the right deviation, or whatever else might be the slogan of the day.[15]

It was probably not temporizing that prompted these tactics, but the condescension of the universal philosopher towards those on the lower slopes of Olympus. When Deborin first challenged the mechanists at the beginning of 1925, he stated his point of view quite plainly:

> . . . We demand the re-working of the new data in each field of knowledge from the point of view of materialist dialectics, while various 'critics', often without being aware of it, are inclined towards the 're-working' of dialectical materialism from the point of view of *particular* facts, of a *particular* science. . . .
>
> The method of dialectical materialism is the result of the entire *accumulation* of human knowledge. Therefore it cannot be overthrown by *particular, contingent* facts, which are themselves subject to critical examination from the point of view of the general methodology.[16]

It was for this kind of remark that the mechanists regarded Deborin as an arrogant metaphysician, scornful of empirical considerations in his rationalistic search for '*the universal connection of everything with everything*'.[17] He fed the dismay of his empirically minded opponents by such comments as these:

> The question of the possibility of 'reducing' chemistry and biology to mechanical laws is a question of principle. Its methodological formulation and solution cannot be dependent on whether such a 'reduction' has or has not been achieved already in practice.[18]

Quotations in a similar spirit can be greatly multiplied, apparently substantiating the reputation that Deborin had among the mechanists.

To be sure, one can also accumulate quotations of a strikingly different character. For example, in the same speech to natural scientists in which he described philosophy as a search for '*the universal connection of everything with everything*', he described it further in a way hardly different from the so-called mechanist point of view:

> In what does the power and significance of philosophy consist? Above all in the fact that, by synthesizing the results of the separate sciences, it gives us a unitary world view; on the other hand, philosophy has as its subject the process of cognition, the analysis of scientific concepts, and the elaboration of the method of cognition. . . . A firm union between philosophy and the positive sciences is the sole theoretical guarantee of a powerful development of human knowledge.[19]

Moreover, Deborin could show deference not only to the positive sciences but to other aspects of 'practice', which, in line with the Marxist tradition, he declared to be the criterion of truth.[20] On one occasion he even anticipated a predominant feature of Soviet Marxism since 1930. Describing as 'practical dialectics' 'the practice of class struggle in contemporary society', he remarked: 'If thought is determined by being, then it is natural that theoretical dialectics is determined by "practical dialectics".'[21] But one must bear in mind the extreme pliability of the concept of 'practice' in Marxism. Indeed, in Deborin's very claim that particular facts or sciences cannot overthrow dialectical materialism, there was an appeal to an allegedly empirical criterion of truth. He declared his universal philosophy to be irrefutable because it summed up the whole of human experience.

To a large extent Deborin was of course merely reproducing the contradictions already noted in the 'classics' of Marxism-Leninism. Like them, he replied to the charge of inconsistency with the claim that truth is contradictory or dialectical, and the present writer will not try to decide whether Deborin substantiated the claim. But it is possible to decide, without undue intrusion of one's own philosophical judgments, whether Deborin was more metaphysical than positivistic in his attitude towards the natural sciences. Appropriately enough for a dialectical thinker, he was both, and in order to perceive where his emphasis lay on the most general level it is necessary to examine his thought on some specific issues.

Deborin's ambivalence in defining 'matter', one of the central concepts of dialectical materialism, is illuminating. He could repeat the mainly epistemological definition to be found in the 'classics' of Marxism-Leninism: 'matter is the objective reality that exists in time and space, acts on our senses, and is reflected in them.'[22] But he was not entirely satisfied with this definition, and supplemented it with a markedly Hegelian, ontological explanation:

> In a broader sense, matter is the whole, infinite, concrete aggregate of 'mediations' [*oposredstvovanii*], i.e., relations and connections. And the concrete scientific disciplines—mathematics, mechanics, physics, chemistry, biology, etc.—deal with various forms and stages of the 'mediations', i.e., the processes, relations, and connections of this same matter.[23]

He heaped scorn on the typical mechanist effort to give an ontological definition of matter in terms of current physical theories, and by this scorn he caused the mechanists to complain that he would deny physicists the right to visualize electrons and protons as the building blocks of the universal substance, matter.[24] But, far from denying such a model to physicists, Deborin himself could speak on occasion of the physicist's ether as 'the single universal substance', and could give electrons and protons Hegelian dignity by picturing them as 'nodal points' in the universal substance, matter or ether.[25] His definition of matter depended on his particular audience and purpose. To teach the mechanist faction, and natural scientists at large, a sense of their own inadequacy and a consequent reverence for the dialectical materialist philosopher wrestling with the truly universal problems. Deborin stressed the Hegelian type of definition. To reassure natural scientists that he had no intention of interfering with their proper work, that he was in fact generalizing their work on an inspiring level of abstraction, he stressed the mainly epistemological type of definition, or even translated current physical theory into dialectical terminology.

Deborin showed some awareness of these contradictions, and tried to justify them by his central doctrine of 'quality'. Matter, he taught, the universal substance, might have the quality of the physicist's ether and elementary particles on its lowest level of organization, but higher levels of organization of this same matter had additional qualities or specificities (*spetsifichnosti*); that is, the higher levels exhibited regularities or laws (*zakonomernosti, Gesetzmsäsigkeiten*) that could not be reduced to the regularities of simpler or lower levels.[26] Deborin insisted that his philosophy was still a monistic materialism, for he taught that 'the higher forms arise from the lower forms',[27] and on this basis he approved of the method of reduction, at the same time that he sought to limit its application.

Corresponding to each level of the organization of matter, he argued, was a particular science that used the method of reduction but also sought the irreducible qualities, the specific laws, of its particular level. At the summit of this hierarchy was materialist dialectics, the epitome of all the sciences below but at the same time qualitatively different from them: the universal ontology, studying the most general characteristics of all reality, and simultaneously the universal methodology, showing men how to study all reality. Hegel was considered the chief source for this universal discipline, but Deborin taught that Hegel was to be interpreted in a materialist way, and granted the propriety of enriching dialectics from the positive sciences and 'practical dialectics' or politics. In all this, no less than in the definition of matter, Deborin's stress depended on his particular audience and purpose. Arguing against the mechanists' tendency to make reduction the single universal method of science and to neglect the elaboration of dialectics as an independent discipline, Deborin emphasized the concept of irreducible qualities and the study of Hegel as the chief occupation of the dialectical philosopher. Seeking to reassure mechanistic or uncommitted natural scientists, he stressed his approval of reduction or whatever methods or particular concepts might be convenient in particular disciplines.

'We are striving for this,' said Deborin, 'that dialectics should *lead the natural scientist, that it should indicate the correct path to him. . . .*'[28] In contrast he pictured the mechanists as seeking a philosophy that would 'hobble after the "brilliant successes"' of the natural sciences.[29] Quite naturally therefore he felt compelled to give concrete examples of the dialectical philosopher's leadership of the natural scientist, but the results of his efforts in this direction were tardy, few, and insubstantial. His most ambitious effort was a trilogy of articles called 'Engels and Dialectics in Biology',[30] which he would have been wise not to publish, for they were, as Aksel'rod the Orthodox remarked, 'oppressively boring and essentially inane'.[31] One example will suffice. Biologists with Marxist sympathies were debating the rival theories of 'Lamarckism' and 'Morganism' when Deborin wrote his articles, and he took due note of the debate. He explained that the philosophical root of the trouble was confusion about the twin concepts, 'internal' and 'external', and he gave Marxist biologists this advice:

> The opposition between the internal and external has, it goes without saying, only a relative character. The relationship between them is such that they pass reciprocally from one into the other. The possibility of the transition is explained by their oneness, their 'reality', their reciprocal connection. The external is the appearance of the internal, and the internal is along with that also external. In the internal there is nothing that would not appear in the external, while in the external there is nothing that would not be in the internal. The external and the internal constitute merely elements [*momenty*] of one and the same thing; the external is only *relatively* external, consequently it is also simultaneously *internal*, and vice versa.[32]

To be sure, 'Morganism' was proclaimed a Deborinite position, but only at the very end of the philosophical controversy, and not by Deborin himself. Indeed, at the climactic Conference of April, 1929, after listening to many speakers wrangling over rival theories of heredity and other major issues in the natural sciences, Deborin remarked:

> . . . it seems to me that we have conducted the discussion poorly today, for the discussions have treated really all the problems there are in the world. . . . It was necessary to concentrate attention on some nodal, central problems, but this wasn't done. . . .[33]

If his silence concerning the endorsement of 'Morganism' as the Deborinite position in genetics may be interpreted as acquiescence, he was probably motivated by tactical considerations that will be examined farther on, for there was little or no logical connection between 'Morganism' and Deborin's views on 'the internal' and 'the external'.[34]

Deborin offered physicists as little practicable guidance as biologists. In 1924 and 1925, to be sure, he showed hostility to the theory of relativity;[35] but in 1926, as he was emerging from tutelage to A. K. Timiriazev in the philosophy of natural science, he took a neutral position on relativity.[36] By 1930 he learned from some of his followers to picture relativity as the realization of dialectical materialism in physics.[37] But all these were incidental remarks; Deborin never undertook an extended analysis of Einstein's theories. Nor did he get heavily involved in the problem of contingency, which became a critical issue in discussions of physics following Heisenberg's announcement of the principle of indeterminancy in 1927. Deborin could readily repeat Hegel's general formulas: contingency is an objective category, existing not only in the mind but also in external reality; it is indeed a manifestation of necessity. He could also repeat the Aristotelian definition, which seemed to make contingency an appearance and not a reality: it is the intersection of two or more independently developing lines of causation. But when he tried to show how these views might be reconciled, or how they might help physicists through the philosophical difficulties caused by quantum mechanics, he did nothing but quote at great length from 'bourgeois' physicists, arguing that they were spontaneously approaching a dialectical solution based on the recognition of different types of causality for different types of phenomena. And Deborin completed his circular argument by saying that the solution would be hastened if the physicists would study materialist dialectics.[38]

Thus, if one asks whether Deborin was arrogantly metaphysical or worshipfully positivistic in his attitude towards the natural sciences, the answer must be that he was both. He could argue, as in the case just cited, that scientists were becoming dialectical materialists spontaneously. But he frequently pictured the overwhelming majority of natural scientists as 'bourgeois' specialists, stubbornly resisting the penetration of dialectical materialism into their domain, while the mechanist faction allegedly acted as their apologists.[39] His explanation of the paradox was that the increasingly complex and contradictory nature of natural science had a double result. To solve problems within their specialties natural scientists were becoming unwitting dialectical materialists, while, to comprehend science as a whole, their Philistine (*obyvatel'skie*, *meshchanskie*) minds turned to anti-dialectical, 'bourgeois' philosophies, among which varieties of positivism were the most popular.[40] Accordingly Deborin saw his own role as that of winning natural scientists to a dialectical materialist view of science as a whole, to recognition that the universal philosophical problems were answered by Marxist philosophers digging in Hegel. He showed indifference, or even impatience, towards specific problems of the natural sciences, however pregnant with philosophical issues they may have been. It was his job to elaborate an abstract theory of dialectics out of Hegel and other great philosophers; it was up to the scientists to accept the theory, profess it, and apply it in their work.

On this basis it is possible to understand what Deborin meant when he described his work in the philosophy of natural science as the stimulation of a 'social movement about dialectics':

> Nor is it a secret to anyone that the polemic concerning problems of dialectics . . . has stirred up our natural scientists and compelled them to take a closer interest in problems of the dialectical method, so that on this front we may note significant progress. . . . True, some of the patented Marxists, and especially the Marxist *natural scientists*, turned tail at a most critical moment, but the social movement about dialectics continues to develop in breadth and depth. It has taken hold of the youth, who are turning from their 'authorities' in the field of natural science, for they have already grown a head taller than their teachers in matters of theory.[41]

Against the intractable '"authorities" in the field of natural science', who were unwilling to acknowledge the relevance of Marxism to their disciplines, and especially against the mechanists, whom Deborin pictured as apologists for the 'bourgeois' professors' contumacy, he held up his grand metaphysics in authoritarian fashion. Towards those who professed his metaphysics, or might profess it but for fear that it was a Procrustean bed for the natural sciences, he showed himself to be as permissive as only he can be who is basically disinterested in particular scientific theories.

'The method of dialectical materialism . . .,' Deborin said, 'cannot be overthrown by particular, contingent facts. . . .'[42] This could be the assertion of the philosopher's right to tell scientists what they might or might not accept; it could also be the assertion that anything the scientist might declare to be fact would be consecrated by the philosopher as dialectical materialism. 'We proceed,' Deborin said, 'from the most profound conviction that materialist dialectics . . . can help natural scientists raise natural science to a new, higher level, just as happened in the social sciences, thanks to the efforts of Marx and Engels.'[43] He offered almost nothing in the way of empirical proof to support this conviction.[44] In demanding that natural scientists profess it after him, he was therefore demanding a declaration of faith in the universal, beneficial potency of dialectical materialism. But he did not demand much more. If they were willing to seek the blessing of the dialectical materialist philosopher, to profess belief in the transubstantiation of their disciplines as a result of the blessing, he was willing to leave the rest to them.

'Deborin's students', the nucleus of the Deborinite faction, did little more in the philosophy of natural science than to ruminate, expand, and broadcast Deborin's arguments. Like Deborin, they did their most creditable work in the history of philosophy,[45] and came to the philosophy of natural science largely for polemical reasons. Nevertheless their polemics are a rewarding object of study, if only because of their propensity to simplify and exaggerate, and thus to show the faction's basic trends more vividly than Deborin. Thus, V. P. Egorshin, one of the very few of Deborin's students who were also students of a natural science (physics in Egorshin's case), pictured the natural scientists as already dialectical materialist 'in their special researches'. But he derided as typical of the mechanist faction exclusive stress on this spontaneous dialectical materialism, allowing natural scientists to continue their old way of life undisturbed: '*Materialism and dialectics not only in the workaday laboratory but also in the holiday speech and in the generalizing book*—such must be the slogan we cast to the natural scientists of the Soviet Union.'[46] Another young Deborinite who was studying a natural science, B. N. Vyropaev, straining for a succinct description of the desired relationship between natural scientists and Marxist philosophy, said that Hegel, the chief source for the philosophy, had to be read through a materialist prism. 'Here,' he concluded, 'a division of labour is necessary. The philosophers must become such a prism.'[47]

Both these authors evidently took it for granted that the relationship between scientists and Marxist philosophers would be based on what might be called mutual respect for sovereignty, and it is perhaps significant that both these young Deborinites were of the select few who were also fledgling natural scientists.[48] More common in the writings of Deborin's students was a figure of speech that represented philosophers as *ruling* natural scientists.[49] This startling metaphor had its origin in Engels' posthumous *Dialectics of Nature*, where it was a derisive comment on the self-deception of natural scientists who claimed to need no philosophy: 'Natural scientists may take what stand they please,' went the redaction current in the 'twenties, 'but they are ruled by philosophers.'[50] Engels had hardly been in a position to mean this as anything but an historical comment, but in the writings of Deborin's students it took on the tone of a high-handed programme or prescription, which might have been interpreted as threatening the sovereignty of natural scientists even 'in their special researches'.

Some of Deborin's students exaggerated not only their master's superciliousness towards natural scientists, but also his philosophical justification for it. The following, for example, might be taken for a caricature of Deborin's manifesto to the mechanists, though it was actually part of an intensely serious Deborinite tract for members of the Young Communist League (Komsomol):

> If dialectics as a science studies the universal laws of motions [*sic*], then the positive sciences study the concrete forms of motion. As the general is the basis for the particular, so also dialectics is the basis for particular concrete facts. . . . Our philosophy rests on the experience of the entire history of cognition of the world, and therefore it is unconditionally truer, more objective, than individual positive sciences in their theoretical constructs.[51]

Against such extravagant declarations it is of course possible to set repetitions of Deborin's formal adherence to an empirical criterion of truth, especially the formula that 'practice' has priority over theory. But the repetitions of this ambiguous formula by Deborin's students must be considered in context if their meaning is to be understood. For example, in order to refute the mechanists, who gave the formula a positivist meaning, two leading Deborinites (Karev and Sten) recalled Marx's contempt for an understanding of 'practice only in the dirty-Jewish form of its appearance'. (They changed 'dirty-Jewish' to 'dirty-swinish' [*griazno-svinskoi*], just as the newest Soviet Russian translation makes it 'dirty-hucksterish' [*griazno-torgasheskoi*].) 'Practice', the Deborinites insisted, in order to be the criterion of truth, must be 'considered from the vantage point of universal history'.[52] But in that case, one cannot help noting, 'practice' had become indistinguishable from the universal philosophy that was declared to be 'truer, more objective, than individual positive sciences. . . .'

Marx and Engels had derided 'systems' of philosophy often enough for the Deborinites to deny the mechanist charge that they were trying to build one, but at least one of Deborin's students boldly agreed that he advocated the elaboration of a 'philosophical system'.[53] What is more significant is the inescapable fact that the Deborinite nucleus, even though it usually shunned the forbidden term, was indeed beginning to create a system of metaphysics, as the period under review was drawing to a close. At the very beginning of the controversy a Deborinite projected such a system, while defending the cultivation of philosophy as an independent discipline:

> To express one's self in Hegel's language, the *subject* of dialectical materialism is *the most general determinations* [*opredeleniia, Bestimmungen*] *of being*: matter, quality, quantity, measure, causality, subject-object, etc. All the concepts we have indicated are applicable *in equal* degree both to natural and to social science, but at the same time they *do not merge with either one*.[54]

The studies that this Deborinite seemed to be promising began to appear only late in the 'twenties, and then only in very small number. Moreover they never lost the character of researches in the history of philosophy, which never ceased to be the forte of the Deborinite faction. For example, a paper on 'Hegel's Doctrine of Contingency', read to the Communist Academy in March, 1929, was a purely historical study, except for its initial declaration that Hegel's doctrine was acceptable to materialist dialectics virtually without change.[55] It is significant that, with no mechanists participating in the discussion, the only criticism of the paper came from young Deborinites who complained that the speaker was too abstract, too hard to understand. But no one objected to the virtual identification of Hegelian and materialist dialectics on the subject of contingency.[56]

Behind the increasingly ponderous, Hegelian terminology of Deborin's students, it is possible to perceive the same willingness to endorse current theories of natural science that has already been noted in the case of Deborin himself. Examination of the so-called Deborinite positions in biology and physics will reveal this most clearly. But it may be discerned also in the fact that Deborin's students were remarkably moderate in their treatment of the concept of irreducible qualities. Indeed N. A. Karev, the Deborinite who was most notorious for arrogance towards the positive sciences, formulated the Deborinite position on quality and reduction in a way that moved A. K. Timiriazev to exclaim that the dispute was ended. Karev granted

> that every higher form of the motion of matter rises from a lower one, and its appearance is not due to the intervention of any special forces besides those that exist in the lower form. It rises as a result of a complication of the motion of the lower form, a complication that engenders new qualities. In this case there cannot be any thought of vitalism. Every higher form can be created under appropriate conditions out of the interaction of the lower forms—this is indubitable.[57]

But such endorsements of reduction as a method in the natural sciences did *not* end the dispute, for the Deborinites were not quarrelling about the methods actually used in natural science. They stressed the concept of irreducible qualities in order to check scientists from expanding special theories and methods into a universal ontology and methodology, in order to make the scientists look to 'Deborin and his students' for the universal philosophy.

12

DEBORINITE NATURAL SCIENTISTS

FOR a while even some Deborinites followed the mechanist habit of describing the controversy as a clash of Marxist natural scientists (the mechanists) and Marxist philosophers (the Deborinites).[1] Before long, however, the Deborinites dropped this usage, in part because they perceived the handicap that it placed upon them in winning the allegiance of scientists, but also because natural scientists began to join Deborin's 'social movement about dialectics' in significant numbers. The bulk of these scientific Deborinites were young people, mostly graduate students in the universities and institutes of Moscow and Leningrad, whose views during the period under review were virtually indistinguishable from those of 'Deborin and his students'. But the same cannot be said of several mature natural scientists who gave their support to the Deborinite faction. To a large extent they complaisantly repeated the general slogans and formulas of the Deborinite nucleus, but they also revealed some significant differences.

Alexander Aleksandrovich Maksimov was prized by the Deborinites as their first physicist (until he quarrelled with them in 1929).[2] On the other hand, an eminent physicist once commented that 'Comrade Maksimov is not a physicist but a philosopher'.[3] The facts are that Maksimov graduated from the University of Kazan in 1916 with a major in physics, joined the Bolsheviks just after the Revolution, and served the new régime as a provincial educational official and soldier.

On demobilization he got a post at the centre (in the Commissariat of Education's division of secondary education), and in 1922 went to help A. K. Timiriazev win scientists to Marxism at Moscow University.[4] There he found his career as a prolific and persuasive propagandist. He became chairman of the University's Department of the History and Philosophy of Natural Science, which grew out of the informal study circle that he and Timiriazev organized in 1923, and by the end of the 'twenties he was considered an eminent Marxist physicist by mechanist as well as Deborinite philosophers.[5] Let us say that Maksimov became a physicist to philosophers, a philosopher to physicists.

His first contribution to the Deborinite cause was a polemic against the popularizations of natural science that had been coming from the Timiriazev Institute, the stronghold of the mechanist faction. These popularizations, he complained, put too much stress on the method of analysis, the explanation of complex phenomena in terms of simple constituent elements; they neglected the method of synthesis, and the importance of Marxism as a *Weltanschauung* for natural scientists was thereby diminished. As Maksimov saw it, the mechanist faction was furnishing a theoretical justification for such vulgarizations of natural science.[6] In short, he repeated, with variations appropriate to his particular interests, the Deborinite theme that the mechanists were blocking 'the penetration of dialectical materialism into natural science'. And, like the other Deborinites, he understood this slogan mainly as a demand for a profession of faith from natural scientists.[7] Indeed, he quarrelled with the Deborinite faction when it finally took a stand on a specific issue in physics, the theory of relativity.

To understand Maksimov's disgruntlement it is not necessary to delve into the variety of Soviet Marxist positions on relativity that had been developing since A. K. Timiriazev attacked Einstein's physics in 1921.[8] For Boris Mikhailovich Gessen (or Hessen, as he has been known in the West since 1931 for his notorious paper on Newton)[9] manœuvred the Deborinites into the simple assertion that relativity was the realization of dialectical materialism in physics. In 1927, when he began this manœuvre, Gessen was a forty-four-year-old *dotsent* or Assistant Professor in Maksimov's Department of the History and

Philosophy of Natural Science at Moscow University.[10] A specialist, like Maksimov, in the borderland between physics and philosophy (but one with a greater knowledge of physics), he quietly observed a Congress of Physicists where Timiriazev wrangled with eminent non-Marxist scientists over the theory of relativity. When he gave a neutral report of the wrangle in *Under the Banner of Marxism*, Timiriazev protested: Marxists could not be neutral to relativity for it contradicted materialism.[11] Gessen replied that it did not, since it did not question the existence of matter as the source of sensations and knowledge. He even went so far as to say that Newtonian physics was in need of a supplement. 'Whether Einstein's theory or some other theory will serve as that supplement, physics will reveal; and furthermore, no theory will destroy materialism.'[12] Thus Gessen's début as a Marxist was typically Deborinite; discreet neutrality on a concrete issue was combined with sweeping assertiveness on a metaphysical level. But he pressed Timiriazev vigorously on a practical matter, picturing him as a dogmatist baiting non-Marxist physicists:

> Here we come to conclusions in which theory is already interwoven with practical politics; and we, along with the Party and the Soviet régime hold to the unshakeable conviction that we need to work in unison with the representatives of contemporary science, that we can make dialectical materialists of them only by means of *joint* work with them, while the point of view of nihilism, of a peculiar scientific 'On Guardism,'* would do irreparable harm to the Revolution and to Marxism.[13]

This was a startling inversion of combat positions that had become almost conventional; a Deborinite was charging a mechanist with endangering harmonious co-operation between Marxists and natural scientists by thrusting an alien and dogmatic philosophy upon the scientists. By 1928 Gessen was ready to take the final step in his manœuvre, presumably after he had made sure that Timiriazev was almost entirely alone in opposition to the theory of relativity. Gessen then announced that 'in the field of physics the views of the theory of relativity on space and time basically coincide with the views of dialectical materialism on the relationship of space, time, and matter'.[14]

* For an explanation of this term, see p. 159.

When Gessen's manœuvre was completed in 1928, the Deborinite faction was on the verge of its complete triumph over the mechanists, and Maksimov had material incentive to stay with the winning side. Why then at the end of 1928 or beginning of 1929, did he call Gessen a 'Machist' and 'Right deviationist'?[15] The first epithet is not hard to account for. Gessen had concluded that Einstein's physics coincided with dialectical materialist views on space, time, and matter; Maksimov believed that 'through Lorentz, in the person of Einstein, physics has finally reached ideas that *approach* the adoption of dialectical materialism's viewpoint on space and time'.[16] In other words, Gessen put less stress than Maksimov on the 'idealist' elements in Einstein's writings, and Maksimov therefore accused him of accepting Einstein's 'Machist' philosophy along with the valuable physics in the theory of relativity. On that basis one can account for the second epithet, 'Right deviationist'. At the end of 1928 and the beginning of 1929 this term was not yet associated with mechanistic materialism, but was used to describe, in Stalin's words, 'people in our Party who try, perhaps without themselves realizing it, to accommodate the work [*delo*] of our socialist construction to the tastes and needs of the "Soviet" bourgeoisie'.[17] Apparently, then, Maksimov pictured Gessen's position on the theory of relativity as evidence of a larger effort to accommodate the 'social movement about dialectics' to the tastes and needs of the 'bourgeois' specialists in natural science.

In effect, Maksimov was accusing Gessen, and by implication the Deborinite leaders who approved Gessen's manœuvre, of being too soft and permissive in their efforts to convert natural scientists to Marxism; he was turning against the Deborinite faction one of its main charges against the mechanists. It is therefore easy to understand why his quarrel with Gessen did not take him to the mechanist faction. He became an outsider. Within a new, Deborinite Institute of Philosophy at the Communist Academy he was made a subordinate of Gessen's,[18] and he was given no place at all on the editorial board of *Natural Science and Marxism*, which the Academy's Section of Natural Science began to publish in 1929.[19] He stayed away—because of illness, he later said—from the Conference of April, 1929, which, while crowning the Deborinite victory, announced

Gessen's view of relativity as the Deborinite view.[20] (Actually, it had been the mechanist Semkovskii's before it was Gessen's.) Maksimov would come to the top in the 'great break', when the Deborinites would be condemned for their lack of partyness (*partiinost'*) in relation to natural science.[21]

In 1928, about the same time that Gessen was placing the Deborinite label on the equation of Einstein's theory and dialectical materialism, Israel Iosifovich Agol, a 37-year-old geneticist, undertook a similar manœuvre. He claimed that true dialectical materialism (i.e., the Deborinite version of it) and 'Morganism' (roughly, the theory that combinations and mutations of genes are the chief cause of evolution) were mutually corroborative; and he declared the philosophy of the mechanist faction to be inseparable from 'Lamarckism' (roughly, the theory that adaptation to environment and inheritance of acquired characters are the chief cause of evolution).[22] On that basis he demanded the suppression of 'Lamarckist' work at the Communist Academy and the Timiriazev Institute.[23]

This seems at first to be basically analogous to Gessen's manœuvre, but the crucial difference is apparent in Agol's demand for the suppression of 'Lamarckism'. Gessen had no need to make an analogous demand, for he was giving a Deborinite blessing to a scientific theory that was almost universally accepted; his main purpose was to enhance the Deborinite appeal to natural scientists. Agol tried to use the Deborinite ideology as a club against a biological theory that enjoyed considerable support among Soviet Marxists. In order to strike a blow at his opponents in biology he was willing to risk a loss of strength to his philosophical faction. Other Deborinites were unwilling to take this risk, but Agol had the zeal of a new convert—to uncompromising 'Morganism', one must specify. In 1926, when he had become a Deborinite but not yet a militant 'Morganist', he had pictured the Austrian Lamarckist Kammerer as a persecuted hero of science, seeking a haven in the Soviet Union.[24] But in April, 1929, Agol demanded that Kammerer's followers should be expelled from the Communist Academy and the Timiriazev Institute, since they were enemies both of the true science of genetics and of the true, Deborinite version of Marxist philosophy.[25]

Similar anomalies point up the basic motivation of the other 'Morganist' biologists who spoke as Deborinites at the Conference of April, 1929. Their efforts to link the biological and philosophical factions were as strange and new as Agol's, and were similarly directed much more against the flourishing school of 'Lamarckism' than against the decimated mechanist faction in philosophy.[26] Solomon Grigorevich Levit, a 35-year-old physician who had also been recently converted from 'Lamarckism',[27] came the closest to demonstrating a logical connection between 'Lamarckism' and the mechanist faction's philosophy. 'Lamarckism', he argued, by its stress on environmentally induced changes and gradual adaptations, denies the 'self-movement' and 'development by leaps' of living organisms. He also charged the 'Lamarckists' with failure to appreciate the dialectical 'interconnectedness' of the germ plasm and the soma, the organism and its environment.[28] But the 'Morganists' were themselves vulnerable to such arguments from dialectics. If, at the Conference of April, 1929, they displayed a slightly greater facility in this sort of argument than the 'Lamarckists', the reason was probably that they had a head start, and not that their position was inherently more 'dialectical'. In the judgment of the present writer, the unquestioning reductionism of both biological factions made them both more akin to the mechanist than to the Deborinite philosophy. But genuine intellectual affinity was hardly at issue in Agol's manœuvre and the arguments that it provoked.

A similar effort to use the victorious Deborinite philosophy was made by some of the Marxists engaged in the psychological controversies of the late 'twenties.[29] For example, three psychologists who spoke at the Conference of April, 1929, ostensibly to describe the penetration of dialectical materialism into their discipline, did not let slip the opportunity to hint that their own theories of psychology were part of the victorious Deborinite philosophy, and to insinuate that the theories of their opponents were linked to the 'revisionist' philosophy of the mechanist faction.[30] In speaking this way they came into conflict with each other, for they advocated different theories; they agreed that 'reflexology' was 'mechanistic', but they could not agree on what should supersede it. From the leading

Deborinites there was only silence on this question, and one of the three psychologists complained that 'Comrade Deborin . . . has not demarked himself from those who . . . speculate in dialectics [*spekuliruiut dialektikoi*], and there are not a few of them'.[31] The Russian phrase does not have the charitable ambiguity of its English translation, which may mean either that some people were using dialectics for speculation, in the sense of theorizing with insufficient evidence, or that some people were trying to profit from the rising market in Deborinite dialectics. The Russian phrase unambiguously intends the latter meaning; the speaker, A. N. Zalmanzon, was complaining that unscrupulous people were cynically using Deborinite talk to advance special interests in scientific disputes, and Deborin's silence was being taken as support for such people.

Actually it is not necessary to put Zalmanzon's sordid interpretation on the efforts of rival scientists to climb on the Deborinite band-wagon—as, indeed, he himself was trying to do, through grumbling that the best places had been taken. People climb on band-wagons for many reasons, not all of them sordid. Moreover, the silence of the leading Deborinites in regard to conflicting theories in psychology was part of a pattern of Deborinite reluctance to become involved in the concrete issues of the various sciences, whether social or natural. The appearance of the bickering psychologists at the Conference of April, 1929, and the concomitant silence of the Deborinite leaders, have therefore only indirect significance for an understanding of the Deborinite philosophy of science. They show once again that the Deborinite faction had created a 'social movement about dialectics' without being ready or willing to give authoritative answers to the theoretical questions that agitated Marxists in various scientific fields. Paradoxically, one may also note that the Deborinite philosophy of science made possible (or even probable) a scramble for authoritative answers.

The clearest evidence that the Deborinite leaders were not trying to become arbiters of scientific controversies is the activity of the one truly eminent scientist among them. Otto Iulevich Shmidt (or Schmidt, as he wrote his name in Western scientific journals) was only twenty-eight in 1918, when he joined the Bolshevik Party, but he already had a reputation both as a mathematician and as a revolutionary (he had been a prominent Left Menshevik). During the 'twenties he was honoured (or burdened) with offices as important as those of A. K. Timiriazev, the only Bolshevik scientist of comparable stature during this period.[32] For a time Shmidt was head of the State Press (GIZ) and chief editor of the *Large Soviet Encyclopedia*. As Vice President of the State Council of Scholarship (GUS) and President of its Scientific and Technical Section, he played a central role in the continuing transformation of higher education and research. When the Communist Academy in 1925 finally established a Section of Natural and Exact Sciences, he organized and then headed the new Section, and accordingly became a member of the Presidium that governed the Academy. Still he managed to continue his work in mathematics, publishing both in Soviet and in foreign journals, and already in the 'twenties he began the geographical explorations that were to make him something of a popular hero. But perhaps the most remarkable aspect of Shmidt's versatile personality was his attitude towards the problems of a Marxist philosophy of science. He had probably the most crudely practical, the least 'philosophical', attitude of all the major figures in both factions.

Shmidt's essential position was vividly revealed as early as February, 1924, even before the philosophical controversy began, when he made one of his rare contributions to the Communist Academy's continual discussions of Marxism and natural science. Having heard A. K. Timiriazev reject the theory of relativity in the name of dialectical materialism, and Bogdanov and Bazarov defend the theory from a 'Machist' philosophical viewpoint, Shmidt anxiously insisted on the possibility of supporting both the orthodox Marxist philosophy and Einstein's physical theory. Shmidt argued that the physical theory must be judged independently of its creator's philosophy; 'the ideology of a certain professor living in Berlin is of no interest to us'.[33] And he quite frankly, almost cynically, explained his pragmatic attitude towards the philosophical aspects of the theory:

Relativist philosophy was the first to make use of Einstein, the first to begin extracting capital out of him. It does not follow that we must reject him. Let us also try to extract capital from him and take what answers to our world view.[34]

It is hardly surprising, therefore, that Shmidt conceived his task as head of the Communist Academy's Section of Natural and Exact Sciences to be 'essentially political'. His main purpose, he said, was 'to intensify the [political] division among natural scientists', by encouraging those with materialist sympathies to make some demonstration of sympathy with the Soviet régime and its ideology.[35] Nor is it surprising that for a long time he kept out of the philosophical controversy, though evincing some sympathy with the mechanists. (His difference with A. K. Timiriazev over the theory of relativity was not for Shmidt any more than for Tseitlin or Semkovskii a reason to oppose the mechanist faction.) He felt that most natural scientists who had materialist sympathies were closer to the position of the mechanists than to that of the Deborinites, and he wanted not only to avoid offending them but to draw them into his Section of Natural and Exact Sciences, or at least into one of the societies of materialist scientists that his Section organized. On the other hand, he did not want to offend the growing number of Deborinite graduate students in the natural sciences, in part because he agreed with some of their views. But as late as the academic year, 1927–1928, he invited a leading mechanist, Aksel'rod the Orthodox, to give the seminar on dialectical materialism in the Communist Academy's Section of Natural and Exact Sciences.[36] Evidently he still believed, at the beginning of this academic year, that it was more important to propitiate mature natural scientists who had mechanist sympathies than fledgling natural scientists who were Deborinites.

At a conference in March, 1928, Shmidt, goaded into a public comment on the philosophical controversy, showed that he was being pulled towards the Deborinite faction but was still reluctant to commit himself entirely to it.[37] In order to rebuff 'bourgeois' ideologies among natural scientists, to make sure that the rising generation of natural scientists would be thoroughly devoted to the Soviet cause, in a word, to establish the Party's tentacle (*shchupal'tse*) in all branches of natural science, Shmidt felt it necessary to assert the primacy (*primat*) of philosophy and social theory in the elaboration of the Marxist philosophy of natural science. This was of course a, or even, *the* central principle of the Deborinite faction, though Shmidt did not identify it as such. In effect, he was saying that if he had to choose between the mature natural scientists with mechanist sympathies and the Deborinite militants among the graduate students, he was ready to choose the latter. But he did not feel it necessary to choose; he hoped to reconcile the two. He followed the mechanists' custom of deploring the philosophical controversy as an internecine struggle distracting Marxists from their proper work of combating 'bourgeois' ideologies. (The Deborinites generally represented the controversy as an unfortunate but ultimately beneficial purge of 'revisionists' from the body of Marxists.) And he tactfully repeated the most common mechanist criticism of the Deborinites, without identifying it, by saying that metaphysics must be avoided, that the necessity of concreteness in the philosophy of natural science must be recognized.[38] Even in January, 1929, when *Natural Science and Marxism* was launched as a quarterly publication of the Section of Natural and Exact Sciences, with Shmidt as the chief editor, the editorial manifesto revealed Shmidt's adherence to the Deborinites only glancingly.[39]

Only at the Conference of April, 1929, did Shmidt attack the mechanist faction and declare himself a Deborinite publicly, strongly, and in detail. Indeed, he gave one of the two main speeches denouncing the mechanist faction. But even there he could not entirely repress his *arrière-pensées*, which broke into his speech continually, a discordant refrain clashing with the predominant Deborinite theme. He drew a laugh from his audience by quoting a bit of anonymous doggerel that described the Conference as a meeting of '"Catholics"' trying to '"'convert botany and zoology into Marxotheology"'.[40] He told mockingly of

> one graduate student, a very sweet girl, a student of a man who considers himself a Marxist, [who] submitted an essay in which the first lines began: 'The entire great multiformity of the universe is a continuous transition from quantity into quality.'[41]

He imagined a physicist charged with an absurdity who would counter: '"How is that absurd? That is the unity of opposites."'[42]

Even some parts of Shmidt's theses were found objectionable by the Deborinite committee selected to write the 'Resolution in Accordance with Comrade Shmidt's Speech'. Some of the committee's emendations were trifling, but indicative all the same. Where Shmidt wrote that 'the union [of Marxist natural scientists] with Marxists working in the social sciences, especially philosophy, is very important', the committee changed 'union' to 'close union', and described such union as 'absolutely necessary' instead of 'very important'.[43] Shmidt deplored

> the primitive [*kustarnye*] attempts of the so-called mechanists to build 'their own' philosophy on the basis of contemporary natural science in isolation from the development of dialectical philosophy and social theory in general,

and the committee struck 'so-called', Shmidt's one conciliatory reservation in an otherwise thoroughly Deborinite condemnation of the defeated faction.[44] But one of the committee's emendations was not trifling. An entire thesis was dropped, especially since the mechanists at the Conference claimed that it conceded the justice of their hostility to the Deborinites:

> The struggle of the dialecticians [i.e., Deborinites] with the mechanists within our own ranks . . . has weakened us: it has distracted attention from the common enemy (idealism triumphant in the West); it has temporarily led some philosophers to underrate positive knowledge, which portended scholasticism; the whole discussion was at times conducted on an insufficiently high level. . . .[45]

These points had indeed been among the mechanists' central criticisms of the Deborinites. Apparently Shmidt hoped that this thesis would reassure natural scientists who feared the metaphysical arrogance and partisan unscrupulousness that the mechanists attributed to the Deborinites.

It must not be imagined that Shmidt always objected to polemical methods of 'an insufficiently high level'. At the Conference of April, 1929, he gave a review of Soviet Marxist discussions of the theory of relativity that was entirely in accord with Gessen's manœuvre. He completely ignored not only the support given Einstein's theory by Soviet Marxists who were not Deborinites but even his own earlier comments.[46] In other words Shmidt gave his wholehearted support to the legend that the Deborinite faction had saved Einstein's theory from the attacks of the mechanist faction. To be sure, he resisted Agol's manœuvre, and promised that 'Lamarckists' would not be expelled from the Communist Academy and the Timiriazev Institute, but his motivation was probably not an abstract desire for fair play.[47] The Gessen manœuvre was designed to extend the appeal of the Deborinite philosophy to as many natural scientists as possible, while Agol's manœuvre was intended to destroy a school of thought that was fairly widespread among Soviet Marxists. Shmidt, in supporting one and condemning the other, showed once again his main objective in the philosophy of science: to bring as many scientists as possible to ideological sympathy with the Soviet régime.

This objective made Shmidt conciliatory, but only on condition that certain Deborinite fundamentals be accepted. In March, 1928, he had already been inclined towards the Deborinite faction because he had sensed the need of a Communist 'tentacle' in the natural sciences, and he had recognized that unless the 'primacy' of philosophers and social theorists were granted, it would be impossible to establish such a tentacle. But he had feared then that the bold assertion of this primacy would alienate the natural scientists whose friendship he had been cultivating: those with a preference for some kind of mechanist philosophy. By April, 1929, he had changed his mind on some essential matters. 'Now,' he declared, 'it is no longer a question of the Party's having some sort of tentacle in this field [of natural science] . . . . We must concentrate [our] forces in this field of work.'[48] Vaguely sympathetic natural scientists had proved insufficient; after all, the overwhelming majority of mature natural scientists were still followers of 'bourgeois' philosophies. The coming generation of natural scientists, Shmidt warned, including even many Communists, were being 'confused' by their teachers of natural science. It was necessary to mobilize these young people in a struggle for the supremacy of Marxism as the philosophy of natural science; the students would have to teach their teachers. Philosophers and social theorists were to show the students what to teach, and the theoretical justification of the entire process was 'the primacy of philosophy'. If this basic position would be granted,

Shmidt argued, natural scientists had nothing to fear from the Deborinites:

The ridiculous dispute between philosophy and natural science does not exist for us dialectical materialists. We recognize the leadership of philosophy in the sense of a general methodology and a general viewpoint [*ustanovka*]. On the other hand, the philosophers, through the mouths of Deborin and his students who have spoken here, have refuted the absurd accusation sometimes made against them, that they scorn concrete natural science and prescribe their own laws to nature. While continuing their own work in the development of dialectical materialism, the philosophers are in need of facts and concrete applications of the method from us [natural scientists]. We, in our turn, need the leadership of dialectical materialism for the solution of the tremendous task before us, that of mastering all natural science as a whole, and we shall utilize that leadership.[49]

The Deborinite philosophers at the Conference of April, 1929, did not take exception to this peroration. And they had no reason to, for Shmidt had neatly expressed the practical side of their main position: natural scientists could do as they pleased in natural science, if they recognized the primacy of Marxist philosophers and social theorists in questions of *Weltanschauung* and abstract methodology.

Shmidt, who was probably the most estimable natural scientist *qua* scientist in either faction, had the least 'philosophical' attitude towards the issues. Perhaps this was the result of his preoccupation *qua* government official with the Cultural Revolution in natural science. Perhaps, paradoxically, it was the result also of his genuine profundity as a scientist: maybe the feeble light of the Deborinite philosophy (or of dialectical materialism altogether) could not penetrate the depths of his mathematical mind. Or, assuming the Deborinite philosophy and dialectical materialism to have been more than a feeble light for science, perhaps he had the kind of scientific mentality that was insensitive to all but technical and political issues. An attempt to choose the most likely possibility would raise the larger problem of determining what are the meaningful questions in the philosophy of science that are neither technical nor political. For present purposes it is sufficient to note that Shmidt's view of the controversy was largely political. He perceived the Deborinite and mechanist positions in terms of the relations between natural science and its adepts, on the one hand, and Communist ideology and its interpreters, on the other hand. What is more, something similar can be said of nearly all the mature scientists who joined the Deborinite faction. (The exceptions were those who tried to use the Deborinite philosophy as a weapon in scientific controversies.) The Deborinite faction appealed to Marxist natural scientists to join a 'social movement about dialectics', and that apparently is what those who responded did. It is therefore hardly surprising that the semi-philosophical issues dividing natural scientists *qua* scientists were rather artificially involved in the writings of the Deborinite natural scientists.

13

SOCIAL THEORISTS IN THE DEBORINITE FACTION

MUTUAL indifference prevailed between the Deborinites and the social theorists of the 'twenties,[1] but there were two exceptions to the rule. A few social theorists gave support to the Deborinite faction, and the Deborinites began to extend their philosophy into the field of social theory in the second half of 1929, that is, when they had finished their dispute with the mechanists and were on the verge of being attacked by a third faction. These exceptions deserve examination, for they point up some of the essential features of the Deborinite philosophy.

Without doubt the most notable of the social theorists who gave articulate support to the Deborinites was David Borisovich Riazanov (*né* Gol'dendakh in 1870), a sardonic old revolutionary, crusty enough to decry the cult of Lenin,[2] but nevertheless kept on as the chief of research on Marx and Engels until 1931.[3] There is no place here to review his impressive work as an historian of Marxism, for his support of the Deborinite faction was a peripheral incident in that work, very nearly the result of a misunderstanding. Among the manuscripts that he wheedled out of the archives of the German Social Democratic Party was Engels' fragmentary *Dialectics of Nature*, which he deciphered and finally published in 1925, just as the controversy between the mechanists and Deborinites was getting fully under way.[4] This new 'classic of Marxism-Leninism', as the Deborinites hailed it, contributed to their victory over the mechanists. But Riazanov did not regard the work as a 'classic', in the sense of an authoritative standard for the Marxist philosophy of science; he agreed with Einstein's opinion that the manuscript had only historical interest as a source for the intellectual biography of Engels.[5] If Riazanov nevertheless argued publicly that Engels' new work helped to prove the Deborinites right and the mechanists wrong, his reasoning was simply that the mechanists denied the universality of Marxism, while Marx and Engels declared, in many 'classics' as well as this new work, that dialectical materialism was applicable to natural no less than social science.[6] Such an argument was of relatively little help to the Deborinites, for the mechanists were not positivists explicitly but only in effect; the chief issue was not whether but how the Marxist philosophy applied to natural science. Riazanov probably perceived this as the Deborinites became more and more involved in natural science by their efforts to prove that their version of dialectical materialism rather than the mechanists' was the truly universal Marxist philosophy; for he ceased active support of the Deborinite cause as early as 1926.

By the spring of 1928, though he still felt that the Deborinites were right by comparison with the mechanists, he was sarcastic towards the entire effort of Soviet Marxists to 'penetrate natural science'. Indeed, many aspects of Soviet Marxism must have been troubling the old man, for he disturbed a Conference of Marxist-Leninist Research Institutions in March, 1928, with sarcastic interjections on several subjects.[7] He derided the Communist Academy's efforts to conduct Pavlovian research.[8] He confessed that Deborin was justified in accusing him of indifference towards 'the hunt after natural science', which he characterized by the Russian equivalent of catching a tiger by the tail ('the hunter seized the bear, and the bear will not let him go').[9] 'Recently,' said Riazanov, shifting his figure of speech, 'these symptoms of the infection not of natural science by Marxism, but of Marxism by natural science, have been growing stronger and stronger.' When the mechanist Varjas interjected at this point, 'And what does that mean, infection?' Riazanov gibed at him and specified the mechanist disease of reductionism.[10] But apparently he felt that the Deborinites had also contributed to the infection, or the 'jabberology' (*boltologiia*), as he described the philosophical controversy at one point.[11]

Soviet Marxism, he argued, would not be ready to penetrate and transform natural science in a meaningful way until a great many natural scientists had been trained with Marxism as their general philosophy. In effect, he wanted the universality of Marxist philosophy taught to natural scientists as a self-evident truth, leaving the task of explaining what this truth meant to a future generation of natural scientists that would take it for granted. Thus Riazanov combined the mechanist argument that the philosophy of natural science must be elaborated by natural scientists in the specific context of their disciplines with a peremptory declaration in the Deborinite manner—but embarrassingly naked—that natural scientists must begin by professing dialectical materialism on faith. One may accordingly surmise that his complete absence from the closing stage of the controversy was the result of Deborin's wishes no less than his own.[12]

In the closing stage of the controversy, as Riazanov was withdrawing his original support of the Deborinites, some other prominent social theorists were going through an opposite process: they were relinquishing an original coldness or hostility towards the Deborinites. V. V. Adoratskii, who had shown pronounced mechanistic tendencies in the early 'twenties, retreated into silence when he found himself lumped with Minin and Enchmen.[13] He did not cease his writings in the field of social theory, where he revealed his continuing mechanistic tendency,[14] but the Deborinites kept aloof from this area of Marxist thought. One wonders what discussions occurred between Adoratskii and Deborin, who were colleagues in a number of institutions of Soviet Marxist higher learning.[15] In this connection, there would appear to be some significance in the manner of publication of Lenin's philosophical notes. Deborin published the first fragments in 1925 as a blow at the mechanists,[16] but Adoratskii, who apparently became the editor of the notes in 1926, when Stepanov became head of the Lenin Institute,[17] delayed publication of the rest. Some Deborinites began to complain publicly and to recall Adoratskii's 'Mininism'.[18] When the first volume of the notes finally appeared early in 1929, Deborin was the author of the Preface. Apparently Adoratskii had agreed to support the Deborinite faction, though he had required some unspecified alterations in Deborin's Preface.[19] Perhaps Adoratskii, who had been a friend of Lenin's, was converted to grudging support of the Deborinite 'struggle for Hegel' by poring over Lenin's admiring comments on *The Science of Logic.*

Whatever the causes may have been, Adoratskii's support of the Deborinites could hardly have been more niggardly, for he never publicly and explicitly endorsed them or condemned their opponents. This was an especially difficult feat, for Adoratskii's report on Lenin's philosophical works was one of the principal speeches at the climactic Conference of April, 1929.[20] In one portion of this speech he implied agreement with the Deborinites, but even there he could not forbear a little irony at their expense:

> Dialectics is the logic and the theory of knowledge of Marxism, dialectics that is of course purged of idealism, that reflects the external motion of the real world. And so, until the new Leibniz promised by Comrade Iurinets appears and writes us a theory of dialectics, until then, Hegel's book, *The Science of Logic,* studied with the guidance of Lenin's notes, is the most important guide in this theory of dialectics.[21]

The reference to 'the new Leibniz promised by Comrade Iurinets' was a fairly plain sneer at Deborin, for Deborin's students (Iurinets included) tended to picture him as the successor to the 'founders' (*osnovopolozhniki*) of Marxism-Leninism who would soon write the definitive work on materialist dialectics. Moreover, Adoratskii's twin stress on purging Hegel's dialectics of idealism, and on the indispensability of Lenin's notes in the process, distinguished his remarks from the usual profession of the Deborinite creed (and anticipated a later school of Soviet Marxism), for the Deborinites usually stressed neither the purging nor the use of Lenin's notes. One supposes that Adoratskii was trying to be loyal to Lenin's admiration of Hegel, and to the 'social movement about dialectics'. But his old belief that materialist dialectics must not be considered a philosophy, which he had formerly dismissed as 'ideology', was apparently still working in his mind; perhaps along with suspicion—perhaps jealousy—of Deborin's growing eminence.

Adoratskii had been supported in his mechanistic tendencies

of the early 'twenties by I. P. Razumovskii, a specialist in historical materialism and the philosophy of law. Razumovskii became a supporter of the Deborinites as early as 1926, and was a good deal more outspoken and less stinting in his support than Adoratskii.[22] Perhaps his motives were similar to those already noted in the case of some psychologists, for he used charges of 'mechanist revisionism' as a weapon in the disputes concerning social theory.[23] In the judgment of the present writer, Razumovskii's social theories were no less mechanistic than his opponents',[24] but it is not necessary to examine them, for the leading Deborinites were aloof from issues in social theory until late in 1929. The only aspect of Razumovskii's social theory that his Deborinite colleagues commented on was his definition of 'ideology'. The memory of his earlier condemnation of philosophy as an 'ideology' evidently rankled, especially in view of his continuing relutance to grant that ideologies might, like sciences, be judged true or false in a 'theoretico-cognitive sense'. He regarded ideologies as significant only in a 'class-useful' (*klassovo-poleznaia*) sense, and this could be interpreted as a belittlement of philosophy, which was generally considered an ideology.[25] But the leading Deborinites took an indulgent attitude towards Razumovskii's aberrations even on this touchy issue, probably because there was only one other notable social theorist who gave more than perfunctory support to the Deborinite faction.

The one other was I. P. Podvolotskii, 'one of Bukharin's pupils' in the field of social theory,[26] who gave no indications of changing his views on social theory when he became a Deborinite. In his first contribution to the Deborinite faction in 1927 he stressed the usual theme that 'methodology [has] the task of guiding the work of natural scientists'; indeed, he was tactless enough to imply that the *humbling* of natural science was one of the Deborinite aims.[27] As late as April, 1929, he still failed to connect the Deborinite philosophy with issues in social theory.[28] But in the fall of 1929 he produced the first Deborinite attack on Bukharin's social theory as the correlate of mechanistic materialism in philosophy and of right deviationism in politics.[29] And when the Communist Academy developed this theme in a series of meetings at the end of 1929, Podvolotskii was the continuing chairman and a principal speaker.[30]

One cannot exclude the possibility that Podvolotskii's view of Bukharin's social theory had been changed before the end of 1929 by the logic of Deborin's philosophy; he may simply have kept the change to himself until it was permissible to criticize Bukharin without risk of being called a Left deviationist. Nor can one rule out the possibility that Podvolotskii had been unaware of a conflict between his Bukharinite position in social theory and his Deborinite position in philosophy until the highest officials of the Party demanded that the Deborinites join in the criticism of Bukharin.[31] It is also possible that Podvolotskii was a time-server. But whatever the cause of his abrupt appearance as an opponent of Bukharin's social theory, he was apparently somewhat uncomfortable, and tried hard to show that the attack on Bukharin's social theory was rooted in the previous activity of the Deborinites. In doing so he took a step that does credit to his logical ability but raises considerable doubt of his political acumen. The main conclusion to be drawn from the discussions of Bukharin's social theory, he stressed, was that the Deborinite philosophers had been the Party's prophets:

The struggle against the theoretical foundations of the Right deviation is a direct continuation of our struggle against the mechanists and a verification of the correctness of the theoretical, the philosophical positions that we held in our struggle. It is the political verification of our theoretical positions.[32]

Podvolotskii may not have realized it, but he had taken the Deborinite belief in the primacy of philosophy to the verge of a dangerously Platonic position on the relationship of philosophers and kings. He could have been interpreted as implying that the philosopher, since he was the master of the universal dialectics, was to guide not only the work of the practical man as natural scientist, but also the work of the practical man as political leader and statesman.

But a radically different conclusion could be drawn from the Deborinite philosophy when applied to social theory. The Deborinites repeated the usual Marxist formula that practice has priority over theory, but usually interpreted 'practice' in such a way that it became indistinguishable from the universal theory of dialectics, thereby justifying once again the primacy of philosophers over practical men.[33] The mechanists had continually used the same formula to a contrary end, and quite naturally one of them (Malyi) did so again at the meetings late in 1929, when the Deborinites extended their philosophy into political and social theory. *Stalin*, he claimed, had discovered the error of Bukharin's social theory, because Stalin was a practical man and therefore a better dialectician than the abstract, speculative philosopher, Deborin.[34] A young social theorist of Deborinite sympathies, P. A. Shariia, thereupon objected:

> Comrade Malyi opposes the practical dialectician Stalin to the theoretical dialectician Deborin. This is once again, in essence, an attack on theory. Stalin is a practical dialectician because he is a good theoretical dialectician. Stalin's correct practical standpoint results from a correct theoretical standpoint. . . . *It is impossible to be a practical dialectician without being a theoretical dialectician.* The old argument, when Lenin was called a practical man and not a theorist, is now being repeated by Comrade Malyi. *I categorically deny that the chiefs* [*vozhdi*] *of our Party are only practical leaders* [*rukovoditeli*]. If they were not theorists they could not determine the Party's general line. To say that there are practical dialecticians is to insinuate the mechanistic view that the theorist has no significance for the practical struggle. It is once again the struggle against theory, against theoretical philosophy, the separation of practice from theory, and the preaching of crawling empiricism and tailism.[35]

With a few months similar arguments would be used by a third faction, neither mechanist nor Deborinite, to justify the subjection of philosophers to kings.

Most likely neither Shariia nor Podvolotskii realized that they had come to the verge of sharply different conclusions in arguing that there was harmony between the Deborinite philosophy and the Stalinist social theory. Their thoughts as philosophers were still oriented towards the controversy with the decimated mechanist faction, a controversy that had been focused on the relations between philosophy and natural, rather than social science. In dutifully undertaking a philosophical rebuttal of the right deviation, they unwittingly revealed once again the essential ambiguity of the Deborinite philosophy, its mixture of arrogance and deference towards less exalted studies than the elaboration of dialectics from Hegel. But they were taking the Deborinite philosophy into a field, as the controversy of 1930 was soon to reveal, where such an ambiguity would not be tolerated.

14

CLOSING THE CONTROVERSY, 1926-1929

By 1926 it was already apparent that Soviet Marxists were reluctant to defend the mechanist faction publicly. This reluctance was especially notable in the case of anti-religionists, social theorists, and natural scientists with Marxist sympathies, for they were predominantly mechanistic in the tendency of their thought. On the other hand, several Marxists whose views differed from the philosophy of 'Deborin and his students' nevertheless made some public show of support for the Deborinite faction. This anomalous weakening of the mechanist and strengthening of the Deborinite faction, with apparently scant correspondence between public behaviour and inner thought, continued beyond 1926. At the Conference of April, 1929, which closed the dispute by condemning the mechanist faction as 'revisionist', a Deborinite could boast that only a negligible handful of diehards was left in the mechanist faction.[1] At the same time another Deborinite warned that it was deceptive to judge the strength of the mechanist faction by the eight or nine who still stubbornly spoke up for it; there were, he cautioned, many silent sympathizers.[2] And neither of these Deborinites was far wrong. The mechanist faction was depleted even though mechanism as a tendency remained strong; the Deborinites gained their victory on a nearly deserted field.

One of the chief reasons that the mechanist faction withered away was that important Party organs showed favour for the Deborinite faction as early as 1925. *Bolshevik*, the theoretical journal of the Party's Central Committee, had carried the original polemic between Sten and Stepanov, but had closed its columns to the controversy in apparent neutrality at the end of 1924.[3] However, less than a year later *Bolshevik* began publishing occasional pieces by Deborinites, and none by mechanists.[4] Other important journals and newspapers (for example, *Pravda*, the Party's chief daily, *The Young Guard*, journal of the Communist Youth, and *Chronicles of Marxism*, published by Riazanov's Institute of Marx and Engels) took similar notice of the philosophical controversy in occasional reviews and articles that were uniformly Deborinite.[5] The journals that regularly carried philosophical articles were, with one exception, under Deborinite control from the start of the controversy; they allowed space for mechanist polemics, but appended deprecatory editorial footnotes and Deborinite rebuttals much longer than the articles rebutted. This was true of the chief philosophical journal, *Under the Banner of Marxism*, long before the end of 1926, when Deborin was formally elevated to the post of 'responsible editor', or editor in chief, as we should say.[6] And though Deborin never held the same position on *TheHerald of the Communist Academy* or *The Militant Materialist* (published by the Society of Militant Materialists), as early as 1925 and 1926 these periodicals were following a Deborinite policy in their philosophical offerings.[7] By 1928, if not earlier, the State Press (GIZ) was in Deborinite hands, as far as philosophical publications were concerned.[8] It is thus small wonder that Soviet Marxists, most of whom no doubt wanted to be in the swim of Party thought, increasingly shunned the mechanist faction.

However, the Party's Central Committee made no formal pronouncement on the philosophical controversy, and it was possible to stand against the current. As early as August, 1926, the Deborinite V. V. Sorin appealed, from the pages of *Pravda*, for a decree to end the philosophical controversy; but the Central Committee did not respond.[9] Perhaps this silent rebuff caused Sorin to adopt a liberal view. At any rate, until mid-1927, when he and the leading anti-religionist Iaroslavskii were dropped from the editorial board of *Red Virgin Soil*, this famous review gave space to Aksel'rod's polemics against the Deborinites.[10] And there were other cases of important periodicals occasionally granting space to the mechanists: *Izvestiia*, which was edited by Stepanov, did so,[11] and as late as 1928 Sarab'ianov was allowed to present the mechanist case in *The Young Communist*, a publication of the Communist Youth.[12] To be sure, these were isolated cases, as also was *Dialectics in Nature*, the thoroughly mechanist organ of the Timiriazev Institute, which appeared five times between 1926 and 1929. All these were indications that the philosophical discussion was formally held open until April, 1929, even though increasingly few writers or editors were willing to show favour to the mechanist faction, or even to appear neutral. We are accordingly faced with this singular pattern: high-placed support for the Deborinites contributed signally to the rapid weakening of the mechanist faction almost from the start of the controversy, yet the same high places seemed in no hurry to shut off the dispute altogether.

Any effort to explain this singular pattern must take note of the tactful self-restraint displayed by the Deborinites, and the corresponding self-restraint of those who might well have opposed the Deborinites but did not. The Deborinites confined their attack on mechanistic thought to certain fields, mainly the history of philosophy and the philosophy of natural science, and even in the latter field they tended to ignore mechanistic thought as long as it was not explicitly associated with the mechanist faction. The literature of the anti-religionists, for example, laid heavy stress on natural science, interpreted in a pronouncedly mechanistic way; yet, with the single exception of Sten's initial attack on Stepanov, the Deborinites ignored the anti-religious literature. The anti-religious movement returned the favour by refraining from throwing its weight, which became considerable in 1927, behind the mechanist faction. Indeed, by 1927 and 1928 a few anti-religious writers were showing faint signs of Deborinite influence,[13] and towards the end of 1928 the chief anti-religionist, Emilian Iaroslavskii, fleetingly acknowledged the accumulating defeat of the mechanist faction. But he did so somewhat ambiguously, in the course of a tribute to Stepanov, who had just died:

> This discussion [which Stepanov's book precipitated] is not yet finished, and however mistaken the views of the group of Marxists to which I. I. Stepanov adhered may have been, his direct and precise

formulation of the problem has aided the work of giving Hegelian dialectics a materialist explantion.[14]

Apparently Iaroslavskii was a bit uneasy in regard to the gathering victory of the Deborinites; not until the discussion was formally 'finished' by the Conference of April, 1929, did *The Anti-religionist*, which Iaroslavskii edited, publish a Deborinite article.[15] Even then the predominantly mechanistic tendency of the journal was only slightly altered; and it is noteworthy that in 1930, when the Deborinite ascendancy was attacked by a third faction, Iaroslavskii not only helped mount the attack, but tried to make it a resurgence of mechanism.[16] Thus, it seems fair to assume that Iaroslavskii and the anti-religious movement were one of the forces that kept the controversy from being 'finished' until 1929, even though they were unwilling to give active support to the mechanists.

The Deborinite faction overlooked mechanistic tendencies also in the field of social thought, perhaps intentionally, thereby making it possible for a few social theorists like Razumovskii and Podvolotskii to become active Deborinites in the philosophical controversy without altering the mechanistic tendency of their social thought. But the active support that these few gave the Deborinite faction was probably of less moment to the outcome of the controversy than the abstention (or simple indifference) of most social theorists. It may well be that many Soviet Marxists did not realize the similarity between mechanistic tendencies in social thought and the mechanist faction's disfavoured philosophy of natural science. But at least one eminent social theorist, S. S. Krivtsov, realized the similarity, and alluded to it in a passing comment that is noteworthy for its perceptive distinction between mechamism as a tendency and as a faction. 'The dispute still goes on, it is not yet decided,' he wrote in 1928,

> for which reason this is not the place to give an evaluation of this still uncompleted process, all the more since each of us belongs, if not to one or the other faction, then to one or the other tendency.[17]

Krivtsov's comment, with its stress on the formal openness of the dispute, is as noteworthy as Iaroslavskii's for it too was made late in 1928, when the disfavoured mechanist faction had lost the active support of all but a few Soviet Marxists. Krivtsov was no more willing than Iaroslavskii to join this hardy band, especially since his own work was not being criticized by the Deborinites; but, like Iaroslavskii, he appears to have been anxiously aware of the similarity between his own philosophy and that of the failing mechanist faction. On the assumption that other social theorists, perhaps including Bukharin and Trotsky themselves, felt as Krivtsov did, they must be considered another force holding back the formal termination of the controversy.

There was a most revealing episode at the First All-Union Conference of Marxist-Leninist Institutions, which was convened by the Communist Academy in March, 1928. The philosophical controversy was deliberately excluded from the agenda, and one of Deborin's former students, S. L. Gonikman, who was then, at thirty-one, head of the Leningrad Institute of Marxism, sharply protested at the exclusion and at other signs of what he called 'a certain neutrality of the Communist Academy as an institution in the philosophical dispute':

> In this discussion, not with the 'so-called' mechanists, [the Conference's chief speaker, V. P. Miliutin, had used that expression] but with genuine philosophical revisionists, how has the Communist Academy shown itself . . .? If one looks at the journal of the Communist Academy, one can find in it a paper of Comrade Bogdanov, 'A Reasoner's Reasonings Concerning Reasoning', ...[18]

As Gonikman proceeded with his sarcastic indignation, V. P. Miliutin, an economist and a member of the Party's Control Commission who shared the leadership of the Communist Academy with the historian Pokrovsky, called out from the floor: 'Are you suggesting that the list of the Academy's members should be revised?'[19] Gonikman backed away from this logical conclusion of his argument, and Miliutin subsequently took the speaker's stand to explain the attitude of the Academy's leaders towards the philosophical controversy. He said that the Academy was 'an institution where the clash of opinions that exist among us Communists on one or another theoretical question is permitted and must be permitted'.[20] At that point an unidentified voice from the floor resumed Gonikman's line of argument by calling out: 'Including the revision of Marxism?'[21] Miliutin did not answer directly, but reminded his critic that freedom of discussion had been granted even to Preobrazhenskii's deviationist views on economics, and asked again whether the expulsion of the mechanists—he mentioned Stepanov and A. K. Timiriazev specifically—was desired.

Miliutin felt that the mechanists should not be expelled, but he was not consistent in explaining himself. On the one hand, in appealing to the Deborinites for moderation, he implied recognition that the Deborinite philosophy was right and the mechanist wrong. 'Comrade Deborin cannot complain; we have given the preference to him.'[22] (At this point Riazanov, ever sensitive to what would now be called the cult of the personality, called out: 'Not to *him*, but to the trend [of thought]", and Miliutin accepted the correction.)[23] On the other hand, Miliutin justified toleration of the mechanists by the argument that the truth was not clearly on either side. He hoped that the two factions might be reconciled, but in any case,

> it cannot be said that some comrades or other—Timiriazev, Stepanov-Skvortsov—have abandoned Marxism. It is not known whether, as a result of elaboration, discussion, in the final analysis a single line will be laid down here, or whether the same situation will remain; we cannot tell this beforehand, and therefore we are going to allow the elaboration of these problems within the walls of the Communist Academy.[24]

Pokrovsky also spoke in neutral fashion,[25] though he too must have approved of the 'preference' being given to the Deborinites. Presumably the historian Pokrovsky, the social theorist Krivtsov, and the anti-religionist Iaroslavskii had the same peculiar mixture of feelings towards the philosophical controversy as the economist Miliutin. Acquiescing in the increasing ascendancy of the Deborinites, they appear at the same time to have been somewhat anxious to allow still a little room for defence of a mechanistic philosophy.

The 'preference' that Miliutin inconsistently boasted of giving to the Deborinite faction in the Communist Academy had begun to show itself clearly in 1926, just after the debates in the Institute of Scientific Philosophy crystallized the opposing factions. Miliutin had then proposed the creation of a separate

Section of Philosophy, characteristically avoiding any show of partiality, arguing merely that Marxist philosophy could not for long endure the existence of two factions, which might harden into two permanent schools.[26] But preference for the Deborinites was implicit in the proposal itself—did not most mechanists oppose the study of philosophy by itself, apart from the concrete material of the sciences?—and became evident in the implementation. The Section of Philosophy, designed to eliminate the division in Marxist philosophy, was put under the direction of A. M. Deborin.[27]

Nevertheless, until the end of 1928 the mechanists in the Communist Academy could still function within the Institute of Scientific Philosophy, and in O. Iu. Shmidt's Section of Natural and Exact Sciences *they* probably enjoyed a slight 'preference'. Shmidt tried, to the very end of the controversy, to conciliate the mechanist faction; in 1927 he even invited Aksel'rod to give the seminar on dialectical materialism in the Section of Natural and Exact Sciences.[28] He probably joined the Deborinites only at the end of 1928, at a time when further 'preference' was being given them: the Institute of Scientific Philosophy was fused with the Section of Philosophy to form a single Institute of Philosophy headed by Deborin.[29] Evidently, it was at the end of 1928 that the authorities at the Communist Academy decided to suppress the mechanist faction altogether, regardless of any offence to the natural scientists that Shmidt had been patiently wooing. But until that time, Marxist and potentially Marxist natural scientists had probably exercised an influence on the controversy similar to that of the social theorists and anti-religionists. Unwilling for the most part to defend the mechanist faction, they had nevertheless evinced enough sympathy for it to hold back its formal condemnation.

Very likely the conflict in the Timiriazev Institute was one of the main reasons that Shmidt, and Soviet Marxists generally, took it for granted that the mature natural scientists who might be won to Marxism were predominantly sympathetic to the mechanists. This Institute, which was the centre of attraction for Marxist natural scientists before Shmidt's parallel institution was organized in the Communist Academy, voted a formal resolution of support for Stepanov's book early in 1925 and became the mechanist faction's stronghold. It remained to the end of the controversy the only institution of Marxist higher learning that the Deborinites did not control, although they appear to have tried repeatedly to win it. There is an unconfirmed report that a Deborinite resolution was pushed through in 1927 but was not implemented.[30] In the second half of 1928 Stepanov complained that a mechanist had been removed from the management of the Institute's seminar on dialectical materialism.[31] But the biological laboratory and the Institute's philosophical publication remained in mechanist hands to the end of the controversy. Perhaps this continuing difficulty in gaining control of the Timiriazev Institute was the chief reason that many of the Deborinite leaders were permissively silent when Agol linked the philosophical controversy with the dispute over genetical theories. Agol was trying to oust the mechanists from their one remaining centre by charging them with false science ('Lamarckism') as well as false philosophy. At the Conference of April, 1929, he boasted that he had achieved control of the Institute, and promised a Deborinite reorganization.[32] But he spoke prematurely, for Shmidt defended the right of Marxist biologists to continue 'Lamarckist' research, and the Conference resolved that the Timiriazev Institute should be fused with Shmidt's Section of the Natural and Exact Sciences at the Communist Academy.[33]

The proposed fusion did not grind its way through the bureaucratic mill until the latter part of 1929 (the Institute was subject to the Commissariat of Education of the R.S.F.S.R., while Shmidt's Section of the Communist Academy was subject to the Council of People's Commissars of the U.S.S.R.), and in the meantime the mechanist diehards at the Timiriazev Institute got out two more issues of their philosophical journal.[34] When the Communist Academy's Presidium finally obtained approval of the projected merger, it ordered that the Timiriazev Institute's press be absorbed by the corresponding part of the Communist Academy, thereby closing the mechanist faction's access to print. But the Presidium qualified other aspects of the merger:

> Transfer this Institute's section of methodology into the [Communist Academy's] Section of Natural and Exact Sciences, but preserve it in the form of a separate group. The biological laboratory of the Communist Academy shall be fused with [that of] the

Timiriazev Institute, but the continuation of Professor Kammerer's [i.e., 'Lamarckist'] experiments shall be provided for in it.[35]

Thus, even at the end of 1929, when mechanism had been condemned not only as a philosophy of natural science but also as the philosophical foundation of the right deviation, the Communist Academy's authorities did not entirely relinquish Shmidt's efforts to conciliate the mechanistic tendencies of Marxist natural scientists.

This combination of official hostility towards the mechanist faction with indulgence of mechanistic tendencies, especially among natural scientists, was clearly revealed at the climactic Conference of April, 1929, which condemned the mechanist faction as preaching 'revisionism'. Indeed, the Conference was arranged in such a way that the chief conciliator, O. Iu. Shmidt, who had only just become a Deborinite, appeared to be the equal of Deborin in the leadership of the victorious faction. At separate sessions each one, the philosopher and the natural scientist, gave a 'main speech' (*doklad*); and two separate resolutions were adopted, 'The Contemporary Problems of the Philosophy of Marxism-Leninism (In Accordance With the Speech of Comrade A. M. Deborin)', and 'The Tasks of Marxists in the Field of Natural Science (In Accordance with the Speech of Comrade O. Iu. Shmidt)'.[36] Shmidt made the intended lesson explicit by trying to insert in his resolution a rebuke to those philosophers who 'have underrated positive knowledge, which has portended scholasticism'.[37] The other Deborinites excised this point from Shmidt's resolution, but several Deborinites echoed Shmidt's vigorous assurances that 'the philosophers, through the speeches of Deborin and his students, have refuted the absurd accusation sometimes made against them, that they scorn concrete natural science and prescribe their own laws to nature'.[38] And the Deborinite editorial committee did approve, as part of the Resolution 'In Accordance With the Speech of Comrade O. Iu. Shmidt', a thesis that stated one of the central views of the mechanists: 'The Marxist methodology of the natural sciences cannot be built on general reasoning, isolated from the concrete problems of science and from participation in concrete, experimental work.'[39]

In view of this conciliatory attitude towards mechanistic

tendencies among natural scientists, social theorists, and anti-religionists, one may well ask why authoritative Party organs favoured the Deborinites in the first place, why Soviet Marxists of mechanistic tendencies shunned the mechanist faction, and why the decision was finally made to suppress that faction altogether. The answer is to be found partly in posthumous works of Engels and Lenin, but much more in living exigencies of the Cultural Revolution.

15

'CLASSICAL' AUTHORITY AND THE CULTURAL REVOLUTION

It is hardly surprising that the Deborinites claimed victory for themselves when Engels' *Dialectics of Nature* was published in the summer of 1925, for this new 'classic' was an extended polemic against positivist and mechanist interpretations of natural science.[1] Luppol, for example, complacently declared that

> the publication of *Dialectics of Nature* in Riazanov's *Archives* [*of the Marx-Engels Institute*] should put an end to the disputes that have filled the pages of our theoretical journals for many months. To all the questions of principle that are involved in these disputes *Dialectics of Nature* gives clear and unequivocal answers.[2]

And *Bol'shevik*, the theoretical journal of the Party's Central Committee, published a review that described Engels' new work as 'naturally the decisive blow in the struggle, the blow that . . . achieves complete victory for the camp of the supporters of the philosophy of dialectical materialism', i.e., the Deborinite faction.[3]

Of course, these claims and predictions were proved wrong by events; the controversy continued for nearly four years after the appearance of Engels' book. One reason was that the mechanists would not surrender the new 'classic' any more than the old ones, but stubbornly claimed it as support for their own position. Stepanov, for example, argued that *Dialectics of Nature* showed Engels' development through two periods, from

Hegelian metaphysics to a philosophy of science that agreed with Stepanov's own.[4] To be sure, Deborin turned this argument against Stepanov by demonstrating that the mechanistic passages in *Dialectics of Nature* corresponded to no chronological pattern; Deborin concluded that there were not different periods in Engels' development but dialectically contradictory elements in his philosophy.[5] Stepanov thereupon turned to the simpler, and much less effective, argument that *Dialectics of Nature* is a jumble of raw notes and hasty first drafts, from which conclusions about Engels' philosophy can be drawn only with the greatest caution.[6] Other mechanists followed a similar practice, adducing congenial citations from the book and brushing aside plainly Hegelian passages as confusions that Engels would have eliminated from the final draft.[7] On balance, the Deborinites profited more than the mechanists from the new 'classic', for it contained more Hegelian than mechanistic sentiments. Moreover the Deborinites never made the mechanits' mistake of describing an uncongenial passage as the result of Engels' error or unfortunate expression. The mechanists could and did quote against the Deborinites such passages as Engels' warning that 'in theoretical natural science one must not construct connections into the facts, one must disclose them in the latter and, upon disclosing, prove them, so far as possible, empirically'.[8] The Deborinites shrugged off such citations not only with talk of dialectical contradiction, but also with declarations of their own deference to empirically established scientific theories.

Lenin's *Philosophical Notebooks* was also claimed by both factions, but this new 'classic' did not excite as much contention as Engels'. Throughout the 'twenties Engels was generally regarded as a more important source for the philosophy of natural science than Lenin; it took the 'Bolshevization of philosophy' in 1930–1931 to put Lenin next to Engels in the standard list of 'classics'.[9] Moreover, Lenin's new 'classic' was much closer to the state of raw notes than Engels', which had considerable stretches of argumentation and exposition. Finally, the full text of Lenin's new work was not published until the controversy was nearly over.[10] Nevertheless, it will be recalled that Deborin, early in 1925, published sections of the notes as support for the crusade that he was then proclaiming against the mechanists, and throughout the controversy his faction constantly recalled the fact that Lenin had studied Hegel's *Science of Logic* long and carefully. The mechanists were obliged to moderate their opposition to the study of Hegel or give up their claim to orthodoxy. But Lenin's new 'classic' did not, any more than Engels', induce a fundamental change in the mechanists' position, if only because they were able to find support in Lenin's new work as in Engels'. They quoted the *Philosophical Notebooks* to prove that the Marxist philosophy of natural science must be primarily empirical in origin, must be drawn out of the natural sciences and used to transform Hegel's dialectics, rather than *vice versa* as the Deborinites argued.[11] In short, Lenin's new work, like Engels', helped the Deborinites to victory, but it cannot be pictured as an irrefragable text that obliged the mechanists to capitulate.

Supporting the mechanist refusal to capitulate was a general vagueness with regard to the force of 'classical' authority as a criterion of philosophical truth, and supporting *that*, the absence of an institutional authority that might have issued binding interpretations of disputed texts. Lacking a presbytery (until April, 1929) or a pope (until December, 1930), Soviet Marxist philosophers of science could not agree on the limits of private judgment.

The closest thing to a generally recognized rule concerning the binding force of 'classical' authority in the philosophy of natural science was an understanding that it applied to 'fundamentals' and not to 'particulars'. 'We distinguish,' wrote a Deborinite,

> two sides in the system of views of the founders of Marxism: in the first place, a number of particular propositions, entirely determined by the concrete historical conditions and the level of science of their time, and second, Marxism's methodological foundation, its philosophy. This is the unchanging side of Marxism, and along with that, its essence.[12]

Concrete propositions on natural science were generally regarded as 'particulars'; the 'classics' were not binding in such matters. For example, B. M. Zavadovskii, a 'Morganist' in the biological discussions and a Deborinite in philosophy, was quite vexed when a 'Lamarckist' cited Engels' argument that a

European child learns arithmetic more easily than a Bushman. 'The facts that I have been studying these past years,' said Zavadovskii, 'dictate to me the necessity, in the problem of the inheritance of acquired characteristics, of abandoning the views of Darwin, and Engels, and [K. A.] Timiriazev, and Marx.'[13]

There is no record of Zavadovskii's being charged with 'revisionism' for this forthright declaration.[14] Indeed his point of view was used to put the 'Lamarckists' on the defensive; they were charged with trying to subordinate empirical science to outmoded philosophical texts.[15] And a 'Lamarckist' who replied to this accusation showed the spirit of the times by arguing that he and his colleagues used empirical evidence to prove their views, while they cited the 'classics' merely to show that these views did not contradict Marxism.[16] Of course, it is possible to find Deborinites asserting the absolute truth of Marxist philosophy in such a sweeping manner as to imply the primacy of 'classical' authority even in concrete issues within the natural sciences. Karev, for example, once denied the charge that the Deborinites thought truth could be established by quotations from Marx and Engels rather than by empirical proofs; but then Karev went on to assert that 'agreement with a theory that has been confirmed by the experience of the class struggle of millions is of immeasurably greater weight than very many so-called "proofs"'.[17] Nevertheless, there is no evidence that Karev or the other Deborinites meant such assertions to apply to concrete issues within the natural sciences.

In discussions of abstract philosophical issues, both factions displayed a basic ambivalence towards 'classical' citations as evidence of truth. The Deborinite Luppol has been quoted in a quite dogmatic vein, declaring that Engels' new 'classic' must end the philosophical controversy. But the same Luppol on another occasion ridiculed a mechanist for withdrawing his opposition to the study of Hegel simply because Lenin said that Hegel must be studied.[18] Even if Luppol and the other Soviet Marxists had been wholeheartedly inclined to use *magister dixit* as sufficient evidence of truth, the intellectual atmosphere of the 'twenties palpably inhibited them. They were, after all, trying to win natural scientists to Marxism.

Once, when Deborin was arguing that his faction was 'defending the orthodox point of view', someone in his audience called out: 'That does not correspond to science!' Deborin's first reaction was to sneer at 'the very "learned"' person who 'shouted to me that Marxism contradicts science', but he quickly regained his composure and tried to show by extended quotations from German physicists that non-Marxist natural scientists unwittingly shared the philosophical views of Engels and Lenin. Running out of time, he concluded:

> But I beg those who say that dialectical materialism, Marxism, contradicts contemporary science to tell us where, in what, in which parts? Then we shall discuss with them. But I affirm beforehand that Marxism, both in its general philosophical assumptions in the field of general methodology, and in its sociological part too, is entirely confirmed daily by life and by science.[19]

This ambiguous mixture of superciliousness and deference towards empirical criteria of truth had certain Deborinite peculiarities, but in one form or another it was shared by nearly all the participants in the controversy. Bogdanov was virtually alone in his objection on principle to philosophical argument by 'classical' citation, and even Bogdanov occasionally quoted Marx to prove a point.[20]

The effort to win mature natural scientists to Marxism was not the only cause of ambivalance in regard to 'classical' citations as evidence of philosophical truth. Perhaps a more important cause was the vagueness of the distinction between 'revisionism' and legitimate criticism of the 'classics'. Aksel'rod the Orthodox once apologized for repeating elementary truths when she defined 'revisionism' as disagreement with the fundamental views to be found in the 'classics', and legitimate criticism as the alteration of the 'founders'' views on particular subjects, a procedure actually required by 'the eternally creative theory of dialectical materialism'.[21] The Deborinite Karev replied that her distinction was indeed 'a rudimentary truth'; but, as one might expect, he found her guilty of disagreeing with the 'classics' on '*fundamental* problems, problems of method and world view', and therefore repeated the accusation of 'revisionism' that had prompted Aksel'rod's distinction in the first place.[22] A similar exchange occurred between Sarab'ianov and Deborin. Sarab'ianov sneered, as Aksel'rod did, at the tendency to repeat this great thinker [Engels] complacently, and to consider every newly announced idea as an insult to his memory, as a profanation of his doctrine, and even more so anything that disagrees with him to one or another degree!

We understand perfectly well that criticism of such teachers of Marxism as Marx, Engels, Plekhanov, Lenin, easily leads into the camp of Marxism's enemies; but here one needs merely to follow carefully after the critics in order to admonish in time both the critics themselves and also their readers, in case a critic enters the slippery path.[23]

Deborin did not dispute this general point, but, again as one might expect, he argued that 'the mechanists with their "criticism" *have* "entered the slippery path" of the revision of Marxism'.[24]

It is noteworthy that both factions, with the exception of the unique Bogdanov, agreed in principle on the difference between 'revisionism', which both abjured, and legitimate criticism. The mechanists tended to be only somewhat less hedging, somewhat more emphatic, than the Deborinites in their approval of the right to criticize the 'classics'. This difference in emphasis rather than principle was evident not only in the rare comments on the right of criticism, but also in the equally rare exercise of it. 'Morganist' biologists, as already noted, expressly disagreed with the 'classics' on the inheritability of acquired characteristics; A. K. Timiriazev did not object when it was pointed out in a friendly manner that his Newtonian view of space differed from Engels';[25] and in other scientific issues of less theoretical significance obsolete views in the 'classics' were quietly ignored. On a philosophical level express disagreements with the 'classics' tended to follow the pattern of the philosophical differences between Plekhanov and Lenin. Deborin and his students brushed aside as lapses or oversights Plekhanov's statements of the 'hieroglyphic' theory of knowledge, and other formulations of Plekhanov's that appeared to deny the objectivity of all but a few primary qualities.[26] On the other hand, some mechanists openly criticized those statements by Engels and Lenin that denied the possibility of reducing all phenomena to a few primary qualities.[27] In doing so, these mechanists not only quoted Plekhanov in their behalf, they also quoted judiciously selected passages from Lenin and Engels against the passages stressed by the Deborinites. Thus, neither side ventured very far in its criticism of the 'classics', and the willingness of some leading mechanists to go slightly farther than the Deborinites was probably not a major reason why the reputation of 'revisionism' fastened on the mechanist faction.

The most striking aspect of the disputants' attitude towards the 'classics' was their silence with regard to questions that seize the attention of the outside observer. The 'rudimentary truths' concerning the difference between 'revisionism' and legitimate criticism were the object of only fleeting comment on rare occasions. No serious effort was made to define the authority—in a philosophical *or* institutional sense—that would decide whether disagreement with the 'classics' touched the inviolable 'fundamentals' or the ephemeral 'particulars' of Marxism, whether such disagreement 'revised' the basic method of Marxism or creatively 'applied it to new circumstances' and 'developed it further'. The reason for this silence was that almost all the participants in the controversy were not outside observers but orthodox Marxists. They were intent on proving their agreement with the 'classics', not on establishing their right to disagree. Disagreement with the 'classics' was nearly always a crime charged against one's opponents; *they* were 'revisionists', however crowded their writings might be with misinterpreted quotations from the 'classics'. The effort of the disputants was therefore not to decide when, in general, disagreement with the 'classics' was permissible, but to define for present purposes the fundamental principles with which there could be no disagreement.

In this effort, both factions quoted the 'classics' with ease; the two new 'classics' gave the Deborinites an advantage, but not a decisive one. Accordingly the ultimate appeal of both factions was to the objective reality that the 'classics' were supposed to reflect, the 'practice' that, both factions agreed, had priority over 'theory'. And if the Deborinites were more successful than the mechanists in their claims to orthodoxy, the main reasons must be sought in that 'practice', in this case the demands of the Cultural Revolution.

Seen in retrospect, the clear tendency of official policy during the 'twenties was to undermine the ideological autonomy of natural scientists. But it was possible during the 'twenties to imagine otherwise, not only because such influential people as Trotsky, Lunacharsky, and Kalinin told natural scientists that they might have what philosophies they chose as long as they did their work as natural scientists conscientiously;[28] but also because most mature natural scientists, their institutions, and organizations took advantage of the profferred right, and were almost as little Bolshevik or Marxist at the end of the 'twenties as they had been at the beginning. On the eve of the First Five Year Plan even the Union of Scientific Workers, which had abstained almost completely from the dissemination of Marxism as a philosophy of natural science, could claim barely six per cent of all scientific workers as its members.[29] VARNITSO, the newly established Association of Scientific and Technical Workers for Support to Socialist Construction, had enrolled even less.[30] The Communist Academy's societies of materialist scientists, which *had* campaigned for the acceptance of Marxism as a general philosophy, had a combined membership of a few hundred by the most generous estimate.[31] Obviously, peaceful persuasion was a slow method of transforming mature scientists into dedicated supporters of the régime. Yet precisely such scientists, the government was convinced, were essential to the success of the Plan.[32]

The progress of the new 'red' specialists was hardly more encouraging, as the Plan began in 1928. The law required that the sons and daughters of manual workers and peasants be favoured in admissions to higher schools, and there was an impressive network of *rabfaki* or 'workmen's faculties' to give them the academic pre-requisites for higher education. But this system had been in effective operation only a short time, not enough for even the first contingent of predominantly proletarian students to have received degrees.[33] They were reaching the universities as undergraduates, but the children of non-proletarian elements were still the majority even of first-year graduate students in 1928.[34] To the Bolshevik authorities this seemed the main reason that most graduate students in the natural sciences either were uninterested in Bolshevik ideology or shared their 'bourgeois' professors' scepticism.[35] Beginning in 1927 all graduate students were required to pass an examination in Marxism, but when an examiner asked a future mathematician for an appraisal of dialectical materialism, his question was answered with another one: '"Why should I bother with such nonsense?"'[36] Nor was this 'contradiction between the ever-growing role of the scientific worker in socialist construction and his ideological and socio-political backwardness' the only cause of deepening Bolshevik anxiety.[37] The elementary problem of numbers threatened to get out of hand; the rate of production of new technicians and scientists seemed to be falling hopelessly behind the staggering increase in the Plan's demand for them.[38]

'Military measures' designed to achieve 'maximum results in the shortest time' seemed the only way to achieve the 'scientific change-over' (*nauchnaia smena*), as Andrei Vyshinsky, then an important official in higher education, wrote in 1928.[39] The Party's Central Committee ordered a detachment of one thousand Communists to be enrolled in higher schools in the fall of 1928 with scant regard for academic pre-requisites.[40] In June, 1927, the Council of People's Commissars had decreed a new statute for the Academy of Sciences, which extended the right to nominate a greatly increased number of Academicians to organizations outside the hitherto autonomous Academy, and laid it down that an Academician, however nominated or elected, might be deprived of his chair for activity 'harmful to the U.S.S.R.'[41] For a time the new statute lay dormant, but following the 'Shakhty affair' in the spring of 1928 (a number of mining engineers were convicted of 'wrecking' in an intense glare of publicity), a spirited public drive was set in motion to 'renovate' the Academy of Sciences. The Press clamoured for the election of new, pro-Soviet members, crying to the existing Academicians, who were to do the final voting: 'Yes or no, for or against, together or apart—these are questions that demand a clear and unequivocal, an operative and not a declaratory answer.'[42]

In the universities professors of ten or more years' standing were ordered to undergo 're-election', as required by statutes that had lain dormant since 1918. At public meetings their fitness both as specialists and as 'social men' [*obshchestvenniki*] was to be examined by colleagues, students, and representatives of the Party.[43] Of course, membership in the Union of Scientific Workers or in VARNITSO counted heavily in a candidate's favour, especially since the Union, early in 1929, formally established the Marxist-Leninist *Weltanschauung* as a requirement for membership.[44] In short, efforts to push forward new 'red' specialists and pressure on mature specialists to give up 'neutralism' or 'the so-called simply legal relationship to the Soviet régime' were considerably intensified in 1928 and the first half of 1929.[45] The vision that Kalinin had given a conference of physicians in 1925, of socialism being built even 'by the man who says, "I am against Communism, decidedly against Communism,"' was evidently being dispelled.[46]

Such were the forces that produced the condemnation of the mechanist faction by the Conference of Marxist-Leninist Research Institutions in April, 1929. Pokrovsky put the matter very clearly when he called on the Communist Academy to convoke the Conference. He explained that

> the moment has arrived when it is necessary to put an end to the peaceful collaboration still existing in some scientific fields between Marxists and scholars who are far from Marxism or even hostile to Marxism; . . . it is necessary to begin the decisive offensive on all fronts of scientific work by creating our own Marxist science, and in particular to take more decisively in hand the work of preparing the scientific change-over.[47]

Pokrovsky pointed out that Marxists had already won a number of disciplines, chiefly in the social sciences, 'pressing into the background scholars that are hostile to Marxism'.

> The most backward in this respect is the field of the natural and exact sciences, where Communists have not yet entirely rid themselves of that fetishism before bourgeois scholars which Communists working in the field of the social sciences long ago shook off. Therefore, the most important current task must be a heightened offensive in the field of these sciences.[48]

This was the recurrent theme of the speeches at the Conference, which condemned the mechanist faction as proponents of continued autonomy for 'bourgeois' natural scientists.

It must not be imagined that the mechanists accepted this condemnation. They argued, and very likely with complete sincerity, that the *Deborinites* were hindering the celebrated 'penetration of dialectical materialism into natural science'. 'The natural scientist,' warned one of the last mechanist declarations, is transmitting his science to young comrades in the university classrooms and his technical instruction in the laboratories of factories and plants. It is impossible that he transmit only this, and not his general world view, his ideology, along with it. We must try to draw him to our side, to convince him that he is in practice a dialectician, though often a poor one precisely because he does not know the theory of dialectics. But if this is presented to him in works that treat of everything, that are like goods in a scientific department store where there are all sorts of things on the shelves, only of a very bad quality, in a word rejects—then the natural scientist will throw up his hands and take care not to enter such a collection.[49]

But this sort of argument made little impression on Marxist students who joined Deborin's 'social movement about dialectics', and on the Communist authorities who tacitly approved of the movement. Probably the main reason was that the mechanist programme of winning natural scientists by proving dialectical materialism in the context of the natural sciences was not nearly as sure of swift success as the Deborinite demand that natural scientists accept dialectical materialism ready-made from the 'classics' and Hegel. The mechanist faction wanted to argue with 'bourgeois' natural scientists on ground where *they* were far stronger than both Marxist philosophers and Communist students. Even if the mechanist programme might have succeeded in the long run in winning significant numbers of converts among mature natural scientists, the demand in 1928 and especially in 1929 was for the speediest possible 'change-over' from 'bourgeois' to 'red' specialists. The wager was placed on the young.

On this basis one can understand the causes of the Deborinite victory over the mechanists. At the Conference of April, 1929, S. L. Gonikman, the Deborinite who had criticized the Communist Academy's authorities a year before for their protracted toleration of the mechanist faction, expressed impatience with the conciliatory gestures still being made to natural scientists of mechanistic sympathies. No one, he said, had ever doubted that

we must construct a demonstration of the victoriousness of our method in individual concrete problems. However, I think that this is not the high road along which we can proceed to the fulfilment of the task before us. The point is that natural scientists at present are demanding above all, not a demonstration of the method's superiority in one or another *individual* problem, they are demanding a *world view*, they are demanding a *system*, they are demanding *integral views*.[50]

Gonikman's attribution of this demand to natural scientists, who were predominantly non-Marxist, is hardly credible, especially in view of his own warning, in the same speech, that 'we are going in to attack on the most difficult sector of the ideological struggle: we are going in to attack on natural scientists'.[51] But his description of the main goal sought by the Deborinite faction reveals on analysis the basic reasons for that faction's triumph, and also for the anomalous pattern of that triumph.

The Deborinites had insisted all along that natural scientists must accept a ready-made *Weltanschauung* and methodology from Marxist philosophers, and the mechanist faction had all along challenged this basic principle. The mechanist faction had thereby exposed itself not only to the suspicion of 'revisionism' (here, the denial in principle of the applicability of Marxism to natural science), but also to the more serious suspicion of justifying the 'bourgeois' scientist's resistance to Marxist ideology. The reductionist arguments by which the mechanists had supported their opposition to the Deborinite thesis struck responsive chords in a variety of Soviet Marxist hearts, including many high-placed ones. But these sympathizers would not speak up for the mechanist faction, because the faction's central argument—that the Marxist philosophy of natural science could not be accepted ready-made from Marxist philosophers, but must be worked out in the context of the natural sciences—seemed a defence of permanent ideological autonomy for natural scientists, who were predominantly 'bourgeois'. Paradoxically, this fatal defect of the mechanist faction's philosophy was probably the chief reason not only for the rapid atrophy that the faction suffered but also for the extended toleration that it enjoyed. The mechanist faction was frowned upon but tolerated as long as ideological autonomy for 'bourgeois' natural scientists was frowned upon but tolerated; the faction was formally condemned when an all-out attack was mounted against the scientists' autonomy.

But the formal condemnation of the mechanist faction was

also the formal coronation of 'Deborin and his students'. The Party seemed to have delegated to them the authority to define the philosophy that was to penetrate natural science as part of the 'scientific change-over'. The business meeting that followed the adoption of the 'Resolution in Accordance with Deborin's Speech' ended with a paean to Deborin, 'the only one of the old generation of Marxist philosophers who followed Lenin's instructions, who understood the task that stands before the philosophy of Marxism in our extraordinarily complex and difficult circumstances, in this situation in which we are struggling'.[52] In the list of philosophical readings required for admission to graduate training in the Communist Academy, several works of Deborin's followed the 'classics' of Marx, Engels, Plekhanov, and Lenin, and preceded a single work of Stalin's.[53] Until the end of 1929 the explosive potentialities of this situation were not realized. To be sure, one of the chief Deborinite philosophers was Sten, who urged Communist Youth in the spring of 1929 not to repeat blindly the dicta of the Party's chiefs, but to 'criticize and verify, on the basis of your own experience, the whole policy of the Party'.[54] But this scandal was taken as evidence of Sten's deviationist politics rather than his Deborinite philosophy—until 1930, one must specify, when the Deborinites were attacked for their superciliousness towards 'practical' leaders. Until that time the authority of Deborin and his students was not only tolerated but tacitly approved by the highest Party authorities. After all, it was natural scientists over whom Deborin and his youthful admirers sought to exercise 'the ideational hegemony' of Marxism, and these scientists were largely 'bourgeois *spetsy*'.

The formal condemnation of the mechanists was also the formal establishment of conciliarism, as we may conveniently call the practice of deciding philosophical truth by resolutions of conferences. V. P. Miliutin, a member of the Party's Control Commission, pointed out the significance of this new departure. 'We are adopting—for the first time in history, I dare say—a philosophical resolution. It will be, so to speak, a definite platform in regard to dialectical materialism and a formulation of all the work that will be done in this field.'[55] And he predicted that the practice of establishing 'clarity' and 'definiteness' in theoretical work by resolutions adopted in conferences would be extended to other fields than philosophy and natural science.[56] He was of course bringing into the philosophy of science an attitude that had long been characteristic of Bolshevik political thought: if truth is to be made effective in human affairs, it must enjoy uniform acceptance among a well-disciplined group. Soviet Marxists were conditioned to accept this attitude; only a couple of mechanists challenged 'the propriety of deciding scientific questions by a majority of votes'.[57] Most of the mechanists complained that the issues of the dispute had not been thoroughly examined, that a resolution ending the controversy was premature rather than improper. This complaint assumed the propriety of deciding what the truth is by resolutions of conferences. Deborin, with a similar assumption in mind, declared that the issues had been thoroughly examined, the controversy exhausted, the time come to close it; unity was necessary for the accomplishment of the Party's tasks.[58]

Neither Deborin nor anyone else felt it necessary to examine in any detail the philosophical presuppositions of this conciliarism. A Deborinite editorial simply brushed aside the conventionalist theory of knowledge that may be inferred from it, in conflict with the empirical theory espoused by materialism. 'It goes without saying.' the editorial snapped, 'that no kind of voting by itself is the last instance in settling philosophical arguments.'[59] The Conference had simply made clear what no Bolshevik of right mind could doubt. Neither were the practical consequences of this conciliarism explored. M. B. Mitin, a Deborinite militant, declared that 'essentially, the decrees of the All-Union Conference have put an end to the situation when [*sic*] problems of Marxism have been problems for discussion. Now great positive work rises before us'.[60] This seemed to rule out for good the possibility of questioning the Deborinite version of Marxist philosophy; yet only a year later Mitin himself would denounce that version as a revision of Marxism.

What warrant could he produce mightier than the Conference's 'decrees', as he chose to call its resolutions? And what sanctions were proper against recalcitrants like Tseitlin, who told the Conference that the passing of a hundred resolutions would not change the facts that made him a mechanist, or against the other mechanists who defied decreed truth?[61] And those who renounced their 'errors'—how could their sincerity be tested?[62] Would the same warrants, sanctions, and tests be applicable to natural scientists as well as Marxist philosophers, when the 'scientific change-over' was completed? In April, 1929, these questions were left hanging. When the mechanists complained that it was premature to close the controversy, Karev exclaimed:

> We declare merciless struggle on all deviations from Marx, Engels, and Lenin. Your defence of spinelessness and freedom of criticism is advocacy of the undermining of the dictatorship of Marxism in our country. We stand for the unity of the Marxist-Leninist front, while you want to disrupt it by giving 'freedom' to all the critics of Marxism and all the deviators from Marxism.[63]

But almost a breath before this outburst against the 'critics' and 'deviators', Karev had inconsistently boasted of the mechanist works that 'we [the Deborinites in charge of the State Press' philosophical offerings] have printed'.[64] And Luppol, an official of the Union of Scientific Workers who could also speak of the 'ideological dictatorship', at the Conference of April, 1929, distinguished between deviant Communists, who must be proscribed, and balky non-Communist specialists, towards whom a patient, tactful attitude must be adopted.[65] Evidently, the ambiguities that had characterized the Cultural Revolution during the 'twenties were not yet entirely dissipated.

PART FOUR

THE GREAT BREAK

1929-1932

16

THE GREAT BREAK FOR NATURAL SCIENTISTS

MIDWAY through the fury of his first Five Year Plan Stalin singled out 1929 as 'the year of the great break [*perelom*]', the year of shattering transformation, 'on all fronts of socialist construction'.[1] He had in mind the beginning of 'the decisive *offensive* of socialism on the capitalist elements of town and country', and of course he did not mean to suggest that the offensive would be completed in 1929. The shattering and transforming, he made clear, had only begun. Academic historians, who like to speak of this as a 'watershed' or 'turning point' or 'spinal year',[2] ought to concede that Stalin's more violent image is more appropriate for the crisis of forced industrialization and collectivization, though they are probably right in shunning his effort to dramatize the great break by assigning a particular year to it. On 'the scientific front of the cultural revolution' the great break took about two and a half years, from the middle of 1929 to the first part of 1932, which is short enough, considering the magnitude of the changes accomplished, to require no further dramatization. In this brief period 'the scientific change-over' from 'bourgeois' to 'red' specialists, and the accompanying search for a suitable philosophy or ideology of science, reached a crisis, a breaking point, by which past trends were selected, some for destruction, others for dominance over a generation of Soviet scientists and philosophers of science.

The 'scientific change-over' had been accelerated already

in 1928 and the first part of 1929. A thousand Communists had been sent to higher schools without regard to academic prerequisites; professors of ten or more years' standing were ordered to undergo 're-election'; the Academy of Sciences was 'renovated' by the imposition of many new members; and highly publicized trials of 'wrecking' specialists were staged. But still, as the Soviet régime approached its supreme test towards the end of 1929, the drive for solid collectivization of agriculture, it showed growing dissatisfaction with the progress of the 'scientific change-over.' Professors were reportedly sneering at the dispatch of 'the thousand' as an effort '"to prove the theorem that any illiterate can become a university student"'[3] And it appeared that such professors usually had little to fear from the 're-elections'. In most institutions there had as yet been none; in others, all candidates, regardless of their 'social physiognomy', were being 're-elected' by a formal ritual of meetings and eulogies. When there were genuine 're-elections', the Communists on the spot (students for the most part) tended towards one of two extremes, equally denounced as deviations in the central press. Either they attacked the professorial candidates as if they were *lishentsy* (a Soviet neologism for such people as priests and former gendarmes, who were deprived of civil rights), or, more often, the Communists succumbed to the mysteries of the academic guild and agreed to use professional competence as the sole basis for judging the candidates.[4]

The 'renovation' of the Academy of Sciences also moved forward haltingly. In January, 1929, A. M. Deborin and V. M. Friche, the chief Soviet Marxists in the fields of philosophy and art criticism, though nominated to membership by many Communist institutions, were voted down by the Academy, while the mathematician N. N. Luzin, an intuitionist in the theory of mathematics, was elected *qua* philosopher.[5] To be sure, the Academy quickly reconsidered this affront to the Bolsheviks, and elected the two Marxists at an extraordinary session in February.[6] At the public celebration that followed, the Academy's Permanent Secretary, the sixty-seven-year-old orientalist S. F. Ol'denburg, who had been one of the Provisional Government's Ministers of Education, reassured the Bolsheviks with pathetic earnestness: 'We feel still more our close connection with public opinion; we feel that there is no "we" and "you", but only "we"'.[7] Apparently not all shared his feeling. A meeting of the Academy's graduate students [*praktikanty*] resolved that compulsory training in Marxism should *not* be made a part of their programme.[8] And in the fall of 1929 a special investigating commission descended on the Academy and fired at least one hundred and twenty-eight people, some of whom appear to have been prosecuted subsequently in secret before administrative tribunals.[9] For Soviet scientists the great break had begun in earnest.

Without access to the archives one cannot know much about secret arrests and punishments, but the public record does reveal some things. In the first place it shows that mass terror, as a means of pressing the Cultural Revolution, made its appearance in the fall of 1929.[10] From the spring of 1928, to be sure, there had been intensely publicized trials of specialists accused of 'wrecking', that is, of activity, 'bringing economic and political harm to the Soviet state with the purpose of sapping its power and preparing for an anti-Soviet intervention'.[11] Such loud admonitory shouts at the 'bourgeois' specialist did not cease at the end of 1929, but they were joined by a more poorly defined threat, by obscure, unexplained acts of terror. For example, the most serious published allegations of malfeasance in the Academy of Sciences concerned some historical documents found in the Academy's library. They were supposedly 'of political importance', and the Academy's Librarian, the historian S. F. Platonov, was 'relieved of all administrative posts' for failing to turn them over to the proper agency and for allowing unauthorized people to see them.[12] Platonov's further fate was not publicized, nor was light shed on the misdeeds or the further punishment of those fired with him. A similar obscurity, an ominous omission of specifics, characterized most of the references in 1930 and 1931 to 'methods of terrorizing the accomplices of counter-revolution', methods that were being used, if we can believe a speaker in December, 1930, against 'that whole upper-echelon bourgeois intelligentsia, which, though not caught *flagrante delicto*, fosters wrecking activity by its sympathy or by its neutrality'.[13]

There was no longer any question whether *antipathy* to Communism was permissible. The great outcry was against 'apoliticism' and 'neutralism'. VARNITSO took scientific workers out on the streets of Moscow to demonstrate against these sins,[14] and a 38-year-old Bolshevik mathematician warned against the most refined kind of 'apoliticism': verbal endorsements of Marxism-Leninism unaccompanied by deeds to prove sincerity.[15] 'Never,' he exclaimed,

> has the class struggle in science been carried on with such bitterness as just now. Never has the demand for *our* science, a science that really serves socialist construction, been as great as today. Whoever now is not with us, whoever is still neutral, is against us.[16]

Nor were lesser Bolsheviks the only ones to erase the former distinction between political loyalty and ideological solidarity. Lunacharsky, the former Commissar of Education who, early in 1928, had given scientists one of the last assurances of their right to reject Marxism, in 1930 spoke to them with a new toughness, saying nothing of rights but only of obligations.[17] Perhaps Karl Radek's exhortation, 'On One Side of the Barricades or the Other', which appeared in August, 1930, was the most revealing. '*The broad mass of specialists,*' he wrote,

> stunned by shootings and arrests, dash off in various directions, and, frightened by the hostile atmosphere that events have created about them, do not know where to submit, and try in the meantime to hide their heads in their wings, in expectation of better times.[18]

Radek warned such 'Philistines' that 'the mistrustful attitude of the working class cannot be assuaged by correct declarations of loyalty or by silence'.[19] Still, declarations of loyalty abounded, culminating at the end of 1930 in a birthday greeting from the Unions of Scientific and Educational Workers to the Star Chamber itself: the Unions thanked the OGPU, on the occasion of its thirteenth anniversary, for purging their ranks of those not worthy of the honourable title of Soviet scientific worker.[20]

One can also get a vivid sense of the 'hostile atmosphere' surrounding the mature specialist in 1930–1931 from the plays and novels of the time. Gorky's irritation with the Russian *intelligentsiia* (the Russian word runs together what English separates as 'intellectuals' and 'white-collar class') was given new expression in a play about a counter-revolutionary engineer, whose villainy is inspired by arrogant pride in his special knowledge and by contempt for the working man's ignorance.[21] 'The workers have seized state power,' he explains to his wife,

> but they can't manage. . . . In general, the dictatorship of workers, socialism, are fantasies, illusions, which we, the *intelligentsiia*, involuntarily support by our work. . . . Machinists, house-painters, weavers, they're not capable of state power: it must be taken by scientists, by engineers.[22]

When his troubled wife asks whether he is not two-faced, he exclaims: 'Am I two-faced? Yes! Any other way is impossible! . . . The role of the defeated, the prisoner's role, is not my role!'[23] This was one of Gorky's exceptional individuals (gone wrong, to be sure), but a similar sense of outraged pride was presumed to be hidden behind the mask of complaisance put on by the ordinary 'Philistine' [*obyvatel'skii*] specialist, the type who had no stomach for plots against the Soviet state but might, if the proletarian dictatorship relaxed the threat over him. Anyone educated before 1917 was a 'bourgeois' specialist unless he proved himself otherwise. He belonged to that

> long-winded, weak-kneed intelligentsia . . . which met the October Revolution with passive sabotage or with active, armed resistance, and which in part continues to struggle 'in word and deed' against the Soviet régime even to the present day, wrecking consciously and unconsciously. . . .[24]

Was the suggestion in Gorky's final phrase (that a specialist might *unknowingly* commit the capital crime of 'wrecking') simply an extravagant flight of rhetoric? Probably not, for the central problem of a very popular play written in 1930 was precisely a scientist's unwitting 'wrecking'.[25] The play, *Fear*, dealt with an eminent professor of human physiology (was Pavlov the author's original inspiration?), who was not one of the Soviet state's conscious enemies, but played into their hands by developing a theory that fear was the essential motive in the behaviour of Soviet people. His theory was 'exposed' in a public meeting by an old Bolshevik woman, whose rich political experience compensated for her lack of formal education. When she concluded her indictment of the professor with a cry to the audience on stage to be vigilant and merciless towards the class enemy, the real audience in the theatre responded with loud applause.[26] The response lends verisimilitude to the dramatist's picture of 'public opinion' working with the OGPU to make the professor a Soviet patriot or break him. Certainly the *vydvizhentsy* in the audience, a Soviet coinage for proletarians 'pushed up' into scientific or other responsible work, must have felt a tightness in their throats at the symbolism of the play's ending. The professor, converted, promises to give a public criticism of his 'wrecking' theory and to hand over the keys of all the offices to the *vydvizhentsy* in his institute.

One naturally wonders whether the great break of 1929–1932 was actually the triumphal completion of the 'scientific change-over', as this play suggests. Were Russia's mature scientists forced to choose between full-throated Bolshevism and self-destruction by 'wrecking'? Did the management of scientific institutes and university departments pass in this brief period to the 'pushed-up' new generation of 'red' specialists? Certainly there is non-fictional evidence that seems to suggest an affirmative answer to both questions.

'The thousand' of the academic year 1928–1929 (that is, the detachment of Communists pushed into universities and institutes with little regard for academic pre-requisites) were joined by two thousand more in 1929–1930, and by still more in the next two academic years.[27] Even without such special detachments the staggering overall expansion of higher education between 1928 and 1932 (the student body trebled and the teaching staff doubled) suggests that the older specialists were being 'dissolved in a sea of new forces', as a report of the State Planning Commission put it in 1930.[28] 'A young man who studies our science,' said the mathematician O. Iu. Shmidt, who helped write the report, 'has every chance of becoming a professor at twenty-five.'[29] At the same time drastic measures were taken to reduce the sense of need for highly trained professors. In November, 1929, the Party's Central Committee ordered 'continuous productive practice' for students in higher technical education, with the result that abstract, theoretical subjects, the stronghold of the old specialists, were pushed into the background.[30] Some institutes even abandoned courses in theoretical physics and chemistry altogether, brushing aside as reactionary—or worse—the professors who protested that technicians rather than engineers would be the result.[31] New methods of teaching and grading in 'brigades' were designed to get around the need for individual *expertise* in students as well as teachers.

Even in research it seemed that the masses might break down the *tsekhovshchina* or guildlike seclusion of the old specialists. When T. D. Lysenko, a virtually unknown thirty-one-year-old seedman from the Caucasus, failed to impress a scientific convention in 1929 with a report of his experiments in plant physiology, he got dirt-farmers to try them out, and the Commissariat of Agriculture was so impressed that in 1931 it ordered collective farms to experiment with Lysenko's allegedly new methods 'on a mass scale. . . . Only in this way', the Commissar of Agriculture declared, 'will the business be set up in a really scientific way, in a really revolutionary way'.[32] Such things seem to confirm the playwright's vision of the great break as a time when the old specialists surrendered the keys of their institutes to the *vydvizhentsy*, to those 'pushed up' from bench and plough.

Moreover, one can find real analogies to the fictional indictment of a scientist for unwitting wrecking, in apparent confirmation of the notion that the older scientists were actually forced to choose between full-throated Bolshevism and self-destruction by 'wrecking'. A physician's pamphlet on the control of venereal disease, for example, aroused intense anger in a Bolshevik reviewer because of its calm, objective tone. The pamphlet did not sufficiently extol Soviet accomplishments in this field nor sufficiently berate bourgeois failures. 'In our time', the reviewer lectured,

> the time of the socialist offensive, when all hostile class forces are resisting desperately, the pen is obliged to shoot just as accurately and truly as the revolver. Paper, the printed word, speech—all are weapons that must guard our life, our construction, our philosophy from all sides. . . . We say loudly and clearly: it is not only useless to write such 'scientific' works [as the pamphlet under review], but also harmful, and criminal.'[33]

It may be that the 'criminal' physician survived this rhetorical fusillade, but more than verbal guns seems to have been used against statisticians in the State Planning Commission. Apparently they did not satisfy the Party's insistence that the Plan's goals must be the scientific prediction of mathematicians no less than the passionate desire of 'shock brigadiers' (*udarniki*).[34]

To be sure, such statisticians, like the physician writing on public health or the fictional professor in the play, *Fear*, were on the dangerous frontier where the natural and social sciences met, but one finds an attempted crusade against wrecking in the mathematical theory of statistics too. A fifty-three-year-old insurance specialist, who had vainly criticized the established authorities in statistical theory before the Revolution, won a following in the time of the great break. He helped 'unmask the wreckers' in the Planning Commission, and then teamed up with two young Bolshevik mathematicians to produce an allegedly revolutionary *Theory of Mathematical Statistics* in 1930.[35] If other mathematicians were obliged to support this book in a milieu where the distinction between pistols and pens was lost, then it might seem reasonable to suppose that 'the proletariat on the front of the Cultural Revolution', to quote a young Bolshevik mathematician in 1930, was indeed 'storming heaven itself,' forcing the mature scientist 'to place not only himself as citizen [*obshchestvennik*] but also his science in the service of socialist construction, to reconstruct it'.[36]

Yet there is considerable evidence that requires major amendments to these simple conclusions. Mathematicians apparently were *not* forced to accept the new theory of statistics. At the end of 1930 a supporter of the theory complained that O. Iu. Shmidt, the most important Bolshevik 'on the scientific front' and a mathematician himself, was indifferent to it, and within a few years it appears to have died altogether.[37] Lysenko's programme for boosting yields did not turn into a full-scale attack on geneticists until 1936. The pistol-waving review cited above was not at all typical of the journals of natural science; they continued throughout the great break to print specialized articles, and showed the influence of the times only by occasional editorial declarations of Soviet loyalty and by considerable transformations of their editorial staffs.[38]

Looking through them one begins to understand why Radek reported that 'the broad mass of specialists' responded to the clamorous demands for positive proofs of patriotism in a 'Philistine' [*obyvatel'skii*] way, with little more than 'correct declarations of loyalty, or silence'.[39] In fact this conclusion is suggested by the Bolsheviks' exasperated repetition of the warning that such a flaccid response was not enough, that 'Philistine' specialists could not escape history, or, to use Radek's industrial version of Lincoln's phrase, that they would be 'cast aside by the flywheel of history'. Perhaps the 'Philistine' specialist had a firmer grip on the dizzily spinning Russian flywheel—than Radek himself? (He was condemned only a few years later.) To ask this in sneering malice is to take sides, and not with the liberal opponents of regimentation, but with the type whose only cause was self-preservation, who would not *sacrifice* himself in any cause.

Direct evidence of the extent and geological force of this 'swamp' [*boloto*]—if the metaphor may be shifted to the standard Bolshevik pejorative for the passive, adaptive, self-centred type—is provided by the scientific conventions of 1930. They were the highpoints of a drive to capture the scientific societies, whose virtually complete autonomy had aroused only headshakes and grumbling before 1930.[40] Now, the Bolsheviks proclaimed, this autonomy was to be destroyed. But for all such loud talk the conventions of 1930 were much the same as those of previous years: a great mass of special papers were read, most of them trivial or repetitive, as in scholarly conventions the world around.[41] The 'great break' at each convention was a keynote speech, with a corresponding resolution, on science's role in the construction of socialism, and, when the convention broke into its many sections, perhaps two or three papers on the dialectical materialist reconstruction of science.

The pettiness both of the victories claimed and of the rebuffs lamented is the most striking characteristic of the Bolshevik reports of these conventions. The Congress of Physiologists, for example, was pictured as making history in its handling of the *affaire* Pavlov. He stayed away from it as he consistently did from all but foreign conventions, to demonstrate his disapproval of the Soviet régime. Previous congresses of physiologists had elected him honorary chairman *in absentia*, but the Congress of 1930 passed him by, and elected to its 'honorary presidium'—the entire Political Bureau of the Communist Party. Against this victory a defeat: the famous Professor A. F. Samoilov, whose report on electrical methods in physiological research was the convention's most memorable event, dismissed dialectical materialism 'with genial irony'.[42]

In Baku, to take another example, a Congress of Pathologists adopted the proper resolutions and elected to its presidium the thirty-six-year-old Dr. S. G. Levit, one of the leaders of the drive for a dialectical materialist reconstruction of science. But then a foreign professor, who told the convention that science should be free both of religion and of Marxism, was duly applauded at the end of his speech. Perhaps, the Bolshevik reporter noted hopefully, the audience did not understand him, for he spoke in German.[43] In Odessa, where the physicists had a pleasant August meeting, there seem to have been no such contretemps. The fifty-year-old Academician A. F. Ioffe, a universally respected physicist and genuinely enthusiastic supporter of the Soviet cause, though not of dialectical materialist reconstructions of physics, gave a keynote appeal for planned research to aid industrial expansion.[44] Clearly Ioffe and Levit belonged to a tiny band of prophets in a heathen land, where principled and outspoken opponents were even rarer, and certainly much hardier. Reading through the reports of the conventions one senses a vast flaccidity silently, perhaps indifferently, absorbing a few brave Bolshevik speeches.

There was one illuminating exception to this rule, the rebellion of the Congress of Mathematicians in June, and the related trouble of the Moscow Mathematical Society in December. The turmoil in statistics had nothing to do with these events,[45] and other substantive issues relating to mathematics were only slightly involved. The 'Moscow school' was famous for its otherworldly absorption in pure theory, and D. F. Egorov, the sixty-one-year-old chief of the 'school', would not criticize this tradition and declare some interest in serving the Five Year Plan. As this fact suggests, the main source of trouble was a general stiff-necked nonconformity in Egorov and an equally stiff-necked liberalism in his colleagues. He scandalized the Bolsheviks by refusing to join the Union of Scientific Workers, while remaining an elder of the Church, and the Moscow Mathematical Society not only kept him on as president but listed *émigrés* in its membership.[46] Already in the 're-elections' of 1929 Bolshevik graduate students at Moscow University singled him out for attack, and it seems that he was removed from control of the Mathematical Institute.[47] Still he was a leading figure at the All-Union Congress of Mathematicians in June, 1930, and it may well be that his example was a contributing cause of the Congress' rebellion: it refused to send greetings to the contemporaneous Sixteenth Congress of the Communist Party. A complete revolution was not attempted; the mathematicians did resolve to aid socialist construction, though adding the caution that theoretical work should not be neglected in the interests of immediate practicality.[48] But 1930, the year of savage class warfare in the villages and of frantic 'shock brigades' in the towns, was hardly a time even for a limited rebellion, which was continued moreover by the Moscow Mathematical Society's refusal to expel Egorov.

The climax was reached at the end of 1930, when Egorov told a meeting of the Society that 'nothing else but the binding of a uniform *Weltanschauung* on scientists is genuine wrecking'.[49] It is significant that the Bolsheviks present could not agree on the appropriate reaction. The one who took the floor tried to smooth over the clash with talk of a misunderstanding, for which he was subsequently accused of 'rotten liberalism' and 'Maecenasism'.[50] Bolsheviks with the proper 'irreconcilability' [*neprimirimost'*] took action after the meeting was over. Egorov was arrested. But his colleagues in the Society, including a member of the Communist Youth, silently defied the terror by holding a regular business meeting. (They were expected to condemn the arrested man and engage in 'self-criticism' for resisting Bolshevization so long.)[51] Thereupon five mathematicians, styling themselves an 'Initiating Group for the Reorganization of the Mathematical Society', published a truculent denunciation of the Society's belief that '"one can be a Egorov by conviction yet work honourably with the Soviet régime"'[52] They could hardly have expressed more succinctly the government's assurances to 'bourgeois' scientists during the 'twenties, but they lashed this belief as 'Philistinism [*obyvatel'-shchina*], hiding in its corner from the class struggle, and decorating this corner with scientific aestheticism instead of the canary of the rank-and-file Philistine. . . .'[53] But the other mathematicians would not yield to revolutionary appeals any more than to terror. The wretched affair had reached a climax without issue. The Society was not reorganized but simply ceased functioning for more than a year, in the course of which Egorov died. The place and cause of his death are not in the public record, nor are the steps leading to the revival of the Society in 1932.[54] One supposes that the locked opposition of intransigent Bolshevizers and unyielding liberals gave way to some such complex adjustment of principle and reality as had already made the functioning of the other scientific societies possible.

Until the end of 1931 it was not clear that an adjustment would be made even with the complaisant majority of old specialists. The Bolshevizers kept up the struggle for something more than complaisance, and looked beyond the scientific societies to the places where scientific work was done. 'However strange it may be'—a young Bolshevik biologist told a meeting of the Communist Academy in January, 1931—

> in the fourteenth year of the Revolution, though we have at our disposal a colossal apparatus of scientific establishments, museums, laboratories, observatories, etc., in essence we do not possess them at all. It would seem to me . . . that the Association of Natural Science [of the Communist Academy] should set itself the organizational and ideological task of entering, of penetrating these institutes, these observatories and laboratories, through the cells of atheists that exist there, the sections of VARNITSO that exist there, all the circles of political or other character that exist there, so that we will have at each institute some cells on which we can rely in our work.[55]

As if in response to this suggestion, the Party's Central Committee decreed, on March 15, 1931, that the Communist Academy should establish its 'methodological control' over the most important non-Communist (*vedomstvennye* was the actual term) scientific research establishments.[56] They were to submit their research plans to the Communist Academy for approval and admit the Academy's representatives to the drawing up of future plans.

The old distinction between the special network of Communist institutions and that of non-Communist (*vedomstvennye*) ones was thus to be erased; the Communist Academy, which had previously been the directing centre only for the former, was now apparently to become the centre for all. Clearly there was a conflict here with the competence of the Academy that Peter the Great had founded, the Academy of Sciences of the U.S.S.R. One wonders with what wild surmise Stalin's young men, newly triumphant at the Communist Academy, gazed upon the order to establish 'methodological control'; they had the most rudimentary knowledge of natural science, or worse. But their chief was then declaring that the most important goals of the economic Plan, audacious enough on a five-year scale, could be reached in *three*, if only the Bolsheviks would nerve themselves (and the masses) to it.[57] 'If there is a passionate desire to do so, every goal can be reached, every obstacle overcome. . . . We lag behind the advanced countries by fifty to a hundred years. We must cover that distance in ten years. Either we will do it, or they will crush us.'[58] Inflamed by such desperate encouragement, the Bolshevizers of science must have grown suspicious of their recent victory over the scientific societies, for it had been too easy. As they established their 'methodological control' of scientific research establishments—in essence it was ideological control of the personnel[59] —they launched a new drive against the societies. Bolshevik speakers and writers 'completely exposed the protective colours with which the societies have redecorated themselves, using Marxist terminology for this and also some change of leadership'.[60] Forcibly converted, mature scientists were coming under the entirely logical suspicion of hypocrisy. Soviet Marxists seemed about to follow the example of Spanish Christians; converting Moors and Jews by force, they felt a natural complusion to lay open the hidden beliefs of Moriscos and Marranos.

At this convulsive moment the first signs of relaxation began to appear. Opening the First All-Union Conference on the Planning of Science in April, 1931, Bukharin brandished a '*physical or moral guillotine*' over scientists, and, raising the question whether the planning of science meant the imposition of Marxism as 'a state doctrine, the doctrine of the proletarian dictatorship', he answered forthrightly: 'Yes, that is true.'[61] Closing the Conference a few days later, Molotov cautioned young Bolshevizers of science to treat distinguished scholars respectfully, and even implied in one fleeting sentence that the latter might take some time yet to be converted to Marxism.[62] In June, Stalin lectured Bolshevik executives on a change in the dominant political attitudes of senior industrial specialists. Within the past two years, said Stalin, active wreckers had been 'routed' or driven 'deep underground'; most industrial specialists would no longer have anything to do with them. 'It would be incorrect and undialectical to continue the old policy under these new conditions. It would be stupid and senseless now to regard just about every specialist and engineer of the old school as an undetected criminal and wrecker.'[63] In August *Pravda* suggested that this absolution might be extended from industrial engineers to professors, by printing an appeal from Academician Bukharin to all Communist organizations: as they selected people for graduate training at the Academy of Sciences, let them remember that the Academy had been transformed into a thoroughly Soviet institution, performing great services for socialist construction; let them cease dumping on the Academy their '"superfluous" people or . . . second-raters'.[64] In October a leading Bolshevizer of science, still stressing the dialectical materialist transformation of natural science as the only guarantee of loyal specialists, nevertheless included scientists in Stalin's absolution of older specialists from automatic suspicion of wrecking.[65]

And then, in 1932, the frenzy of Stalin's first Five Year Plan was spent. Criticisms and cautions that had formerly been denounced as subversive promptings of the class enemy, now came from the Central Committee itself. In June A. I. Stetskii, the head of the Central Committee's Propaganda and Agitation Department, called off the ardent young Bolshevizers who had been establishing the Party's 'methodological control' even in engineering and ichthyology.[66] He voiced the familiar complaint of 'a purely verbal, formal, declaratory' endorsement of Marxism, but he did not conclude, as he had a year before,[67] that the pressure on the non-Party scientist must be increased, the struggle intensified. Now he concluded that Bolsheviks must work long and well on the specific material of the sciences, so that the non-Party specialist might be convinced of Marxism in terms of his own specialty. Foreshadowing the fate of the Communist Academy itself, Stetskii called on Communist scientists to leave their special societies at the Communist Academy and dissolve into the societies of non-Party scientists, which had been brought under Bolshevik control.

Three months later the Central Executive Committee of the Soviet Union called for an end to 'shock brigade' methods of effecting the Cultural Revolution in higher education.[68] The intervention of student and Party organizations in the management of the higher school was to stop—in part, we may suppose, because the political reliability of the rectors had been tested in the fire of the past few years. The overemphasis on 'continuous productive practice' was to end, for this practice had been producing technicians rather than engineers. Of course, the professors who had been called reactionary for predicting as much were not vindicated, very likely because the Soviet economy needed technicians anyhow, and because professors needed to learn respect for the Party's decrees. The 'brigade' methods of teaching and grading were to be dropped, and the individual responsibility of students to individual teachers was firmy re-established. Professional competence was to be the only basis for filling vacancies and giving promotions in faculties, and graduate students were to be appointed only by the faculties concerned, only from graduates of higher schools. Evidently the day of the *vydvizhentsy*, the 'pushed-up' ones, had passed.

In 1933, indeed, the tables seemed to be turned altogether: the Communists in higher learning were subjected to 're-elections'. More precisely, in the course of a general purge, the Party organizations in institutions of higher learning examined their members at public meetings, to which non-Party professors were urgently invited in order to help weed out unworthy Communists.[69] The actual results were hardly revolutionary. Many of the professors at the meetings were discreetly silent, and only a few Party members suffered expulsion (e.g., a laboratory assistant for having his son circumcised with religious rites, a graduate student for giving flippant answers to questions on Leninist theory). Most, including the leading Bolshevizers of higher learning, were merely given a forum in which to tell the story of their lives as model Communists. (I. I. Prezent, for example, who had abandoned the Deborinites and 'Morganists' when they lost favour and was now becoming Lysenko's chief advocate, was the star of the meeting that expelled the religious laboratory assistant and the flippant graduate student.)[70] But however trifling the actual purge of the new 'red specialists' may have been, its staging gave ceremonial recognition to the continuing importance of the older, non-Party specialists.

Of course, it would be quite wrong to imagine that the great break had passed without trace. The older generation of scientists, still the possessor of essential knowledge, was still in charge of many university faculties and scientific research institutes; and most were still felt to be ideologically alien, though all but a very few indomitable spirits now refrained from the expression of any but the correct ideology.[71] What was new was a fundamental transformation in the intellectual autonomy of these old specialists. In principle they had lost it altogether; to use a favourite expression of the day, they had 'disarmed themselves' (*razoruzhilis'*) before the Party's Central Committee. In practice they still enjoyed almost unimpaired autonomy in their subject matter, and immeasurable autonomy in ideology—immeasurable because of the mask of silence and possible hypocrisy that covered it. How long this incongruity of principle and practice would continue depended on the Central Committee's assessment of changing necessities and possibilities. Aside from the 'disarming' of the old specialists (at least in principle), it had gained an enormous number of new scientists in training, most of them from social classes that would, the Central Committee hoped, produce great specialists who would also be genuinely Bolshevik in ideology.[72]

The reader may feel inclined to conclude that the Bolsheviks had became fatally embroiled with the law of enforced belief, as we may call the truism that hypocrisy and unthinking zealotry spring up where heresy and freethinking are cut down. Ultimately, it would seem, one of the two would have to die, either creative thought or thought control. But we cannot rest with this generality, for the Central Committee's silence on the substantive issues of natural science was one of the most striking characteristics of the great break. They had razed the walls of academic autonomy, but one senses a moat of irrelevance still lying between their ideology and the natural sciences, bridgeable perhaps, but hardly by the primitive zealots who assumed control of Bolshevik ideology during the great break. Moreover, on the scientists' side, enforced belief might become genuine in time; the Church that Egorov refused to quit had been forced on his ancestors by an earlier state.[73] Whether the new ideology would enjoy an inner conversion following its outward establishment, and what might come of such a conversion, can only be discovered by further historical research. But first one must discover what happened to the ideology and its adepts during the great break.

17

THE GREAT BREAK FOR PHILOSOPHERS

THE yellow neoclassic building in Moscow that once housed the Communist Academy bears no memorial of its former function. An enormous bronze plaque commemorates instead a visit of Stalin's; not his first (lacking other than biographical interest, that would be the 'cult of the personality'), but his appearance in December, 1929, to address a Conference of Agrarian Economists.[1] The choice is apt, given the Bolshevik effort to forget that the first phase of their assault on 'bourgeois' learning was headed by a somewhat autonomous group of Marxist scholars centred here.[2] For Stalin's speech in December, 1929, began the destruction of that group and the absorption of its functions by 'this areopagus', as he would call the Central Committee two years later, 'in which the wisdom of our Party is concentrated'.[3]

There is no evidence that the end results of Stalin's speech were clearly in his mind when he made it. Indeed it is possible to doubt that he originally intended to speak at the Conference of Agrarian Economists at all. It was opened by V. P. Miliutin, the economist and member of the Party's Control Commission, fulfilling the prediction he had made in April that the use of conferences to establish 'clarity' and 'definiteness' in theoretical work would be extended from philosophy to other sectors of the theoretical front. He spoke on collectivization, decrying the lag of economic theory behind agrarian practice, rejecting the agrarian theories of some economists and laying down the one that best fit the Party's current policy.[4] The discussion that followed turned up only a few diffident dissenters, and the Conference duly voted a resolution endorsing Miliutin's theses.[5] Only then, as the Conference was about to close, Stalin appeared.

Perhaps he wanted to show Communist Academicians that he deserved the encomiums he was then receiving, on the occasion of his fiftieth birthday, as 'the chief [*vozhd'*] of the proletariat and simultaneously the most outstanding theorist of Leninism'.[6] But whatever may have been in the back of his mind, his primary motive was quite apparent. He was irritated. Just when the drive for collective farms was reaching its supreme crisis, and success or failure hung on the absolute dedication of the 'practical workers' pushing collectivization against a world of enemies, a few theorists still dared to doubt and question. They must realize that the Party had already found the practical solution to Russia's agricultural problems; their job was to propagate the theory implicit in this solution, to give 'practical workers the power of orientation, clarity of perspective, confidence in their work, faith in the victory of our cause'.[7] Hammering the theme in his usual iterative style, he seemed at one point to be irritated not just with economists but with all theorists: '. . . it must be admitted that theoretical thought is not keeping pace with our practical progress, that there is a certain gap between our practical progress and the development of theoretical thought.'[8] Whether or not he intended it, these few words were caught up in all fields of Marxist thought, in a cry for a 'turn to the actual problems of socialist construction'.

In philosophy Stalin's fleeting complaint precipitated a storm of controversy. Very likely it was gathering before his speech, for the Institute of Red Professorship was undergoing a fundamental transformation. Though it had been a Communist institution since its founding in 1921, it had troubled the authorities much as the non-Communist higher schools had: it turned out too few specialists, nearly all of middle-class extraction (a grand total of 236 through June, 1929, only twenty-seven of them workmen or peasants in origin).[9] In the fall of 1929 its student body was swollen to 545, the majority of them workmen in origin, many without any higher education. For the latter a Preparatory Section was set up, making the Institute as much an undergraduate as a graduate school.[10] The Philosophical Section, where Deborin's curriculum concentrated on original readings in Spinoza, Kant and Hegel, was now filled with 'young comrades, free of any "specific" philosophical tradition, educated in conditions of our country's civil war and class struggle, who have no special medals in philosophy but in the political field, on the other hand, have served the Party and the working class well'.[11] Such, to take the most notable example, was the thirty-year-old Paul Fedorovich Iudin, who had been a provincial newspaper editor and Party official before he came to the Institute, where he would become head of the Philosophical Section as a result of the great break.[12] Their Deborinite teachers seemed to them 'a sort of philosophical sect, very small, indrawn and barred against those who are not initiated in the secrets of their philosophical guild'.[13]

It is small wonder, then, that there were hot-tempered recriminations when the Party organization of the Philosophical Section, which included all the students and most of the teachers, began to discuss Stalin's complaint of a separation between theory and practice. The Deborinite teachers at the Institute seem to have proposed a resolution dedicating philosophy to 'the actual problems of socialist construction'. Some of the students objected that this was not sufficiently self-critical. Had not Stalin said that *theory*, not just economics, was lagging behind the demands of practice, and was not philosophy a sector of the theoretical front? The philosophers had begun criticizing the right deviation; why had they failed to do the same for Trotskyism? Was there a connection between this negligence and the Trotskyism of two teachers, Karev and Sten? Shouting and fist-shaking followed, and apparently no resolution was adopted.[14]

At the beginning of 1930 the young rebels may have wondered whether their flammable devotion to Stalin's slightest 'instruction' was more ardent than wise. The official organs of the Party and the Communist Youth continued to show undiluted favour to the Deborinites.[15] At the end of March the theoretical organ of the Central Committee published a criticism of Deborin's ethical theory, but this was written by an unregenerate mechanist.[16] If publication of his article was a sign of a high-placed decision to shake up the Deborinites, a revival of mechanism appeared to be the intended instrument. Such a possibility was strengthened about the same time, when the Deborinites in the League of the Militant Godless were rebuffed in their demand for the exclusion of unrepentant mechanists from the League's publications. Iaroslavskii, the head of the League and a major Party leader, countered that no Party Congress had declared mechanism incompatible with Party membership. The mechanists, he said, had served the anti-religious movement much better than the Deborinites, who were not free of philosophical errors themselves. The philosophical front was in need of self-criticism, and he suggested that the philosophical conference planned for July would reopen the issue of mechanism.[17]

His call for self-criticism in philosophy gave new heart to the students who had attacked their teachers. (In the Bolshevik vocabulary 'self-criticism' includes not only a person's criticism of himself but also the group's criticism of itself.) But the young rebels protested, in a joint letter to *The Godless*, against Iaroslavskii's effort to revive mechanism; the Party needed a struggle on two fronts, in philosophy no less than politics.[18] And they were joined by the first defectors from Deborin's entourage, most notably by Mark Borisovich Mitin, who had graduated from the Institute of Red Professorship the year before to become, at twenty-nine, the assistant director of the Krupskaia Academy of Communist Education.[19]

At the end of April, 1930, the Communist Academy's Institute of Philosophy and the directors of the Society of Militant Materialist Dialecticians met to plan the philosophical conference scheduled for July. The theme proposed by Deborin was 'a turn to the actual problems of socialist construction'. The philosophers' main job, his resolution specified, was still to elaborate a dialectical methodology, largely from Hegel; but they must apply it not only to natural science but to all the social and political problems that the Party faced in the construction of socialism.[20] In a spirit of self-criticism Deborin's resolution took note of the philosophers' failure to criticize the methodological basis of Trotskyism, but Mitin and his comrades felt such self-criticism was too mild. They pressed Karev and Sten for personal criticism of their Trotskyist past, and, for the group of philosophers as a whole, moved an amendment to Deborin's resolution:

Passivity in exposing the methodological bases of the views of the Trotskyist-Zinovievist opposition is evidence of the well-known separation of philosophical thought from the actual political problems that have faced our Party in the past period of development.[21]

The amendment was voted down. They recast it, and moved it again, and when the new version lost, tried a third time (saying that 'formalism' had crept into the philosophers' works), again in vain.[22]

Thereupon Mitin's little band sent a complaint to *Pravda*, a standard démarche for Bolsheviks with a grievance.[23] In mid-May they made their first public complaint (in the newspaper of the Institute of Red Professorship),[24] and early in June they scored a major triumph. *Pravda* printed their criticism of 'the well-known separation of Communist philosophical thought from the most pressing political problems that have stood and stand before our Party'.[25] And *Pravda* printed it not with the tag, 'For discussion', by which Soviet editors disclaim responsibility for a controversial piece, nor even bare of editorial comment, which is taken for tacit approval in Soviet editorial practice. A footnote dispelled any possible doubt: 'The editors are in full agrement with the *basic* views of the article above.'[26]

Thus the Deborinites were placed in the same position as the mechanists of the 'twenties: they were attacked by a group that had the support of the highest Party organs, and they responded as the mechanists had, by a vigorous counter-attack. But this controversy, occurring in the midst of the great break, was strikingly different from the earlier one. That one had dragged on for six years before the mechanists were formally condemned and denied access to print; this time it took no more than six months to shut up the losing side. Political themes, other than the problem of winning natural scientists to Bolshevism, had been kept out of the earlier controversy; they were the heart and soul of this one. Then, distinct philosophical positions had been clearly opposed and more or less rationally argued; now, there was so much vituperative shouting with so little sense that it was possible to doubt whether philosophical issues were involved at all. The Deborinites, to be sure, were defending their old views, but against what? What did their opponents stand for?

At first they were not sure themselves what they stood for, beyond the proposition that the philosopher's primary task was to serve the Party's immediate political needs. They called for the 'politicalization' (*politizatsiia*) or 'Bolshevization' of philosophy. Originally they took it for granted that the Deborinite version of Marxist philosophy was basically correct—hadn't *it* won *Pravda's* support in 1926? But its chief interpreters must unreservedly support the call to Bolshevize philosophy, and criticize themselves contritely, in the Bolshevik manner, for having failed to support it before it was announced. At first the young Bolshevizers saw little more than wilful self-esteem in the Deborinite resistance to this demand; the triumph over the mechanists, though deserved, had apparently swollen 'the philosophical leadership' with conceit.

They struck at this conceit simply by ransacking the leadership's works for quotations that could be interpreted as 'errors', given a will to disregard the context or even the authors' specific disclaimers of the 'errors'—'tendentious flea-picking', a mechanist victim called the method, when it was used against him.[27] Deborin, for example, in a book that had gone through three editions while Mitin was a Deborinite, had written that 'Plekhanov was above all a theorist, Lenin above all a practical man, a politican, a chief [*vozhd'*]'.[28] In context this was clearly meant as praise; some pages later Deborin established his main point: 'If thought is determined by being, then it is natural that theoretical dialectics is determined by "practical dialectics".'[29] But Mitin (after six years of tacit approval) found a belittlement of Lenin as a philosopher in the first, isolated quotation, and connected it with an openly anti-Leninist essay that Deborin had written in 1908, when he had been a Menshevik student of Plekhanov's.[30] Mitin and his cohorts never gave up such 'tendentious flea-picking', but gradually they ceased to consider Deborin's philosophy 'a correct line with a system of formalist errors'.[31] They came to call it pure and simple 'a formalist deviation'.[32] Willy-nilly they learned enough philosophy to see that genuine issues divided them from the Deborinites.

They and the Deborinites had sharply different interpretations of the plastic Marxist formula concerning theory and practice. Theory and practice, the formula runs, are dialectically united: theory guides practice, but in the long run practice determines theory. Both sides agreed that 'practice' includes sense data, workable techniques in industry and agriculture, and effective policies in social and political affairs. Both sides also included certain *theories* in the concept of practice, theories that are verified by sense data, technology, or effective state policy. On this basis even the universal abstractions of materialist dialectics could be regarded as part of 'practice', since they were supposed to be generalizations verified by the whole experience of mankind. Dialectics turned out to be truly a philosophy in which opposites interpenetrate, become united—and conflict. For, the young Bolshevizers reasoned, if dialectics was generalized practice (and the Deborinites agreed that it was), then the Deborinites were wrong to use Hegel as their primary source in elaborating dialectical philosophy. Not Hegel, said Mitin, but 'the masterful application of dialectics that our Party carries out enters into the development of philosophical Communist thought as the most important component element'.[33] Seen in this light, the priority of practice meant that the resolutions of Party Congresses, the decrees of the Central Committee, and the speeches of Stalin were to be the chief source for the elaboration of dialectical materialist philosophy.[34]

This new view of practice and theory required a revaluation of the history of Marxism. If philosophical theory was to be assimilated to political practice, then the old view of Plekhanov as the father of Russian Marxism could not stand. For the father had lived too long; he and the son, Lenin, had come to blows over political affairs while they remained in substantial agreement on philosophy. The Bolshevizers could not tolerate such a separation of theory and practice even in the historical past. Plekhanov, they argued, *seemed* a solidly orthodox philosopher only because the ex-Menshevik Deborin had dominated the past phase of philosophical education. Lenin had warned that Deborin must be watched, lest he smuggle in his Menshevik politics; the true Bolsheviks had not been sufficiently watchful, from now on they would know better. They would see that

Plekhanov was guilty of serious errors in philosophy, which Lenin corrected. Even before Plekhanov became a Menshevik he was inferior to Lenin as a philosopher. Lenin showed *him* the way in philosophy, but his degeneration into Menshevism prevented him from following it properly. The Bolshevizers did not spurn Plekhanov altogether. They removed him from the list of 'classics'; he was to be read critically. The genuine 'classics', and the current masters of 'applied dialectics' would show one how.[35]

The new view of practice and theory also required a revaluation, or rather, a 'concretization' of Lenin's concept of partyness.[36] At first Mitin recognized that Lenin had used partyness largely as a synonym for the class character of philosophy, but, he argued, such a broad understanding of the term was no longer sufficient:

> One must be able to understand the peculiar manifestation of philosophy's class character in our peculiar epoch. The fundamental peculiarity of the epoch of the proletarian dictatorship consists in the fact that the ruling theory is the theory and world view of the ruling class, the *proletariat*. This peculiarity requires a more concrete formulation, on a new basis, of a whole series of questions, in particular, of the problem of the class character of science, of philosophy.
>
> The philosophy of dialectical materialism is the official point of view, the world view of the Communist Party.
>
> Hence it follows that the partyness of the philosophy of Marxism-Leninism in general, in the conditions of the contemporary stage in particular, signifies and must signify above all *its politically efficacious character*.[37]

In time Mitin learned to write less awkwardly, and the Bolshevizers as a whole learned to avoid his implicit admission that their concept of partyness differed from Lenin's. But the iron logic that led them to their new concept did not change. They began their rebellion against the philosophical leadership by stressing methodological criticism of the Party's enemies as 'the very first obligation of dialectical materialists'.[38] The political struggle on two fronts had to be matched by a philosophical struggle on two fronts. This assumed an organic correlation between philosophical and political tendencies: every political deviation had philosophical roots, every philosophical deviation political offshoots, while philosophical orthodoxy was to be found in the correct Party line and only there. They learned to justify such assumptions by their new view of theory and practice, that is, by assimilating philosophy to the Party's 'masterful application of dialectics' in political and social affairs. They concluded that the Party's centralized organization must be extended to the 'philosophical sector of the theoretical front'. 'There is not', Mitin announced, 'and cannot be a philosophy that wants to be considered Marxist-Leninist philosophy while denying the necessity of ideational-political and theoretical leadership on the part of the Communist Party and its leading staff'.[39]

In October, 1930, when Mitin made this announcement, it was not yet clear that Stalin and the Central Committee would assume the function that Mitin assigned to the Party's 'leading staff'. *Pravda*, the Central Committee's chief organ, had endorsed the Bolshevizers in June, but *Revolution and Culture*, the organ of the Central Committee's Propaganda and Agitation Section, did not. The Presidium of the Communist Academy, which was defined by statute as the directing centre of Marxist-Leninist learning, took a neutral position in this new philosophical controversy, as it had originally in the old one between mechanists and Deborinites.[40] Many of Deborin's former students went over to the Bolshevizers,[41] but he and a faithful few remained in control of the Communist Academy's Institute of Philosophy. The editorial board of *Under the Banner of Marxism*, the leading philosophical journal, was split three ways: A. K. Timiriazev remained a mechanist, A. A. Maksimov, who had quarrelled with the Deborinites over a year before, eagerly endorsed the Bolshevizers,[42] while Deborin, the editor in chief, retained the militant support of Karev and Sten. The journal became the Deborinite organ, vehemently denouncing the Bolshevizers as 'militant eclectics': their call for philosophical service of the Party's political needs was allegedly borrowed from the Deborinite programme, while their denunciation of Hegelian studies, their assimilation of theory to practice, were allegedly a revival of mechanism.[43] In the Deborinite view, publication of the Bolshevizers' articles in the Party press did not give them the right to 'identify their own eclectic views with the general Party line'.[44] For the last time, perhaps, an officially endorsed article in *Pravda* was denounced as 'a spineless article, eclectic by its very purpose. . . . Shocking falsehood is added to astounding confusion and gross political mistakes'.[45]

The exchange of fierce polemics was very brief. In September, 1930, it was suddenly cut off, and in October the dispute was ended as the controversy over mechanism had been, by a conference of Marxist-Leninist specialists. But this was not a large conference representing the whole network of Marxist-Leninist research institutions, as the earlier one had been. It was a meeting of the directing centre of that network, the Communist Academy, and formally, only of its Presidium, though many other people than the Presidium's members were allowed to speak.[46] Nor was it a meeting primarily of specialists in philosophy. V. P. Miliutin, the economist and member of the Party's Central Control Commission, gave the main speech (*doklad*), in support of the Bolshevizers.[47] His post as Vice President of the Communist Academy gave him some warrant to be the main speaker, but his specialty was economics, and Deborin, who gave a 'companion main speech' (*sodoklad*), drew laughter from the audience by sneering at Miliutin's 'great services on the philosophical front'.[48] Iaroslavskii, also of the Party's Control Commission, dropped in during the ensuing debate to endorse the Bolshevizers, and to produce the document in which Lenin had warned against Deborin's Menshevism.[49] Skrypnik, another old Bolshevik who was then a Party chief in the Ukraine, came to denounce the Deborinites for having '"encommissarized" themselves [*zakomissarilis'*] in the field of philosophy. . . .'[50] (He could have boasted that he had been ahead of Lenin in anticipating the narrow concept of partyness, for the Bolshevik Conference of 1909, dominated by Lenin, had rejected Skrypnik's motion to condemn empiriocriticism.[51] But he modestly refrained.) Aside from Party chiefs, a number of Bolshevik social theorists spoke, telling why their sectors of the theoretical front required the rejection of Deborin's 'formalism', his absorption in abstract Hegelian dialectics.

The specialists in philosophy who spoke at the Conference of October, 1930, were either students just beginning their careers, uniformly opposed to the Deborinites, or former students of

Deborin's, all but five of whom had turned against him. A. K. Timiriazev also appeared, to suggest that the mechanists were being vindicated, but he was laughed out of court.[52] At the beginning of the Conference the Deborinites were still full of fire. One of them flailed the Bolshevizers' views as 'liberal, Kautskian, Menshevik, Struvian—I already don't know what kind of rubbish', whereupon a voice from the audience helped him out: 'Right-deviationist!' and he completed his furious incantation: 'Right-deviationist—this nonsense, this is arch-revisionist trash!'[53] But an avalanche of counter-tirades crushed the fighting spirit of the Deborinites. Mitin warned them that 'one must lend an ear to the signal which the Party press is giving'.[54] Another former disciple of Deborin's was more explicit: the philosophical leadership must not presume to instruct the Party's leaders in philosophical truth; 'at the present moment philosophy is a Party weapon, and it's their [the philosophers'] duty to sharpen it'.[55] Changing his metaphor, he grew ominous:

> Comrade Deborin and especially certain others of his group have taken the bit in their teeth and want to gallop very swiftly off to a side from the general Party line. They must be curbed. Whoever has read several Party documents, whoever has read Comrade Stalin's articles, knows that we can curb a swift gallop. . . . And the comrades themselves will be sorry if this curb occurs in the fiercest manner, if they don't understand now what they're being warned about beforehand.[56]

The Deborinites who spoke towards the end of the Conference abstained from invective, but Karev still insisted that the philosophers' main duty was to provide the practical builders of socialism with 'the universal methodological key' to their problems,[57] and Sten warned the Bolshevizers against 'theoretical liquidationism': 'Theoretical struggle has political significance because it is connected with the class tasks of the proletariat, but theoretical struggle is not by any means reducible to political struggle and is not exhausted by it.'[58]

When it came time for the two main speakers to make their concluding remarks, Deborin apologized for having attacked Miliutin and the Bolshevizers. He had, he confessed, missed the whole political significance of the meeting. He was particularly distressed by his separation from 'the young cadres', with whom he had been working for nearly ten years. But he balked at the description of his philosophy as a formalist deviation from Marxism. If his works deserved that condemnation, how could Lenin have recommended him for teaching philosophy?[59] Other times, other customs, his opponents might have replied, as the Conference elected a committee to write a Bolshevizing resolution, and adjourned.

Thus conciliarism, the establishment of a philosophical line by resolutions of conferences, was used again, this time with much narrower representation of interested specialists, and with much greater participation of the Party's higher officials. And again major questions were left hanging, this time more serious questions, since the philosophers condemned by the Conference were in charge of the principal centres of philosophical teaching, research, and publication. What sanctions were to be used against them, if they refused to 'recognize their errors' in a fully satisfactory manner? On the eve of the Conference the young Bolshevizers at the Institute of Red Professorship had resolved that the central Marxist-Leninist institutions should be thoroughly reorganized.[60] The resolution of the Communist Academy's Presidium restated the Bolshevizers' views, but 'in a more "delicate" form'.[61] A strange impasse resulted. The Central Committee's Propaganda and Agitation Section fired the Deborinite editors of its organ, *Revolution and Culture*, and appointed a renegade Deborinite to turn the journal into a mouthpiece of the militant Bolshevizers.[62] And the Party organization at the Institute of Red Professorship expelled Sten for the 'now' in his declaration, 'I now stand on [the Party's] political line'.[63] (They demanded that he give a full account, in writing, of his former association with the 'bloc of "Leftists and Rightists"', and charged a current association between him and Lominadze, who had started a petition for Stalin's resignation.)[64] But who was to remove Sten from his position in the Communist Academy's Institute of Philosophy, while the Academy's Presidium ignored the demand for a thorough reorganization? And who had the authority to remove him (and Karev, who also refused to 'recognize his errors') from the editorial board of the leading philosophical monthly, *Under the Banner of Marxism*? Who was to judge between the chief editor, Deborin, who had recognized all the charges of error that he thought were justified, and Iudin, who complained in *Pravda* that Deborin had not completely 'disarmed himself before the Party's Central Committee'?[65] One can only imagine what was going on in private. *Under the Banner of Marxism* ceased to appear after the Conference of October, 1930.[66]

Early in December the young Bolshevizers of the Institute of Red Professorship went to the chief. Stalin told them that they had been too soft. The Deborinite philosophy was more than a system of formalist errors or a formalist deviation leading to idealism (the Bolshevizers had vacillated between these two descriptions). It was 'Menshevizing idealism', a sly counterfeit Marxism that was surreptitiously moving towards idealism in philosophy and Menshevism in politics. Mechanism, Stalin granted, was still the main danger on the philosophical front, but the philosophical works of the Menshevizing idealists were to be completely 'dissipated'; Lenin's works, and their continuous practical interpretation by the Party, were to be the basis of all future work in philosophy.[67] A few weeks after this conversation, new meetings were called at the Communist Academy, where Mitin triumphantly announced that 'Comrade Stalin's instructions must be placed at the foundation of all further theoretical work on the philosophical front'.[68] The Communist Academy's Presidium revised its resolution to accord with Stalin's harsher judgment,[69] but even then Deborin Karev, and Sten held back. *Under the Banner of Marxism*, of which they were the effective editors, still failed to appear.

On January 25, 1931, Stalin spoke with the voice of formal authority: the Central Committee published a Decree, 'Concerning the Journal *Under the Banner of Marxism*'.[70] Without giving credit to the young Bolshevizers or mentioning the Conference of October, 1930, the Decree repeated their criticism of the Deborinites, substituting 'Menshevizing idealists' for 'formalists' as the name by which these deviators were to be known. They had 'converted the journal, especially during the recent past, into the organ of their own group'; henceforth the journal must consistently put into practice the partyness of philosophy. To guarantee this, the editorial board was reorganized. Karev and Sten were dropped altogether (they were later condemned, without public trial, as 'enemies of the people');[71] Deborin was demoted from editor-in-chief to an ordinary member of the board; and four Bolshevizers were added to the one (Maksimov) already there. The questions left hanging by conciliarism were thus answered; final authority in Soviet Marxist philosophy had been clearly located.

Driving home the lesson that the new partyness ruled out the former autonomy of Soviet Marxist scholars, the Central Committee struck down the most illustrious one, D. B. Riazanov, founder and head of the Institute of Marx and Engels. Less than a year before, in the spring of 1930, his sixtieth anniversary had been celebrated with messages of personal and official greeting from the highest authorities (excepting Stalin); a lavish *Festschrift* had been prepared, and a biennial Riazanov Prize established for the best work in Marxist studies.[72] But in February, 1931, he was abruptly removed from the Institute of Marx and Engels, expelled from the Party, and subjected to further undisclosed punishment.[73] Iaroslavskii came to the Communist Academy to read the lesson of Riazanov's crime: between the Party and its enemies he had tried to establish a 'third force'. He had given jobs in 'his' Institute to Marxist intellectuals fired from other government posts for deviation, and he had tried to help former Mensheviks who were just then (March, 1931) being tried for organizing a counter-revolutionary group. 'You yourselves must understand', Iaroslavskii confessed,

> that expelling Riazanov from our Party was not an altogether pleasant matter. But can we forbear from such a measure? We would not be a proletarian, Bolshevik Party, if we were to forbear from this. Anyone can be a member of our Party only so as long as he executes its will, so long as he sincerely serves the cause of the working class, the cause of this Party. The moment he changes, the moment he takes a stand on the border where he is, on one side, our trusted person, while, on the other side, he is the trusted person of the Menshevik Party, there is no place for him in our Party. We will not tolerate anybody's double-dealing, whoever he may be; for treachery to the Party we will be ruthless.[74]

Riazanov had also refused to renounce his view that Lenin's genius was restricted to political science.[75] A tolerable foible in the 'twenties, well known to those who had honoured him in 1930, this view of the Party's founder was inadmissible now that the separation of politics and philosophy was ended. While refusing to bow to the Party's will in this matter, Riazanov had allegedly honoured agreements with foreign socialists, who gave him letters of Marx and Engels that were embarrassing to themselves on condition that the letters remain unpublished. The bare fact that Riazanov had not yet published certain letters was joined with innuendo and downright falsehood to support charges of dishonest suppression.[76] V. V. Adoratskii, Riazanov's successor at the Institute of Marx and Engels, had a thoroughly Bolshevik understanding of the scholar's commitment: a mechanist in the early 'twenties, he had supported the Deborinites when they won official favour, only to leave them for the Bolshevizers in 1930. Moreover, he had been among the first to proclaim Stalin the theoretical chief of Soviet Marxism as well as the political chief of the Party.[77]

Riazanov's dismissal was by no means an isolated incident, though the maturity of his successor was (Adoratskii was 53). The Central Committee's Decree of January 25, 1931, unfolded a great irony. The Marxists who had prepared and begun the 'scientific change-over' for 'bourgeois' scholars were themselves the victims of a change-over. And it was far swifter and more thorough in Communist than in non-Communist institutions of higher learning, presumably because traditions of academic autonomy were much weaker in the former, and slowly acquired professional competence was considered less important in the social sciences and the humanities, the special field of the Communist institutions, than it was in the natural sciences. In philosophy crabbed age gave way to Bolshevizing youth with astonishing swiftness. One year after the Decree was issued Iudin reported that 'almost all the old philosophical "gods" had to be dismissed from the Institute [of Red Professorship], and it is primarily young comrades, new cadres, that carry on the work'.[78] He complained that many were poorly prepared for scholarly work, but presumably he did not include himself in this number (he was made head of the Institute's Section of Philosophy within a year of graduating from it). Moreover, the Philosophical Section became a full-fledged Institute within the Communist Academy under a second decree of the Central Committee, which effectively terminated the Academy's vanishing autonomy by bringing the renovated Institutes of

Red Professorship into it.[79] By 1936 Mitin could boast that the Deborinite works were being erased from memory: 'Not in a single question of philosophy have we left one stone on another of that which the representatives of Menshevizing idealism created.'[80] About the same time the Communist Academy was dissolved altogether, its Institute of Philosophy absorbed by the venerable Academy of Sciences of the U.S.S.R., where Iudin, who has not produced a creditable philosophical work to this day, sat as Academician Iudin, Director of the Institute.[81]

The cry for partyness, which brought down the old leaders of Marxist-Leninist scholarship and raised the new ones to their places within the space of two years, rose to a new shrillness in the fall of 1931. In a letter to a Bolshevik historical journal Stalin denounced 'archive rats' who made past stages of Bolshevism seem less than absolutely irreconcilable towards deviationists.[82] This time Marxists on the various sectors of the theoretical front did not wonder whether Stalin's 'instruction' applied only to historians. All sectors criticized themselves, exorcising the spirit of 'rotten liberalism'.

In some cases genuine issues were raised—and quickly settled. One author, who had dutifully given the new evaluation of Lenin and Plekhanov as philosophers, was savagely attacked for remarking at one point that Lenin's separation of political and philosophical issues in the pre-revolutionary period showed the primacy of political considerations, of practice, in his mind.[83] It was a slander of Lenin and a distortion of history, the Bolshevizers shouted, to suggest that in certain circumstances his concept of partyness allowed the separation of political and philosophical issues. Another scholar, who had collaborated on Iaroslavskii's condemned history of Bolshevism, was taken to task for this confession:

> When it came to striking out some facts or others, we approached [the Party's history] not with the viewpoint of political expediency, but with the viewpoint of that objectivity which is absolutely not characteristic of our Bolshevik history, but is a belch of bourgeois liberalism, of bourgeois objectivism, in the sense in which we are accustomed to speak of bourgeois historians.[84]

In a second confession he apologized for the vestige of 'bourgeois objectivism' in this first one: political expediency and objective truth were not opposed for Bolsheviks, whose political line was the embodiment of objective truth, and never required 'striking out some facts or others'.[85] Even Iudin and the Bolshevizers had an error to confess. They had given the impression that their campaign against the Deborinites had been undertaken on their own bold initiative. Now it must be known that Stalin and the Central Committee had directed the campaign from first to last, and they retold the dispute to prove it, striking out inconvenient facts without any qualms of 'bourgeois objectivist' conscience, but supplying no new facts to confirm the new version.[86]

In other outbursts of 'self-criticism' the only issues seemed to be personal: had this or that former leader of Marxist scholarship completely 'disarmed himself' before the Central Committee? Deborin, for example, submitted a letter of self-criticism early in 1932, but it was found inadequate since he distinguished between correct and incorrect elements in his former work.[87] At a meeting in March, 1933, he rejected such critical self-criticism in a second confession:

> No, complete and unconditional disarmament is demanded of us. . . . If one stubbornly resists and does not submit to the Central Committee's Decree unconditionally and, at the same time, honourably and conscientiously, without double-dealing trickery. . . . Menshevizing idealism turns into *Menshevist* idealism, and then, in accordance with the inexorable logic of things, its followers fall into the camp of the complete enemies of the Party, of its leadership, fall into the embrace of counterrevolution. And some in fact [notably Sten] have already taken this road.[88]

Even so, he was attacked again for inadequate self-criticism, since he still did not condemn everything he had ever written.[89]

This truculent refusal to accept Deborin's notice of surrender suggested the victors' fear for the places they had won from the vanquished. But mean personal motives aside, suspicion that repentant Marxist philosophers were hypocrites was probably as unavoidable as similar suspicions of the recently converted 'bourgeois' specialists. After all, there were severe penalties for refusal to repent. Nor could the danger of wary time-serving be avoided among the newcomers to Soviet Marxist philosophy. They had no past errors to repent, but they might well read their seniors' fate as proof that present truth could be future error. Mitin unintentionally demonstrated these inherent paradoxes of the new partyness when he generously decried the uncomradely attacks on Deborin's sincerity, but agreed that Deborin had not gone far enough in his self-criticism.[90] Calling for a rally round the incontestable truths disclosed by the 'classics' and their current interpreters in the Central Committee, he implicitly recognized that some regarded these truths as less than permanent: 'We must brand those—and there *are* such comrades—who are hedging, who formally recognize the Central Committee's decision, but think it necessary to stand aside for a while and see whether someone will fall.'[91]

A further paradox emerged in Mitin's prophetic peroration:

> Rallying all our forces, under the leadership of our Party, under the leadership of our dear and beloved teacher, Comrade Stalin, we will indubitably move forward the work on the philosophical front, and will create serious, fundamental works of research![92]

Even assuming the best of motives on the philosophical front, Mitin's prophecy was paradoxical. Mature philosophers who had known some intellectual autonomy might voluntarily and sincerely give it up for the iron discipline that the cause seemed to demand; newcomers to philosophy might enthusiastically dedicate themselves to the execution of Comrade Stalin's instructions, careless of future instructions that might make deviationists of present enthusiasts. But in one case or the other, how could they ask fundamental questions or seek their answers? Disarmed before the Central Committee and executing its instructions, how could they *search* for philosophical truth? Must they not *wait* for it? Already in 1932 and 1933 the lack of 'serious, fundamental works of research' in philosophy was being lamented.[93] Beading quotations into standardized patterns had become the characteristic occupation of Soviet Marxist philosophers.

What, then, were the results of the great break in the philosophy of *natural science*? The denigration of philosophizing out of Hegel and the new stress on 'practice' struck some mechanists as a belated vindication of their programme.[94] Had they not proposed to build the philosophy of natural science out of empirical data and verifiable theories, which, together with technology, were 'practice' in the natural sciences just as the Party's policies were 'practice' in the social sciences and humanities? Indeed there was an unacknowledged kinship between the Bolshevizers' programme and that of the mechanists. 'The proposal' one of the Bolshevizers explained to the Deborinites in October, 1930,

is that philosophical work should grow in the soil of definite, concrete, scientific disciplines, that you learn the concrete experience of practical work in this or that scientific field and get your philosophers to be connected more closely with these fields. That is the crux of the matter.[95]

But the mechanists were sharply rebuked when they claimed vindication in 1931.[96] They had advanced this proposal when the old generation of 'bourgeois' specialists seemed destined to control the natural sciences for a long time to come, and the mechanists had accordingly played down the class character, the partyness, of natural science. The Bolshevizers advanced this proposal *in the name of* partyness, when young 'red specialists' seemed to be storming the special disciplines of natural science. Their programme of developing the Marxist philosophy of natural science within the special disciplines was by no means a concession to the autonomy of 'bourgeois' specialists. Armed with 'instructions' dug out of the 'classics' and the decisions of the Central Committee, the young 'red specialists' would reconstruct the special branches of natural science, thereby forcing the older specialists to reconstruct themselves or get out.[97]

The initial result, while the great break lasted, was a burst of feverish talk about the dialectical materialist transformation not only of such disciplines as mathematics, physics, and biology, but even of such as hospital planning and agricultural engineering. One example has already been given from the field of statistical theory, and more will be given in the next chapters from the fields of physics and biology. Let it suffice here to say that in some cases a few older specialists, whose pet theories had been rejected or ignored by most of their colleagues, were taken up by fledgling red specialists apparently convinced that the minority view must be the revolutionary one. In other cases the fledglings produced wondrous gibberish of their own.[98] In any case these 'reconstructions' of the natural sciences were advanced in the name of partyness, apparently under the aegis of the Central Committee itself. Older specialists who might have resisted or criticized were prudently silent for the most part, and the Bolshevizers began to complain of their silence as evidence of resistance to dialectical materialism.[99] Clearly the Central Committee, having set itself up as the areopagus of dialectical materialism, had to arbitrate, and this time not on abstract issues of Marxist-Leninist ideology but on their application to special disciplines of great practical importance.

The Central Committee's response, when the great break had run its course, was to cry down hasty reconstructions of the sciences and to put off the problem. In June, 1932, A. I. Stetskii, head of the Central Committee's Agitation and Propaganda Section, lectured the Bolshevizers 'On Vulgarization and Vulgarizers':

> These zealous comrades do not take into consideration the fact that the reconstruction of mathematics, physics, and the rest can be achieved only on the basis of a careful, critical reworking of *all the accumulated material of a given science*, on the basis of a knowledge of its history, an understanding of its specific character, without mentioning such an elementary precondition as the necessity of being both a *specialist* in one's own field and a good *judge of dialectics*.[100]

To be sure, Stetskii's published article did not criticize any of the leading Bolshevizers; minor, relatively insignificant figures were the butts of his ridicule. He blamed the continuing influence of Menshevizing idealism for the 'vulgarization' of dialectics in natural science, and he warned 'bourgeois' specialists that they were not being vindicated.

The reconstruction of natural science on the basis of dialectical materialism was not being abandoned, but clearly its tempo was being greatly slowed. (The federated Institutes of Red Professorship lost their Institute of Natural Science, the students of which were dispatched to appropriate technical schools.)[101] And clearly the Bolshevizers were being warned. They must not press the reconstruction too rashly, ignoring the claims of 'practice' in the natural sciences, where non-Party specialists were still dominant; and they must not presume to speak for the Central Committee on specific issues in the natural sciences. On such issues non-Party specialists were once again free to criticize the Bolshevizers, who became strikingly temperate in their replies, at least for the time being.[102]

A new phase in the interaction of Soviet Marxism and natural science had begun, in which the main subject of discussion would no longer be the *Weltanschauung* or universal methodology appropriate for natural science as a whole. Soviet Marxists were supposed to continue such discussion, for their basic philosophy was supposed to be anti-dogmatic, empirical, open to further development. But now only Stalin and his compliant Central Committee had the requisite world-sweeping vision; lesser philosophers would wait to be told when experience required the Marxist *Weltanschauung* to be developed further. As of 1932 they defined it by quotations and paraphrases of the 'classics', carefully selected to support fulminations against reductionist positivism (the mechanist deviation) and against Hegelianizing metaphysics (the Menshevizing idealist deviation).

Ostensibly, the old tension in Marxist philosophy between positivism and metaphysics had been solved—and if any further solving were needed, the chief would provide it by adding a 'classic' or two to the carefully defined list. In fact, the longstanding argument over dialectical materialism as a universal philosophy had become an internal dialogue, fragments of which might burst into the open from time to time. For lesser Bolshevik philosophers were still free, at their own risk, to discuss the application of dialectical materialism to specific issues within the natural sciences; and non-Party specialists would be prodded to give proof of their conversion to the official ideology by participating in such discussions, also at their own risk. The ultimate arbiter of issues, risks, and all—the Stalinist Central Committee—would decide for itself whether, when, and how to intervene.

Some may say that the mechanists had scored a belated, unacknowledged victory, for Hegelian studies were cut back, the Marxist philosophy of natural science would henceforth grow within the concrete disciplines, 'practice' was enthroned as the ultimate criterion of truth. But 'practice' could mean almost anything, from the data and verifiable theories of the natural sciences (and who can be sure that there are no metaphysical elements here?), to the changing techniques of Soviet agriculture and industry, to the Party's shifting ideological needs (and who can doubt that metaphysical vision enters here?). How the diverse, often conflicting demands of such 'practice' were to be reconciled can be determined only by further historical research. One cannot infer from the situation in 1932 a single, unescapable pattern of further interaction between Soviet Marxism and natural science. For the great break did not settle the contested legacy that the 'classics' had left to Marxist philosophers of science, just as it did not effect a final change-over from 'bourgeois' to 'red' specialists.

Ostensibly, the older generation of scientists (henceforth called 'non-Party' rather than 'bourgeois') had been brought within the Marxist community and would participate in its discussions. In fact, those discussions would still derive much of their meaning from the ideological suspicion with which Bolsheviks continued to regard most older scientists. As for Marxist philosophers of science, the great break seemed to have eliminated the issues that had divided them in the past. In fact, it restricted the field and changed the rules of argument, and brought to the fore a new generation of court philosophers, ready to continue the restricted argument under the new rules, in the presence of the close-mouthed, unblinking chief of the revolution.

PART FIVE

PHYSICS AND BIOLOGY IN THE FIRST PHASE 1917-1932

18

THE 'CRISIS' IN PHYSICS

THE first phase of Soviet Marxism, 1917–1932, coincided with triumphs of physics that produced crises of philosophical understanding. From the solar eclipse of 1919 until the late 'twenties, Einstein's modification of classical concepts of space, time, and motion held the centre of philosophical attention. Then came the new quantum mechanics, calling in question the most fundamental notions of causality and the intelligibility of physical processes. To understand Soviet Marxist reactions to these triumphs of physics and concomitant crises of philosophy one must realize that Soviet physicists actively participated in the triumphs, while treating the crises with indifference or non-Marxist philosophizing.

It is a great mistake, though a very widespread one, to imagine that Soviet scientists were shut off from the new, relativistic physics, which was allegedly 'rejected [by] the majority of Soviet philosophers and physicists', or 'tabood [*sic*] by the Bolsheviks for a long time'.[1] Indeed, no study of Russian sources or special history of Soviet physics is necessary to realize the erroneousness of this view. A. A. Fridman's (or Friedmann's) contribution to relativistic cosmology and V. A. Fok's (or Fock's) contribution to relativistic quantum mechanics—to take only a couple of notable examples—were published in Western journals and became widely known among Western scientists.[2] As for active opposition to the new physics, one might argue that there was less in the Soviet community of physicists than elsewhere. It will be seen below that the two Soviet physicists who declared war on the theory of relativity, took their arguments against it from the German Philipp Lenard, the American Dayton C. Miller, the Englishman J. J. Thomson. They found virtually no support among physicists of their own country.

Those who perceive a markedly abstract, theoretical, mathematical character in the history of Russian science, supposedly the result of the long-enduring backwardness of the country's technology, may find further support for their theory in the minimal opposition to relativity among Soviet physicists. Others, conscious of the minimal opposition among the physicists of all countries, would emphasize the ceaseless movement of scientific papers and people across national boundaries. Until the end of the period under review Russian scientists unashamedly continued their tradition of learning from the West. Only in 1931 one finds the first Bolshevik grumbling that 'time and again, as the best demonstration of our achievements, the number of Russian physicists' works printed in foreign journals is adduced. . . .'[3] There was also an impressive amount of travel abroad. A. F. Ioffe (or Joffe), the founder of a brilliant school of physicists in Leningrad, seems to have gone off nearly every summer; to do research with his mentor, Roentgen, until the latter's death, and then to tour centres of physical research and to lecture (e.g. at the University of California).[4] Various young Soviet physicists received Rockefeller grants for study abroad. Fok, for example, worked with Born in Göttingen; Ia. I. Frenkel' not only studied abroad but returned to lecture (at the University of Minnesota).[5] P. L. Kapitsa stayed so long at Cambridge, and so impressed his English colleagues, that he left behind a legendary picture of himself as the single most important physicist in the Soviet Union. Though there was comparatively little movement the other way, it is noteworthy that Philipp Frank read a paper on quantum theory at a Congress of Russian Physicists in 1928 and contributed a couple of articles to the *Great Soviet Encyclopedia*;[6] P. S. Ehrenfest, of the University of Leiden, was a visiting professor in Leningrad, even participating in a philosophical discussion of electromagnetic phenomena in 1929–1930;[7] and in 1933 the presence of Dirac, Frédéric Joliot, and several other foreign luminaries turned a Leningrad meeting on nuclear physics into something like an international conference.[8] But whatever the weight of the contributing factors (the Russian tradition of abstract science, and the international character of contemporary physics), the resulting fact is clear. During the period under review the Soviet community of physicists actively participated in the world-wide development of the new physics.

To this basic fact another must be added, if the background of the Soviet Marxist discussions of the new physics is to be understood. Soviet physical scientists shared not only in the triumphant progress of the new physics but also in the philosophical turmoil that accompanied it. This is not to say that most Soviet physicists were philosophically minded. The majority, who never bothered to record their thoughts on the philosophical aspects of their specialty, very likely shared the crude, somewhat cynical operationalism of one V. R. Bursian, who dismissed a philosophical discussion of electromagnetic phenomena as a quarrel over words, with possible relevance only to teaching: 'Since [lines of force] can be visualized, one must use them constantly, and in that way one inspires faith in their reality, a faith in which I myself have no faith.'[9] But those Soviet physical scientists who took the philosophical aspects of their specialty more seriously than Bursian were overwhelmingly indifferent or hostile to Marxism. During the Civil War, when the censorship was quite lax outside the field of politics, they published freely,[10] and even after the tightening of the censorship in 1922 a diminishing trickle of explicitly non-Marxist appraisals of the new physics continued to flow until the great break cut it off for good.[11]

But repressive measures could not cope with the philosophical problems thrown up by the development of physics. O. D. Khvol'son (or Chwolson), it will be recalled, when subjected to sharp criticism for the endorsement of fideism in his popularization of relativity, withdrew the offensive passage from later editions.[12] But in a survey of the new developments of physics that he published in 1924, he made the inevitable distinction between the phenomenalistic and the realistic approach in physics, pointing out that the former was clearly gaining ascendancy as concepts with a mathematical expression but without a physical explanation gained increasing significance:

> With an instrument that he does not understand the scientist does wonders, he begins to understand a great deal that was formerly not understood, he discovers new phenomena, he unites the divergent, hoping that in time he will understand the sense and significance of his wonder-working instrument. However, all the scientific triumphs achieved in this way arouse, along with feelings of rapture and amazement, a feeling of profound dissatisfaction. We see that a new scientific structure is being raised, but its foundation is not understood by us and arouses the greatest doubts.[13]

Here obviously was a problem for Soviet Marxists, committed as they were to the Promethean belief that all things are intelligible to scientific man.

Such implicit challenges to the Marxist *Weltanschauung* cropped up repeatedly in the textbooks and popularizations of the new physics. A. A. Fridman, for example, the mathematician whose notable contribution to relativistic cosmology was mentioned above, collaborated with the physicist V. K. Frederiks to produce an introduction to tensor analysis. The authors agreed with Hilbert that the general theory of relativity was beginning the 'axiomatization' of physics, that is, the transformation of physics into a formal science like geometry or arithmetic. But they felt that 'physics requires far greater attention than [geometry or arithmetic] to the explanation of its concepts by means of the material world's objects', and therefore they equated the progress of 'axiomatization' with the growth of scepticism. 'Fortunately,' they concluded,

> it is not given to us to see the future, and we do not know whether the epoch of axiomatization, the epoch of skepticism, will be the deathbed hours of knowledge. . . . [*sic*] But even if it should be so, still the logical beauty of the end would force us to welcome the appearance of the principle of relativity.[14]

If Soviet Marxists agreed that the axiomatic method was inseparable from scepticism, which would undermine the Marxist *Weltanschauung* as well as the traditional conception of physical meaning, they would have to become opponents of the new physics. If they wanted to have both the Marxist *Weltanschauung* and the new physics, they would have to prove that

Frederiks and Fridman were misinterpreting the epistemological significance of the axiomatic method. In either case, whether disagreeing with the physics or the philosophy of Frederiks and Fridman, they could hardly avoid involvement in the forbiddingly abstruse theories of the new physics. Yet, when the discussion began, hardly one of them had anything but the vaguest knowledge of physics. Towards the *end* of the period under review, after nearly fifteen years of the Cultural Revolution, Communists were less than two per cent of the community of Soviet physicists, excluding graduate students.[15]

A. K. Timiriazev's entrance into the Bolshevik Party in 1921 must accordingly have seemed a godsend to those concerned with the Party's ideological tasks. Here was a mature physicist, the first one in the Bolshevik ranks.[16] To be sure his special work had been on the kinetic theory of gases, but he showed no hesitation or diffidence in settling accounts with the theory of relativity.[17] He regarded it as a perversion of physics. Whatever truths he was willing to concede it had already been discovered, and could be explained better, by classical physics, in particular by the theories of J. J. Thomson and Philipp Lenard. The data requiring amendments to classical theories could be challenged, as Dayton C. Miller was challenging the invariance of the speed of light and the resulting evaporation of ether.[18]

In article after article Timiriazev doggedly circled the same track: physical phenomena are explained only when they are reduced to, or expressed in terms of, classical mechanics, and when a visual image or model can be formed; physical phenomena that cannot be explained in this manner are erroneously reported. With this argument he was objecting to the theory of relativity before he became a Communist. His discovery of Lenin's polemic against 'Machism' merely furnished him with an additional weapon against an enemy he was already bent on destroying. The theory of relativity, he began to argue, is inseparable from 'Machist' philosophy, and should be rejected for that reason too. Actually, his objection to 'Machism' was equivalent to his initial objection to relativity. 'Machism' would limit the physicist to the logical arrangement of data, and would inhibit the search for material causes responsible for the data. Since he interpreted 'material causes' in the light of classical mechanics, he was thus returning to his basic theme

—and providing an opening for the Leninist defenders of the new physics, who insisted that Lenin had not committed dialectical materialism to the concept of 'material causes' as understood by classical mechanics.[19]

From the start Timiriazev was virtually isolated in his crusade against the theory of relativity. In 1922, to be sure, Lenin commended him for raising an issue that Marxists could not afford to ignore, for Einstein's theory, Lenin warned, was being used to support various idealist philosophies.[20] But this commendation did not include an endorsement of Timiriazev's hostility to Einstein's physics, and V. I. Nevskii, a Bolshevik ideologist commissioned by Lenin to bring *Materialism and Empirio-criticism* up to date, repeatedly indicated his conviction that Einstein was a sound physicist though a poor philosopher.[21] Trotsky would not commit himself that far, but insisted that the question was open.[22] To be sure, Timiriazev's lustre in the first half of the 'twenties as *the* Bolshevik physicist carried some Marxists in his train. Deborin, for example, revealed his acceptance of Timiriazev's judgment in 1924–1925,[23] and *Krokodil*, the Bolshevik magazine of humour, once carried a cartoon captioned: 'Einstein has invented a scientific theory that contradicts physics and geometry, and therefore it is bourgeois, and the proletariat does not need it.'[24] But such ephemeral gestures of emotional solidarity contributed little to Timiriazev's crusade. Until new circumstances brought him new comrades in the 'thirties, his supporters who denounced the new physics at any length were four: I. N. Stukov, editor of *The Atheist at the Bench*; I. E. Orlov, the philosophical chemist who was a leader of the mechanist faction; L. M. Rubanovskii, a graduate student of chemistry; and G. A. Kharazov, a former anarchist publicist who worked out a *mathematical* refutation of relativity.[25]

There must have been a time in 1921–1922 when Timiriazev had hopes of support from A. A. Maksimov, the thirty-year-old physicist to philosophers and philosopher to physicists, who would be a leader first of the Deborinite faction, then of the Bolshevizers. For Maksimov entered the discussion of the new physics with a respectful review of Lenard's blast at the theory of relativity, which had been translated and annotated by Timiriazev.[26] But already at the end of 1922 Maksimov questioned the assumption of Lenard and Timiriazev that the 'mechanical foundations' of physics had been adequately explained by classical theory. The Newtonian concept of space and time, he argued, was metaphysical in its assertion of an absolute space and time existing separately from matter. Engels had pointed out that space and time are inseparable aspects or forms of material being, and the theory of relativity was bringing physics to a realization of this dialectical truth. Moreover, Einstein was to be commended for breaking down the artificial barriers that separated basic concepts, such as mass, energy, and inertia. He was also given credit for unwittingly confirming the materialist view of geometry as a branch of physics, for demonstrating the physical significance of non-Euclidean geometry, which had been a major point of support to idealist philosophies. At the same time, the great scientist was described as a typical product of the decaying bourgeois academic world; his stated philosophy was opposed to dialectical materialism. From free creations of the mind he allegedly hoped to construct a closed system of thought that would express the whole truth about the universe once and for all. Maksimov concluded that proletarian scientists would have to recast Einstein's physics, establishing it on the proper philosophical foundation.[27]

Maksimov had no practicable suggestions for proletarian physicists who might have wished to begin this great task. Very likely, lacking training in physics beyond the undergraduate level, he would have been embarrassed by a demand for such suggestions, but he was spared by the fact that Soviet scientists in the forefront of physical research remained aloof. He was challenged by Stukov, editor of *The Atheist at the Bench*, who charged that Maksimov was a prisoner of relativism, philosophical no less than physical. Maksimov had the simple task of pointing out to Stukov

> the genius' instinctive sense that Einstein possesses in the formulation of questions. However deeply he may be stuck in the Machist-Humist-Poincaréist philosophy, in the last analysis he still obtains formulations that are correct and useful for us sinful materialists too.[28]

A similar distinction between the physics and the philosophy of Einstein was made by another young specialist in the philosophical aspects of physics, Z. A. Tseitlin, who reminded the opponents of relativity of Engels' apology to the platypus.[29] (Shortly before he died Engels recalled an incident of his youth: 'I saw the eggs of the duckbill in Manchester and with arrogant narrow-mindedness mocked at such stupidity—as if a mammal could lay eggs—and now it has been proved! So do not behave in the way I had later to beg the duckbill's pardon for!')[30] Tseitlin's appreciation of the theory of relativity had this further similarity to Maksimov's: Einstein's notion of space and motion as 'modalities' of matter or substance was essentially dialectical materialist. But Tseitlin's reservations about Einstein's physics were not as vague as Maksimov's. He believed that the special theory of relativity, by denying the reality of world ether, denied the objective, material reality of space. The general theory of relativity he interpreted as restoring ether and thereby the objective reality of material space, but he felt that the restoration was not complete. For Tseitlin was convinced that true physics and true philosophy were essentially a continuation of Descartes' vision of the universe as multiform vortices in an ethereal ocean. Cavendish and Maxwell had deepened and refined this basic conception; Einstein's physics grew out of Maxwell's, through the intermediary of Lorentz, and the theory of relativity would be perfected if Einstein would only recognize the great tradition of Cartesian physics and restore a fluid either.[31] Tseitlin himself merely pointed the way; he did nothing to develop his idea in application to the specific concepts and theories of the new physics.[32] Like Maksimov, he may have lacked the ability to do so, but he was never put to the test, for his discussion was with Timiriazev's little school, which would not recognize any merit in Einstein's physics, and with the Deborinite philosophers, who were concerned only to rebut Tseitlin's Cartesian understanding of Marxist philosophy, and ignored the discussion of physics until 1927.

As the reader may recall, the Deborinites did not devise an original position on the new physics; in 1927 they merely endorsed one variety of a view that had been current among Soviet Marxists since Timiriazev opened the discussion in 1921. The proponents of this view combated Timiriazev and his rejection of the new physics, and looked askance at the feeble efforts of Maksimov and Tseitlin to re-work the new physics from the standpoint of dialectical materialism. They argued that the new physics, without the necessity of alteration, already represented an unacknowledged triumph of Marxist philosophy. But beyond this, there was little unity in this school, for its proponents could not agree in their understanding of the Marxist philosophy that had allegedly triumphed. Some were unabashed 'Machists', others orthodox Marxists who refused to support either the mechanist or the Deborinite understanding of orthodoxy; still others were leaders of the mechanist faction, and belatedly, from 1927, some were Deborinites. Finally, when the great break propelled eminent physicists into the orbit of Marxist discussions, another variant of this view sprang to life.

Bogdanov, Bazarov, and Iushkevich, Lenin's opponents in the philosophical debates of the pre-revolutionary period who continued their participation in Marxist discussions during the 'twenties, had little difficulty in reading the theory of relativity as a confirmation of their 'Machist' version of Marxist philosophy. Though Mach had frowned on the theory of relativity shortly before his death,[33] Einstein paid tribute to him for the philosophical inspiration of the theory,[34] and a variety of positivists ('Machists' in the Leninist vocabulary) were busy finding new support for their philosophies of science in Einstein's physics. Bazarov, reviewing the literature of relativity in *The Herald of the Communist Academy*, quoted with obvious approval Professor A. V. Vasil'ev's scorn both for the mystical conclusions that some drew from Einstein's theory and for 'the narrow fanaticism that thinks materialism . . . the last word of human thought'.[35] The intellectual atmosphere of the 'twenties was not free enough for Bazarov to include Professor Vasil'ev's specification ('the materialism of Haeckel and Engels'); indeed, he foreshadowed things to come in his warning against 'arguments of an extra-logical order, to which the partisans of Dantean cosmology so willingly resorted in Galileo's time, and which, unfortunately, have not lost their prestige in the eyes of some partisans of Copernican cosmology'.[36] But until the end of the 'twenties the 'Machist' Marxists were comparatively free to argue their view of the new physics within the community of Soviet Marxists.[37]

The first orthodox Marxist to endorse the theory of relativity without reservation was a publicist and trade union leader, A. Gol'tsman, who had little reason to feel diffident in matters of physics since he was willing to accept the theories of physicists without reservation. The interpretation of Marxist philosophy that he came to in his quarrel with Timiriazev has been described in a preceding chapter, for it foreshadowed the Deborinite interpretation; but he did not join that faction.[38] Nor did the German Communist, August Thalheimer, who was the first in the Soviet Union to argue that Hegel had actually forecast the relativistic denial of absolute time and space.[39] The most extended presentation of this view of relativity came from S. Iu. Semkovskii, who was considered a mechanist until 1929, when he came over to the Deborinites, noting that they had endorsed his interpretation of the new physics. When he first published this interpretation, in 1924–1926, he used it as a criticism both of the 'vulgar materialism' of Timiriazev, who wanted to tie Marxist philosophy to outmoded physical theories, and of the Deborinites, who, Semkovskii argued, were willing to accept Timiriazev's mechanistic objections to the new physics while they turned Marxist philosophy into a sort of neo-Hegelian scholasticism.

Semkovskii felt that great revolutions in natural science coincide with periods of great social revolution, and he recalled Engels' and Lenin's insistence that materialism must take on a new form with every revolution in the scientific understanding of matter. Dialectical materialism must accordingly absorb the new insights provided by the theory of relativity. The fact that Einstein sometimes expressed sympathy with 'Machism', and that 'Machist' philosophers were making capital of his work, should not decide the issue for Marxists; they should consider the theory of relativity itself. (Incidentally, however, Semkovskii defended Einstein as a somewhat inconsistent but basically realistic philosopher.) The only aspect of the theory of relativity that could not be accepted by dialectical materialists, Semkovskii felt, was the notion of a finite universe; but this was a peripheral issue, and still in a highly controversial and unsettled condition. On all essential issues the theory of relativity '*not only does not refute dialectical materialism but, on the contrary, is a brilliant confirmation of its correctness*'.[40]

Semkovskii's reasons for this judgment were essentially similar to Maksimov's: the denial of absolute space and time separate from matter, and their reduction to forms of matter's being; the reduction of geometry to a branch of physics, and the unification of concepts such as mass and energy. To this Semkovskii added a methodological commendation that Maksimov was unwilling to grant: the theory of relativity continued the classical tradition of strict determinism, the reign of objective causality in nature. There can be little doubt that Semkovskii here displayed a more acute perception of Einstein's basic methodology than did Maksimov, even though the latter claimed to be a physicist.[41] In general Semkovskii showed an impressive familiarity not only with the popularizations and philosophical analyses of relativity but even with some of the technical literature concerning it. His book was indeed the only Marxist work of the period under review that was favourably reviewed by a non-Marxist physicist in a technical journal.[42]

Such evidence of the appeal of Semkovskii's argument to physical scientists may well have been a major reason for the Deborinite abandonment of indifference or offhand hostility to the theory of relativity. 'Gessen's manœuvre', as the Deborinite transition to endorsement was called in an earlier chapter, began in 1927, with Gessen turning back against Timiriazev and the mechanist faction the charge of alienating natural scientists by attempting to bind dogmas on science.[43] By 1928 the transition was complete. B. M. Gessen, the forty-five-year-old rival to Maksimov in the borderland between physics and philosophy, published a popularization of the theory of relativity that repeated Semkovskii's arguments, except of course for the charge that the Deborinites were neo-Hegelian scholastics.[44] And at the Conference of April, 1929, which marked the Deborinite triumph over the mechanists, the legend was established that is still current in the Soviet Union: the mechanist faction was equated with hostility to the new physics, the Deborinite with unreserved acceptance.

This was also the time when the new quantum mechanics, with its statistical interpretation of causality and its mathematical description of elementary particles, defying visualization save by contradictory models, was raising philosophical problems that made Einstein's revolution seem, by comparison, a tame addendum to classical physics. Gessen, who replaced the disgruntled Maksimov as the chief interpreter of physics for the victorious Deborinites, turned his attention to these problems. He was far more competent in physics than Maksimov, but, as far as a layman can judge, he did not produce an original interpretation of quantum mechanics. From the works of Smoluchowski and Richard Mises he drew a complex argument that began with a philosophical reconsideration of chance, making it compatible with causality, and, struggling over the abstruse theory of 'ergodic' derived from Gibbs and Boltzmann, concluded that 'statistical mechanics cannot be reduced either to pure mechanics or to pure statistics'.[45] In this way Gessen believed that causality could be saved without arguing that quantum mechanics is a transient first step to a non-statistical theory of elementary particles.[46] (The latter argument, a much simpler method of having the cake of quantum mechanics without paying the penny of causality, was and remains very popular among Soviet Marxists.)

Thus Gessen moved the Deborinites not only to unreserved acceptance of 'bourgeois' scholars' physics, but even towards a serious consideration of their philosophy. A high degree of confidence was implicit in this move: confidence that Deborinite interpretation was sufficient to transform even the *philosophy* of 'bourgeois' scientists into Leninism. Ironically, Gessen's confident move coincided with the drastic failure of Bolshevik confidence in 'bourgeois' specialists even within the field of *science*. To compound the irony, Gessen himself was both a harbinger and an ultimate causalty of the great break. At the same time that he was seriously considering Mises' effort to solve the philosophical problems of quantum theory, he denounced the editors of physical journals for their indifference and hostility to Marxist philosophy, declared inadmissable the publication of Rice's *Theory of Relativity* in Russian translation, and called for the dismissal of the responsible editor in the State Publishing House.[47] But, when the dismissal occurred, in 1930, Gessen himself was under attack for an allegedly uncritical and permissive attitude towards the ideological poison flowing from the new physics.

The dismissed editor was an outstanding sixty-one-year-old mathematician, V. F. Kagan, who must have had some sympathy with Marxism, for he had been participating in the work of the Communist Academy since the early 'twenties.[48] When one considers that his specialty was multi-dimensional geometry and tensor calculus, it becomes surprising that he had not participated in the Marxist discussions of relativity. Yet his was by no means an isolated anomaly. Another participant in the activity of the Communist Academy was the young physicist S. I. Vavilov, one of the first Russian translators of Einstein, and author of a popular presentation of *The Experimental Foundations of the Theory of Relativity.*[49] He too was aloof from the Marxist discussion of relativity, and, in the preface to his book, explicitly excluded consideration of the philosophical issues.[50] The most striking instance of this anomalous abstention of truly competent specialists from the Marxist discussion of the new physics was the mathematician O. Iu. Shmidt, who founded the Communist Academy's Section of the Natural and Exact Sciences. In 1924 he spoke out against Timiriazev's rejection of relativity, and incorporated in the Statute of his Section a call for an open discussion of the subject.[51] Yet he himself ventured no extended analysis of relativity. The Marxist appraisal of the new physics was left to the variety of publicists and undistinguished 'scientists' reviewed in the preceding pages—*before*, one must stress, the great break brought obstreperous young Bolshevizers into the field, impatient of anything but violent denunciations of 'bourgeois' science and passionate declarations of devotion to the Central Committee.[52]

The career of A. F. Ioffe (or Joffe) is one of the most interesting cases of a physicist who was close to Bolshevism but took no extensive part in the Marxist discussions of physics until the end of 'the great break'. He was certainly qualified to participate. One of his earliest works (1907) had been a defence of Einstein's explanation of the photoelectric effect; he was elected to the presidency of a major physical society in 1915, when he was only thirty-five, and to the Academy of Sciences in 1919; he could count among his students some of the most brilliant theoretical physicists of the Soviet period.[53] On the political side, he joined the exodus of non-Bolshevik scholars to the Crimea in 1917, but he was one of the first to change his mind. Returning to red Petrograd when the Civil War was only beginning, he 'firmly resolved once and for all to connect my fate with the Country of the Soviets, and to contribute my share to the future upbuilding'.[54] He became one of the chief organizers of research and advanced training in physics, and a darling of the Bolsheviks, holding a continuous seat in the Leningrad Soviet from 1926, and leading the cry for dedication to socialist construction during the great break. (He did not join the Communist Party, however, until 1942.) Yet he was a virtually silent observer of the Marxist discussions of the new physics until the 'thirties.

In the period under review Ioffe was involved in these discussions only twice, and then very briefly. At the end of 1926, when Timiriazev carried his campaign against relativity from the narrow circle of Soviet Marxists to a Congress of Russian Physicists, Ioffe replied on the spot, and, worried at Timiriazev's influence among the Communists, published a defence of relativity in *Pravda*.[55] It is significant that he confined himself almost entirely to a popularized analysis of the experimental evidence, and only at the end of his article dealt summarily with the philosophical issues, which he reduced to one. Appealing not to Marxism but to common sense, he remarked that Einstein's theory could not be regarded as a blow at materialism since it deepened our understanding of material nature: 'If, from the properties of matter, some want to deduce its absence, then one must struggle with such distortions of common sense, but not with the theories that describe matter'.[56] Timiriazev replied in *Pravda*, charging Ioffe with 'Machism', but the Deborinites, then moving to their endorsement of relativity, defended Ioffe against the charge.[57] Ioffe relapsed into silence, no doubt reassured by the ascendancy of the Deborinites and by the establishment of the legend that the discredited mechanist faction was the special home of enmity to the new physics.

There was another moment when Ioffe and several other eminent specialists were almost drawn into the Marxist discussion of physics. The Deborinite 'social movement about dialectics', encouraging an interest in philosophical issues among students of the natural sciences, in 1929 reached Leningrad's famous Polytechnic Institute, which was closely allied with the research centre headed by Ioffe. Some of the

Institute's students took note of a major difference in fundamental theory between two of their teachers: V. F. Mitkevich, a fifty-seven-year-old electrical engineer who had just been elected to the Academy of Sciences, and Ia. I. Frenkel, a thirty-five-year-old student of Ioffe's who already enjoyed an international reputation as a theoretical physicist. The difference was this. Mitkevich pictured a magnetic field as fine 'tubes of force' in the universal ether; current (the flow of electrons through a conductor according to most physical theorists of the time) was actually, in Mitkevich's view, the result of displacements among the tubes of force. Frenkel had no place in his model of physical reality for a fluid ether spun into Faraday tubes; he gave his students a picture of electrons as point centres of fields of force interacting through empty space; current was the flow of such electrons, and a magnetic field was its by-product. On the students' suggestion a debate was arranged.[58] In three long sessions, at the end of 1929 and beginning of 1930, the Polytechnic Institute witnessed what the Communist Academy had failed to achieve in a decade of effort: a lively, substantial discussion by a dozen truly competent specialists, Ioffe included, of the borderland where physical theory merges into philosophy.

The specialists made no explicit reference to Marxist philosophy, though nearly all shared varieties of the materialist outlook—or the mechanist outlook, as they generally called it, as if completely ignorant of the recent condemnation of mechanistic materialism. Mitkevich's variety of materialism required the rejection of action at a distance through empty space; in his view it was the beginning of mysticism to suppose that the field equations deriving from Faraday and Maxwell made sense without the ethereal model by which Faraday and Maxwell had visualized electromagnetic phenomena. Frenkel, who had Ioffe and nearly all the other participating specialists on his side, argued that belief in an ethereal substance without any experimental evidence of its existence was the beginning of mysticism. He rejected the suggestion that he was a mathematical formalist while Mitkevich was a constructor of visual models. But even while granting the necessity of visual models, he added that convenience of mathematical analysis and prediction is the basic criterion for the acceptance of physical theories.[59] Marxist philosophers had repeatedly called him a 'Machist' for this view, but he had never deigned to reply—and never did, though he lived to 1952.

Frenkel had also been called a 'Machist', or even an outright idealist, for his model of electrons as point centres of fields acting through empty space, but it is significant that the two young Communist philosophers who participated in the discussion at the Polytechnic Institute did not repeat either accusation. Showing a Deborinite willingness to accept current physical theory but an ignorance of what current physical theory was, they diffidently abstained from endorsing either Mitkevich's or Frenkel's ideas. They lectured the assembled scientists on the general tenets of dialectical materialism as the Deborinites understood them.[60] Mitkevich, who was very upset by the unwillingness of nearly all the scientists to support him, claimed that the 'comrade philosophers' had taken his side.[61] Frenkel, Ioffe, and the other scientists had almost nothing to say about the philosophers' comments.[62]

This debate might conceivably have been the beginning of a serious discussion involving truly competent specialists. Instead, coming at the time of the great break, and making Mitkevich suddenly aware of his painful isolation among physicists and of possible support among Communist philosophers, it was the beginning of a bizarre episode. Within a few months the Deborinites were discredited, and Mitkevich became a hero of the young Bolshevizers, who were intent on the dialectical materialist reconstruction of physics. He was taken to London in the summer of 1931 to tell an International Congress of the History of Science that 'every deviation from Faraday's method of study and analysis of physical phenomena leads to painful results. The roots of the modern crisis in physics must be sought to a great degree in this direction'.[63] Back in the Soviet Union Mitkevich lectured his fellow Academicians and the public at large on the identity of Faraday's and Engels' understanding of basic physical theory.[64] He drew up ten questions on action at a distance that were all to be answered according to Faraday and Maxwell, if the physicist queried was a genuine materialist.[65] And the Bolshevizers took the questionnaire seriously, though with some reservations concerning the mechanistic nature of Mitkevich's materialism.[66]

There is not much point in seeking a clear understanding of what the Bolshevizers of physics had in mind, for they were clear themselves only on one point: they wanted proofs that physicists had renounced 'neutralism' towards the Central Committee and its ideology. Maksimov led the outcry for a dialectical materialist reconstruction of physics, with no more concrete suggestions than he had provided in the quieter 'twenties.[67] A few publicists and fledgling physicists were rash enough to try their hand at the reconstruction. In some cases, such as their support for Mitkevich, this meant the revival of classical theory.[68] Generally the reconstruction of physics meant nothing more than the cultivation of Bolshevik hostility towards the ideological outlook of 'bourgeois' physicists, in order to 'educate seasoned proletarian cadres, struggling with every sort of deviation from Marxism-Leninism and from the Party's general line.'[69]

Perhaps the most illuminating effort at the reconstruction of physics was that of V. E. L'vov, a writer in the field of popular science. He may have been motivated by a desire to clear himself of his Deborinite past, for he had been a serious offender. In 1929, describing Rutherford and Chadwick's alteration of the classical version of the conservation of matter and energy, he had treated some of the most venerable tenets of materialism (and of rhetoric) with great highhandedness:

How indeed are we to understand that substance is *not* eternal, *not* indestructible, does not follow the rule *ex nihilo nihil*? May it be that, right after the tongue-tied concepts of macroscopic thought and language, right after 'causality', 'motion', and so on and so forth, the concepts of the 'eternality' and 'being' of substance are now spilling over on to a statistical level? One way or another, before the philosophy of dialectical materialism again and again now rises a task most welcome, and full of high historical responsibility: to break through the envelope of the macroscopic world and *reconstruct the very foundations of thought and philosophical language*, raising their apparatus to those depths of knowledge that the genius of quantum physics is now turning over with a mathematical lever.[70]

In 1931 L'vov abruptly turned his battery of metaphors and began firing in the opposite direction: not the apparatus of philosophical thought but that of 'bourgeois' physics was to be raised to the depths. He found that Einstein's deism was a logical consequence of relativistic physics, in particular, of the four-dimensional time-space continuum.[71] L'vov conceded that this continuum might be a useful technical device in certain calculations, but real space, he insisted, must still be visualized as three-dimensional and Euclidean, and time must be considered as an independently flowing 'universal *quality*'. The dialectical materialist conception of the world process had as much in common with 'the rotten swamp of Einstein's "continuum" as a metallurgical factory with a church, religion with science, a priest with a Bolshevik'.[72]

This sort of obstreperous crudity was hardly likely to precipitate a serious discussion of the new physics, especially when the Central Committee's Decree of January, 1931, was taken as an endorsement of the Bolshevizers. Those who had misgivings about the Bolshevizers' demand for the revision of physics fell silent. But in the middle of 1932 Stetskii, head of the Central Committee's Section of Agitation and Propaganda, denounced 'vulgarizers' of dialectical materialism.[73] I. E. Tamm, an outstanding young physicist who had been pugnacious enough to challenge Mitkevich on technical issues even before Stetskii's article appeared, stepped into the breach. In the chief journal of Soviet Marxism he surveyed 'The Work of Marxist Philosophers in the Field of Physics'.[74] The dialectical materialist philosophers who had been calling for the revision of contemporary physics, said Tamm, simply did not understand contemporary physics. His own understanding of dialectical materialism could have been summed up in Lenin's dictum that 'the *sole* "property" of matter with whose recognition philosophical materialism is bound up is the property of *being an objective reality*, of existing outside our mind'.[75] The Bolshevizers replied to his implicit positivism, but not in such a manner that he and like-minded physicists had to shut up.

A new phase in the Soviet Marxist discussions of physics had begun. It would be distinguished from the earlier phase above all by the participation of some of the Soviet Union's most eminent physicists, defending their subject against the Bolshevizers. As this new phase opened, Semkovskii and Gessen were restored to their place as major Marxist interpreters of physics;[76] the Bolshevizers were rather chastened. Maksimov, for example, reviewing the 'crisis' in physics in April, 1933, sounded very much like Gessen.[77] He took dialectical pride in the conflicting pictures of the electron as wave and particle, and barely suggested his former Bolshevizing by alluding to the 'hypertrophy' of unspecified 'elements' in general relativity, and by calling Faraday's concept of ether and lines of force 'spontaneously dialectical'.[78] His main point on the 'crisis', repeated at tiresome length, was that the great triumphs of physics 'do not serve the mobilization of public opinion' in the proper direction. But the only physicists he accused of betraying their duty of 'organizing public opinion on the theoretical problems of science' were foreigners. (Heisenberg, for example, was accused of aiding the rise of Nazism by disseminating irrationalism in Germany.) For Soviet physicists Maksimov had nothing but praise, both for their social role and for their scientific successes. (As luck would have it, his one example of recent Soviet successes was the explanation of radioactive decay provided by George Gamow, who was on the verge of emigration.)[79] Later on in the 'thirties, when specialists would fall under suspicion once again, the spirit of the great break would be revived. Maksimov would declare Mitkevich's questionnaire a valid test of the physicist's political coloration.[80] Semkovskii and Gessen would perish as 'enemies of the people', and the one Bolshevizer whom Ioffe would praise for proper Marxist criticism of 'bourgeois' physicists would spurn the praise, noting that Gessen had been 'a student of Ioffe's'.[81] But even in such an atmosphere Ioffe and other eminent specialists, speaking as Marxists, would answer back.

One suspects that in this new phase of the Soviet Marxist discussion of physics the underlying issue would be what it had been in the past: not the problems facing physicists in their work as physicists, but the problems facing Bolsheviks in their ideological appraisal of physics and in assuring themselves of the reliability of physicists. That this was true of the discussion in the 'twenties and early 'thirties is evident from the indifference of eminent physicists; even those who were close to Marxism saw no point in participating until the Bolshevizers made such indifference seem treasonable. It is evident also in the positions taken by those authors who did participate: the defence of classical theories by Timiriazev and Mitkevich, the translation of new physical theories into dialectical materialist terms by Semkovskii and Gessen, the confused fluctuation of the Bolshevizers between defence of classical theories and crude translation of new ones. None of these positions had relevance to the professional problems and disputed issues of physicists who took relativity and quantum mechanics for granted, or worked in specialties untouched by them.

Of course, there can be no absolute separation between the scientific and ideological problems of physicists, but there is no need here to attempt a rigorous analysis of the vaguely defined frontier between the two. For the present point is simply that the philosophical appraisal of the new physics could be turned largely inward, concentrating on a close professional analysis of the new concepts and theories, or it could be turned outward, concentrating on the implications of the new theories for epistemology and *Weltanschauung*. Lenin had set the precedent, in *Materialism and Empirio-criticism*, and in his fleeting comment on relativity in 1922, of attempting to draw a sharp line between these two orientations, of taking new physical theories for granted and concentrating on analysis of their epistemological significance, where alone he claimed to find a crisis of physics. The Soviet Marxists of the 'twenties and early 'thirties tended to follow Lenin's example, but they were deflected now and again by two disturbing factors. One was the presence of Timiriazev and Mitkevich, vociferously urging Soviet Marxists to join the defence of classical theories. The other was Communist mistrust of 'bourgeois' scientists, and the consequent fear that any line separating physics from philosophy might become a wall protecting the growth of anti-Bolshevik ideologies. Clearly the second factor was the more powerful one, if only because Timiriazev and Mitkevich were so plainly eccentrics championing a lost cause.[82] But, having only these eccentrics' support among physicists, and lacking much knowledge of physics on their own, the Bolshevizers could not erase the line that Lenin had drawn between physics and philosophy.

One other factor is noteworthy by its absence: an effort to connect the judgment of physical theories with the solution of urgent technological problems. Infrequently the Bolshevizers tried to argue that eminent specialists were so absorbed in 'bourgeois' theorizing that they were neglecting the practical problems of socialist industrialization. The eminent specialists had only to ask what practical achievements could be claimed by the Bolshevizers. Mitkevich had a notable record as an electrical engineer, but few other technologists showed an interest in the Bolshevizers and their theoretical flights. Even if the work of the eminent physicists had been barren of practically useful by-products, it was full of exciting prospects. Tamm, for example, calculated that the hydrogen in a glass of water held as much energy as Dneprostroi, the great dam that was the special pride of the first Five Year Plan, produced in seven hours.[83]

All these factors can be translated into simpler language. If Lenin had not set the precedent of sharply dividing the scientific from the epistemological in his study of the 'crisis' in physics; if physics had been a less ancient and solidly established science, less rigorous and less prolific in theoretical and practical triumphs; if, accordingly, there had been significant blocs of physicists strongly opposed to each other on basic scientific issues; if, eyeing such a turmoil within physics and anxious for the ideological condition of scientists, the Bolshevik authorities had become involved in a crisis of production so desperate as to nurture feverish dreams of rescue by scientific miracles—then the Soviet Marxist discussions of the 'twenties and early 'thirties might well have produced a genuine crisis in physics rather than talk of a crisis on its ideological outskirts. But then, physics would have been biology.

19

THE CRISIS IN BIOLOGY

THERE was little in the Marxist heritage that portended serious conflict over biology; certainly there was nothing like Lenin's quarrel with the 'Machists' over new physical theories. Before the Bolshevik Revolution, when Marxists were not directly involved in biological research or in agriculture, they showed a lively interest in grand theories of evolution, but hardly in such technical issues as the mechanism of heredity. No one regarded the few, casual expressions of 'Lamarckism' in Marx and Engels as an essential element of Marxism. Kautsky, Plekhanov, and Lafargue, the leaders of orthodox Marxist thought at the end of the nineteenth and beginning of the twentieth centuries, felt free to welcome the genetics of Mendel and De Vries as evidence that evolution proceeds by dialectical leaps, without any of their comrades, Lenin included, impugning their orthodoxy for doing so.[1]

The Revolution did not immediately push Russian Marxists into the profession of biology, which was left in the hands of 'bourgeois' specialists, or into agriculture, which was left to the peasants with their ancient mixture of private and communal enterprise. But even before the great break, when a violent attack was mounted against both these autonomous areas of un-Bolshevik activities, their anomalous presence in the land that was building socialism provoked much anxious discussion among Russian Marxists. The discussions of the peasant problem can be passed over here, for it was only in the 'thirties that Lysenko and his followers linked the peasant problem with biological theory. But already in the 'twenties the fear of 'bourgeois' specialists, and the corresponding call for the Cultural Revolution on the scientific front gave a new urgency to Marxist interest in biology. Ideology was still the paramount consideration, but, if a new generation of truly red specialists was to replace the 'bourgeois' professors from whom they were getting their professional training, ideology was obliged to wrestle with issues formerly slighted as merely technical. And the Marxists quickly discovered that specialists in biology were not agreed on the solution of these issues.

Perhaps the earliest sign of this process was the discussion aroused by L. S. Berg's publication in 1922, just as intellectual life was recovering from the atrophy of the Civil War period, of a novel theory of evolution: *Nomogenesis, or Evolution Determined by Law.*[2] (The implied opposite was 'tychogenesis', or evolution determined by chance.) Berg, an eminent non-Marxist zoologist and geographer, asserted that a primary purposefulness (*tselesoobraznost'* or *Zweckmässigkeit*), inherent in living things, was the ultimate determinant of evolution. To some Marxists this seemed a vitalist challenge to materialist philosophy, and Riazanov, the editor of the Marxist classics, warned Soviet Marxists against such newfangled theories of evolution. Objecting to Plekhanov's endorsement of mutations as well as to Berg's nomogenesis, Riazanov cited the defence of simon-pure Darwinism by K. A. Timiriazev, the famous plant physiologist who supported the Soviet cause.[3] (Riazanov's lame effort to save Plekhanov's point was that Plekhanov could have used physical quanta rather than biological mutations as evidence of natural development by dialectical leaps.)[4]

On the other hand, V. N. Sarab'ianov, a Communist propagandist who was trying with difficulty to disseminate Marxism in the academic world, objected that mutations had been established beyond doubt, and that Berg's method of explaining them was essentially valid, though obscured by idealist terminology. Teleology and the religious argument from design, Sarab'ianov protested, could not be effectively refuted by a flat denial of purposefulness (*tselesoobraznost'* or *Zweckmässigkeit*) in living nature. He agreed with Berg that chance variations could not be considered the basis of evolution, since Marxism as well as natural science postulated nature's conformity to law (*zakonomernost'* or *Gesetzmässigkeit.*)[5] Nor was he at a loss for a biologist with the proper political sentiments to cite against K. A. Timiriazev. M. M. Zavadovskii, a thirty-two-year-old zoologist who supported the Soviet cause, had just published a tactful warning:

> The books of K. Timiriazev . . . , though splendid in their clarity of thought, wholeness, and skilful organization, either do not reflect at all or reflect to an insufficient degree the enormous upheaval [*sdvig*] that has been projected and in part accomplished in biology in regard to methods of research on evolutionary problems. . . . At the present time we must recognize that selection does not play the part that was ascribed to it formerly; it creates nothing, but merely finds and educes what existed already. Modifications or simple variations do not play a role in the general movement of the evolutionary process, for they are not hereditary.[6]

But M. M. Zavadovskii himself, however sympathetic to the Soviet régime and anxious to save its ideologists from Timiriazev's errors, was not yet ready in 1923 to speak as a Marxist on the philosophical and ideological problems of biology.[7] And the Marxist ideologists venturing into biology were still rather diffident amateurs looking anxiously for politically sympathetic authorities.

By 1925 two centres of attraction for such sympathetic authorities had been created in Moscow: the 'State Timiriazev Scientific Research Institute for the Study and Propaganda of Natural Science from the point of view of Dialectical Materialism' (mercifully reduced to the 'Timiriazev Institute' in ordinary discourse), and the Communist Academy's Section of the Natural and Exact Sciences, which in turn organized Societies of Materialist Biologists and Physicians. Most of the participating biologists appear to have been simply materialists in philosophy and Soviet sympathizers in politics, and they are called Marxist biologists here merely in the sense that they wanted to be distinguished from 'bourgeois' scientists, who, plainly indifferent or hostile to Marxism, still constituted the overwhelming majority of Russian scientists. The minority were Marxists largely in a demonstrative sense: they were willing to participate in the work of institutions whose announced purpose was to fuse Marxism and natural science. In actual fact the pamphlets of popular science issued by the Timiriazev Institute, the transcripts of discussions staged at the Communist Academy, and reports of the work of the laboratories within both institutions reveal little that was specifically Marxist or dialectical materialist.[8] Rather they reveal about fifty or a hundred biologists and physicians, most of them between twenty-five and thirty-five years old, with materialist inclinations in philosophy and a radical bent in politics, going over the familiar conundrums of biology.[9] Given these circumstances, one would expect that theories of heredity and evolution would be the favourite issue, as indeed they were.[10]

Originally there was argument but no sharp conflict over this issue.[11] Until the end of the 'twenties the characteristic attitude among Soviet Marxist biologists was eclectically broadminded. 'Formal', 'Mendelian', 'Weissmanist', or 'Morganist' genetics —it was called by all these names, singly or in various combinations—[12] was favourably regarded as establishing incontestably materialist laws of individual heredity, but it was also criticized for its alleged inability to explain the evolution of species and the role of environment in such evolution. For this reason the 'epigeneticists' or 'mechano-Lamarckists', who sought to establish environmental influences on heredity (as distinct from the universally despised 'psycho-Lamarckists', who sought to prove the influence of will-power on heredity), were generally regarded with sympathetic favour. In a laboratory at the Timiriazev Institute E. A. Bogdanov, a specialist in the nutrition and physiology of livestock, supervised crude efforts to induce hereditary changes in blue flesh flies.[13] And the Communist Academy's Section of Natural and Exact Sciences in 1925 offered a laboratory to Paul Kammerer, the well-known Viennese zoologist whose 'Lamarckist' experiments and writings had raised a storm in Western Europe.

On visiting Moscow and meeting with universal goodwill, Kammerer agreed to establish the laboratory, and perhaps also to head a research department at the Moscow Zoo and lecture at Moscow State University. But returning to Vienna for his books and equipment, he was confronted with evidence of fraud in one of his most famous experiments. He wrote a letter to the Communist Academy, disclaiming responsibility for the fraud, but all the same expressing his inability to survive the scandal, and shot himself.[14] It is indicative of the eclectic atmosphere of the time (1926) that Soviet Marxists who would shortly be implacable opponents of 'mechano-Lamarckism' were among those who had supported the invitation to Kammerer and now paid glowing tribute to him as a martyr of materialist science. I. I. Agol, for example, wrote that Kammerer's 'consistent, monistically materialist position', which drove mysticism out of biology, was the cause of his persecution in bourgeois countries. 'Where else,' Agol exclaimed, 'but in the land of the victorious proletariat could he find comradely sympathy and support for quiet, objective, scientific research?'[15] Agol was not alone in contrasting Soviet broad-mindedness with the savage illiberalism of capitalist countries; the Scopes trial had just reminded the world that there was a law against teaching evolution in Tennessee, 'one of the most enlightened areas of the United States'.[16]

There is some evidence to support the claim of Alexander Sergeevich Serebrovskii that he brought to the biological discussions of Soviet Marxists a sense of irreconcilable conflict between 'mechano-Lamarckism' and 'Morganism'.[17] In 1925, when he was a thirty-three-year-old geneticist at the State Institute of Experimental Biology, he began to frequent the discussions of biology at the Communist Academy. As a geneticist —perhaps the first to participate—he was appalled not only at the outright 'Lamarckism' of some, but also at the tolerant and compromising attitude of the others.[18] Within a few years, during which he grew famous in his specialty and became head of the Moscow Zootechnical Institute's Department of Genetics, Serebrovskii won to militant 'Morganism' a number of the biologists associated with the Communist Academy. Among them were I. I. Agol (a student or colleague of Serebrovskii's at the Zootechnical Institute), the physician S. G. Levit (a young specialist in internal medicine on the staff at Moscow State University), and the refugee German philosopher M. L. Levin (one of the few full-time staff members of the Communist Academy, past forty, listing his specialty as 'methodology and comparative anatomy').[19] They became with Serebrovskii the leaders of the 'Morganist' school of Soviet Marxist biologists, and two, or perhaps all four, perished as 'enemies of the people' in the late 'thirties or 'forties.[20]

But we are here concerned with the beginning rather than the dreadful culmination of the new intransigence. In 1929 there was still no talk of 'enemies of the people' in the biological discussions, although N. P. Dubinin (another student or colleague of Serebrovskii's at the Zootechnical Institute) unwittingly pointed the way in reply to a plea for a synthesis of 'Morganism' and 'Lamarckism':

It seems to me that between Lamarckism and Morganism there can be no synthesis, for the fundamental conceptions of genetics contradict Lamarckism absolutely. Morganism and Lamarckism—these are two opposed world views; the attempt to unite them can lead only to eclecticism [a pejorative in the Soviet Marxist vocabulary]; the struggle between them must proceed to the end, and one or the other must win: either Lamarckism or Morganism.[21]

It is therefore hardly surprising that, at the Conference of Marxist-Leninist Institutions in April, 1929, Agol and the other 'Morganists' present called for the suppression of 'Lamarckist' work in the Communist Academy and the Timiriazev Institute. They linked it with the mechanist philosophical deviation that was being condemned at this Conference, prompting a twenty-three-year-old graduate student of zoology to shout defiance:

. . . I declare that we [epigeneticists or Lamarckists] do not agree with the propriety [*pravomernost'*] of deciding scientific questions by a majority of votes. Therefore, even if our opponents are preparing Kammerer's fate for us, and drive us into a scientific underground, we will continue our work to the last remnant of our strength.[22]

Actually the young man's alarm was considerably exaggerated. The Conference did not vote a formal condemnation of 'Lamarckism', and 'Lamarckist' work, though under a cloud in the period 1929–1932, was allowed to continue in the Communist Academy.[23] But passionate intransigence had been introduced—by its future victims.

The causes of this ominous turn are to be sought not so much in the issues of the biological controversy as in the changing circumstances intensifying certain issues. It was not the biological issues that were thus intensified. They centred on the perennial question whether acquired characters could be inherited, or, as the 'Morganists' preferred to specify, whether adaptive changes in heredity could be induced by external forces operating only on the soma ('somatic induction'), or by simultaneous and parallel alteration of both soma and germ plasm ('parallel induction'). A. S. Serebrovskii and the militant 'Morganists' categorically denied the possibility of both, and denounced as 'Lamarckists' anyone who would not endorse this denial. Indeed, they extended their denunications further. Even M. M. Zavadovskii, whose deference to modern genetics was noted above, was lumped with the 'Lamarckists' for a paper in 1929 questioning the belief that the genes have 'a monopolistic role' in determining cell differentiation within a growing organism, and for doubting whether genetics could provide a general theory of biology without amendment and supplement by other biological disciplines (embryology was his own).[24] There were other evidences too that the familiar rivalry of special but related disciplines, the dreams and fears of academic imperialism, were at work in the clash of 'Morganism' and 'Lamarckism'.[25]

These rivalries and fears were enormously sharpened by the special circumstances of Soviet history in the late 'twenties, but that period saw no important change in the biological issues considered as such. Indeed, it is the present writer's feeling that there could have been no significant change in this respect. New experimental data appeared constantly without budging either party. Harrison's induction of melanism in moths by feeding them metallic salts characteristic of smoke deposits in industrial areas; Umeya's alteration of voltinism in silkworms by changes of temperature and by transplanting ovaries; Goldschmidt's induction of mutations in Drosophila by heat—these and many other less famous experiments were hailed by the 'Lamarckists' as substantiating their views on the nature of variations, only to be shrugged off as imprecise and inconclusive, or to be subjected to conflicting interpretation, by the 'Morganists'.[26] When H. J. Muller induced mutations in Drosophila by X-rays, proceeding in strict accordance with the initial assumptions of 'Morganism', A. S. Serebrovskii reported Muller's paper in *Pravda* under the headline, 'Four Pages that Shook the Scientific World'.[27] It was the 'Lamarckists'' turn to argue that their opponents' jubilation was founded on question-begging and hasty generalization. Indeed, some 'Lamarckists' claimed Muller's experiment as support for *their* point of view.

Whether or not rigorous analysis would support these mutual charges of question-begging, the historical pattern of the debate suggests to this layman that empirical data could not resolve the argument. At any rate they did not resolve it.

Nor did abstract philosophy. On that level there was no clearcut separation between 'Lamarckism' and 'Morganism', despite Agol's effort to prove that there was.[28] The 'Lamarckists' might charge the 'Morganists' with undermining determinism by dissolving evolution into a chaos of accidents, but the customary 'Morganist' reply—they compared their picture of evolution to the kinetic theory of matter—was simply another version of the deterministic outlook. In the tradition of Epicurus the 'Morganists' argued that their stress on accident eliminated all taint of anthropomorphism from their understanding of functional adaptation in living things. They expressed sarcastic doubt that the 'Lamarckists' could likewise dissociate themselves from 'the shallow teleology of Wolff', in Engels' mocking words, 'according to which cats were created to eat mice, mice to be eaten by cats, and the whole of nature to testify to the wisdom of the creator'.[29]

To such charges of teleology the characteristic 'Lamarckist' reply was that they were trying to reduce evolution not to the purposeful adaptation of wilful beings but to a pattern of physical and chemical interaction between organisms and their environment. In such a demonstration of mechanistic thought they hardly differed from the 'Morganists', who proved their own mechanism by trying to reduce evolution to combinations and transformations of genes, regarded as the material particles or atoms of heredity.[30] Moreover, both 'Morganist' and 'Lamarckist' biologists hesitated to go beyond professions of elemental mechanism; they avoided intricate and extended examination of the philosophical issues implicit in their dispute. They argued the conflicting claims of 'autogenesis' and 'epigenesis', for example, without reference to a strikingly relevant issue in the contemporaneous philosophical controversy between mechanists and Deborinites. That was the issue of 'qualities' and 'reduction', of explaining the emergence of new qualities without abandoning either the view of natural processes as continuous or the method of reducing complex phenomena to the action of homogeneous elements. One searches in vain for an extended examination of the biological controversy in terms of this issue.[31]

If the biological and the abstract philosophical issues do not adequately explain why A. S. Serebrovskii's impatience with 'Lamarckism' was generalized into a reign of angry intransigence at the end of the 'twenties, neither does any single issue in social philosophy or ideology. Such issues were a favourite subject of debate from the beginning of the biological discussions in the Communist Academy. Was eugenics hopelessly bourgeois and reactionary or did it have important implications for Soviet socialism? Was affirmation or denial of the inheritance of acquired characteristics the best rebuttal of aristocratic disdain for the lower classes? Denial, said some, looking back to the generations of peasants and proletarians who had been kept in a starved, diseased, and illiterate condition. Affirmation, said others, looking ahead to the improvement of the environment under socialism, and hoping that this could effect permanent improvements in human beings, the creation of a new man. 'Are we,' exclaimed a 'Lamarckist', 'slaves of the past or creators of the future?'[32] By such an antinomy he invited comparison with the utopian socialists, and he was duly lectured on the Marxist understanding of freedom as the recognition of historical necessity.[33] Such connections between biology and social philosophy were favourite issues, but it would be a mistake to fasten on any one of them as the main reason for the rise of militant 'Morganism' and for its ephemeral ascendancy at the end of the 'twenties. The biological disputants did not. Agol, indeed, when demanding the suppression of 'Lamarckism', explicitly disdained as demagogy efforts to show that 'Lamarckism' or 'Morganism' supported one or another position in social philosophy.[34]

What generalized and intensified and won a brief ascendancy for the geneticist's professional impatience with 'Lamarckism' was its quality of irascible self-assurance. With the onset of the great break intransigence or irreconcilability (*neprimirimost'*) became a prized Bolshevik virtue as never before. In 1929 and 1930, as the drive was mounted for a rapid scientific change-over from 'bourgeois' to red specialists, the very fact that A. S. Serebrovskii's followers described 'Morganism' and 'Lamarckism' as two opposed *Weltanschauungen*, our or the other of which must win out in the struggle for the minds of biologists, gave them the advantage. They had a programme of struggle for the young red specialists who were being pushed to the fore in biology as in other fields, while the 'Lamarckists', conceding the value of genetics within its field and speculating (often at cross purposes) on the best approach to a synthetic theory of biology, appeared by contrast to be feckless compromisers.[35]

As the reader has already seen, Agol was unsuccessful in his attempt to win a formal condemnation of 'Lamarckism' from the Conference of Marxist-Leninist Institutions in April, 1929. But the Timiriazev Institute, the 'Lamarckists'' chief centre, was shut down, and the Communist Academy indicated its preference for 'Morganism', even though it tolerated the continuation of a 'Lamarckist' laboratory within its walls.[36] Indeed, as the great break proceeded, the 'Morganists' acted more and more as if they *had* been formally designated the official interpreters of Marxism in biology. The Party called on theorists for practical aid in this time of crisis, and the 'Morganists' responded with projects for a mass campaign to spread knowledge of genetics among agriculturists, who were overwhelmingly 'Lamarckist' in their point of view.[37] More and more the 'Morganists' pointed with pride to the work of the Institute of Applied Botany, in spite of the fact that its Director, N. I. Vavilov, still kept himself aloof from Marxist discussions.[38] (In 1929 he was induced to publish a brief report of his work in the chief journal of Marxist thought, but there was not the slightest reference to Marxism in this article.)[39] And late in 1929 A. S. Serebrovskii got out a collection of articles on 'anthropogenetics', which he prefaced with a programmatic essay stressing the practical importance of the subject in the construction of socialism.[40]

He called for intensive studies to establish the precise correlation between human genes and hereditary characters, and to map the distribution of good and bad genes in Soviet society. Such knowledge, he argued, would first of all make possible a great reduction in hereditary diseases and consequently the realization of a five-year plan in two and a half years.[41] Serebrovskii disdained negative eugenic measures such as the sterilization of defectives; only 'an Assyro-Babylonian despotism' could carry them out on an effective scale.[42] But he had a programme for selective breeding that could be realized in a free socialist society, and indeed only in such. Soviet society, he reasoned, was destroying the closed, bourgeois family and with it the desire to have children only from a beloved spouse. Of course sexual pleasure would remain, 'and with any structure of society a man and woman entering marriage will have every reason to esteem it. There is of course nothing reprehensible in that'.[43] But women would be educated within a couple of generations to want babies only by artificial insemination with 'recommended sperm'. Indeed, they would be trained to feel that conception in the old manner, 'the disruption of the complex plan [of selective breeding], figured out for many generations in advance, is a deed that is anti-social, amoral, unworthy of a member of a socialist society'.[44] Taking into account man's enormous capacity to manufacture sperm and the excellent technique of artificial insemination, 'which is widely applied just now only in horse and sheep breeding', Serebrovskii forecast that 'from one outstanding and valuable producer it will be possible to obtain up to 1,000 or even 10,000 children'.[45]

Of course, he had overreached himself. In June, 1930, *Izvestiia* carried a long burlesque of his essay by the famous newspaper poet Demian Bednyi (the Bolshevik Edgar Guest, in Professor Simmons' phrase).

> So far, you know, we've been astray,
> The way we've brought our kids forth.
> But soon, you'll see, will come the day
> For socialistic childbirth.
> A social labour it will be
> And shock brigades, I fancy,
> Will rub themselves callosities
> In their productive frenzy.[46]

In other stanzas Bednyi seemed to scoff at the very notion of genes, seeing it as a carry-over from the aristocratic habit of proving virtue by tracing genealogies;[47] but in still others he implied acceptance not only of the science of genetics but even of eugenics, provided only that they aid 'the construction of socialism'.[48] Nevertheless, the net impression that he gave of

Serebrovskii and the 'Morganists' was one of bourgeois professors arrogantly admonishing the Party on the hopelessness of its drive to transform society by Five Year Plans and collectivization, and condescendingly instructing the Party on the only feasible programme of basic social improvement: eugenics.

This defamation of one of the pioneers in the interpenetration of Soviet Marxism and natural science was not an isolated occurrence in 1930. In the same year A. M. Deborin and the rest of the 'philosophical leadership', which had for several years led the campaign to win acceptance for Marxism as a universal philosophy applicable to all fields of learning, were denounced as pinchbeck Marxists who had only seemed to aid the Party but had actually failed and hindered it. The most militant Bolsheviks, both at the top and at the bottom of the Party pyramid, took up arms against the mature Marxist intellectuals just below the top, who had been somewhat autonomous, somewhat liberal leaders of Marxist thought in many fields during the 'twenties. The 'Morganists', it turned out, who had gained a brief ascendancy by their spirit of intransigence, were not intransigent enough.

Agol had fatally compromised the 'Morganism' that Serebrovskii had taught him by asserting that it was the realization in biology of the Marxist philosophy that he had learned from Deborin.[49] Though neither of the principals, Serebrovskii in biology, Deborin in philosophy, had strongly endorsed Agol's equation, they had not disavowed it,[50] and its logic seemed unassailable to the Bolshevizers. Was not Serebrovskii trying to palm off as Marxist a general theory of biology that was characteristic of 'bourgeois' geneticists, many of whom were quite hostile to Bolshevism?[51] And was not Deborin, by his resistance to the rule of partyness in philosophy and science, assisting in just such cowardly efforts to make peace with 'bourgeois' scientists on their terms? Late in 1930 a conference at the Communist Academy, calling for 'the Bolshevization of science,' condemned

> pseudo-Marxist tendencies of the type . . . of Serebrovskii's in biology . . . which are a form of adaptation to Marxism-Leninism in the conditions of the proletarian dictatorship, and reflect in essence the pressure of the class enemy on the ideology of the proletariat.[52]

The young red specialist in biology still had much to learn from 'bourgeois' professors, but, the conference resolved, he must never again make the mistake of accepting as Marxist the professors' general theories of life. 'Lamarckism' was not restored to favour; it too was a 'bourgeois' theory. The truly Marxist theory of biology was yet to appear. It would be known *inter alia* by its belligerent uniqueness and by its practical usefulness to the construction of socialism.[53]

The condemnation of 'Morganism' as a general theory of biology was initially compatible with high respect for the science of genetics as a special discipline.[54] Indeed, the strident demand of the Bolshevik militants that science be of practical service to the construction of socialism probably served at first to enhance the reputation of genetics, whose contribution to the improvement of agriculture was not questioned. For the collectivization of agriculture, plunging the Bolsheviks as never before into the problems of raising food and fibre, was working a basic change in the Soviet Marxist interest in biology. To raise yields or to perish, that was the question that made both 'Lamarckism' and 'Morganism' seem not only deviant but meaningless:

> Instead of concentrating scientific-research thought on the nodal problems of contemporary biology, which have been posed in the course of socialist construction, in the course of the development of science, some Soviet biologists have up to now trailed along after the themes obtruded on us by bourgeois biological thought. Thus, for example, the mechano-Lamarckist and the Menshevizing idealists exalted the question of the inheritance of 'acquired characters' to the role of the nodal problem of contemporary biology; thus at one time, thanks to the efforts of the Menshevizing idealists, questions of eugenics were elevated to the role of the nodal problem whose solution would entail the fulfilment of the Five Year Plan in two- and one-half years, etc.[55]

The Communist Academy's newly established Institute of Biology followed this scornful dismissal of the old theme with an announcement of the new one: 'man as a factor of evolution.'

> Without considering what man does in various socio-historical formations, without considering what the possibilities are of influencing the animal and plant world in the conditions of a planned

socialist economy, it is not possible either to utilize effectively the accomplishments of world science (and above all of evolutionary theory), or to design new researches consonant with a planned socialist economy.[56]

The Marxist general theory of biology would accordingly have to be not only a suitable philosophy for red specialists, enabling them to participate in world science without danger of ideological corruption; it would also have to be an instrument for raising yields on Russia's straitened collective farms, where peasant attitudes towards the new system varied—in proportions we can only guess at—from hostility through resignation to dedicated hope. Such a general theory of biology would necessarily be compounded of elements taken not only from biology and Marxist philosophy, but also from the beliefs and habits of the practical men of Soviet agriculture and politics.

Thus by the early 'thirties Soviet Marxism was far along on the way to Lysenko's Michurinism. Several turns had still to be taken before the destination would come in view. Lysenko had as yet no connection with Michurinism, a mass movement for the improvement of fruit and berry culture, showing no signs of serious conflict with academic biology. Indeed, during the early 'thirties there were signs of harmonious co-operation between the leaders of the Michurinist movement and the great network of biological research institutions headed by N. I. Vavilov.[57] In 1931 Lysenko emerged from obscurity as the leader of an analogous but separate mass movement for the improvement of wheat culture, similarly without signs of conflict with academic biology. He was merely a seed man and plant physiologist who offered no general theory of biology and specifically denied that his famous 'vernalization' altered the heredity of plants.[58] (Here, perhaps, he showed the influence of the recent eclipse of 'Lamarckism'.) But the Commissarist of Agriculture ordered a multitude of collective farms to try Lysenko's process as a 'mass experiment', and within a couple of years, as this rallying of collective farmers under the banner of science was proclaimed a resounding success and extended further, Lysenko's work began to intrude upon the theoretical discussions of Marxist biologists.[59]

It was initially brought forward in discussions of ecology,

which had taken the place of genetics as the crucial discipline for the elaboration of a general theory of biology; and the great significance that Lysenko's earliest advocates claimed for his work was so nebulous as to provoke no comment or criticism.[60] How Lysenko's work was transformed into a general theory of biology that could not be ignored, how the discussions moved round to centre once again on genetics, how Soviet Marxist biologists divided once again into hostile blocs, this time in an arena slippery with blood—these are matters for a future study. Detailed conclusions on the Soviet Marxist involvement in biology are best left to that future study, but it should already be clear that the widespread tendency to regard the texts of Marx and Engels as the chief determinant of this involvement is very much mistaken. It was the programme of Cultural Revolution that launched Soviet Marxism on its quarrelsome search of an appropriate theory for red specialists in biology; and it was the drive for collectivized agriculture, Stalin's cataclysmic 'revolution from above', that raised the storms on which this search has been violently tossing since the mid-'thirties.

20

CONCLUSION

IN its first phase Soviet Marxism did not settle the inherited conflict within dialectical materialism between positivism and metaphysics. For a time one side was ascendant, and then the other, but in the end both were condemned. Philosophers were exhorted to develop dialectical materialism as the empirical philosophy (more precisely: the philosophy of 'practice') that transcends both positivism and metaphysics. But the narrow partyness that triumphed in the great break turned such exhortations into a cruel joke on the philosophers. In effect, only the Party's chief had the right to develop dialectical materialism further, and he was not greatly interested in the refinements and subtleties of epistemology. He was passionately interested in monolithic unity among his followers. The result was a staggering anomaly: dialectical materialism had become a caesaropapist dogmatism acclaimed with enforced unanimity as a collectivist empiricism. Such an anomaly staggers the intellect, but Soviet Marxism no longer appealed very much to the intellect. It had become a ritualistic incantation of the one true *Weltanschauung* that inspired the one true Party in its desperate struggle. Emotional commitment to the Party was the essence of the incantation; its intellectual content was vague enough to accord with any current shibboleth.

It is not hard to find this Stalinist version of dialectical materialism implicit in the original theses of Marxism-Leninism. But it takes the wilfulness of a Stalin to assert that *only* this version was implicit in Marxism-Leninism. For nearly three decades dialectical materialism, partyness included, had been interpreted in different, often contrary ways by Lenin and his followers. Those who hunger for inevitability may fasten on Lenin's most basic idea—the disciplined vanguard leading the socialist revolution in a backward country—and derive from it the necessity of the Cultural Revolution, whose exigencies in turn produced the Stalinist version of dialectical materialism. Such reasoning is persuasive, but inevitability eludes it, for one must specify: the shifting exigencies of the Cultural Revolution, as perceived by changing Bolshevik minds, never omniscient, never completely rational, inflamed beyond reason during the great break—such were the forces that produced the Stalinist version of dialectical materialism.

To recognize the divergent possibilities in Leninism, and still to read the Stalinist version as the inevitable or the most likely outcome, one might turn dialectician and imagine the opposed implications of Leninism at war within Bolshevik minds, unhinging them, producing at last the intellectual catalepsy that would be the hallmark of Soviet Marxist philosophers in their next phase. At an increasingly tumultuous pace the Bolsheviks were using political power, the state, the 'superstructure', to transform economic reality, society, the 'base'. Striking increasingly violent blows at the refractory world to make it fit their ideas, they were striking at their own conviction that revolutionaries are doomed who pit their minds' fragile should-be against the world's brute must-be. Requiring intellectuals to shout encouragement in fervent unison, finally forcing natural scientists to profess Marxism at gun-point, they were doing violence to their own assurance that scientists would spontaneously recognize Marxism as the logical extension of science into human affairs. Pursuing recalcitrant scientists into their special disciplines, demanding the 'reconstruction' of these disciplines as proof of the scientists' conversion, the Bolsheviks were casting doubt on their own faith that dialectical materialism formulates the methods that have brought success to scientists in their cognition of the world. Worshipping science, the Bolsheviks had to raise cries of a crisis in science. To make dialectical materialism an effective fighting creed in a war against ideologically alien scientists, they had to renounce faith in it as an objective description of the way that scientists discover the natural order. The union of revolutionism and scientism (*nauchnost'*), which Lenin had described in 1894 as the chief power of attraction in Marxism, could hardly be maintained in the face of these contradictions. To believe in one part of their doctrine, the Bolsheviks had increasingly to disbelieve another. At the maddening climax of most intense belief and disbelief they shut off further discussion, 'disarmed' their intellects (the phrase was a catchword of the great break), made their minds wax in the hands of the Central Committee and the chief.

Maybe such flights of dialectical fancy dignify a reality that was actually sordid. The young zealots who came to the fore in the great break may have had true monoliths for heads, incapable of comprehending the contradictions just noted, much less of agonizing over them. But it is a plain fact that dialectical materialism was never more contradictory than at the moment when it was removed from public debate. Even the dullest philosopher of science could not have avoided repeated collisions with one of the principal contradictions, though he might have failed to comprehend it. Repeated declarations of faith in a petrified collection of rather vague formulas,which Stalin soon codified in a brief catechism, were not enough. Without making heretical alterations and refinements in these formulas, philosophers and natural scientists were expected to use them for the 'reconstruction' of the particular sciences. It was as if Bossuet, defining a heretic as one who has an opinion, had required opinions of the scholars in his flock.

What followed, the second phase in the interaction of Soviet Marxism and natural science, can hardly be inferred from the bundle of contrary indications that marked the close of the first phase. Fearing senior natural scientists as ideological aliens behind their new Marxist masks, and restricting philosophers to beading quotations in standard patterns lest they undermine the Party's monolithic unity, how could Soviet Marxists *have* further debate on the philosophy of natural science? Calling on scientists to prove the genuineness of their conversion by participating in the 'reconstruction' of the sciences, pointing to 'practice' as the criterion of truth, and defining it so broadly as to include everything from sense data through technology to proven scientific theories and the Party's policies, how could

Soviet Marxists *avoid* further debate on the philosophy of natural science? The contradictions that dialectical materialism inherited from the pre-revolutionary period seem almost trifling by comparison with those that were added in the first phase of Soviet Marxism and handed on to the second.

NOTES, BIBLIOGRAPHY, INDEX

ABBREVIATIONS

BSE	*Bol'shaia sovetskaia entsiklopediia*, edition specified in citataion
DVP	*Dialektika v prirode*
EIM	*Estestvoznanie i marksizm*
FNIT	*Front nauki i tekhniki*
KN	*Krasnaia nov'*
KPSS	*Kommunisticheskaia Partiia Sovetskogo Soiuza v rezoliutsiiakh i resheniiakh s'ezdov, konferentsii i plenumov Ts. K.* 7th ed., Moscow, 1953.
LM	*Letopisi marksizma*
NR	*Nauka i nauchnye rabotniki SSSR* (See Bibliography for significance of the date of each volume.)
PIR	*Pechat' i revoliutsiia*
PZM	*Pod znamenem marksizma*
RIK	*Revoliutsiia i kul'tura*
SK	*Sputnik kommunista*
Sorena	*Sotsialisticheskaia rekonstruktsiia i nauka*
VKA	*Vestnik kommunisticheskoi akademii*
VM	*Voinstvuiushchii materialist*
VSA	*Vestnik sotsialisticheskoi akademii*

The titles of other books and articles have been shortened in citation. The reader who wants to see a full title should consult the Bibliography, where he will also find an English translation of each title cited. The citations from Lenin, *Sochineniia,* are from the fourth edition, unless otherwise specified.

PREFACE

[1] See Feuer, 'Dialectical Materialism and Soviet Science', *Philosophy of Science*, April, 1949, pp. 116 *et passim.*

[2] Baldwin, *Dictionary of Philosophy*, II, 45.

[3] Runes, editor, *Dictionary of Philosophy*, p. 191. Cf. also the relevant articles in H. Schmidt, *Philosophisches Wörterbuch.*

[4] J. J. Thomson, *The Atomic Theory*, p. 3.

[5] Cf. Nagel, 'The Meaning of Reduction in the Natural Sciences', in Stauffer, editor, *Science and Civilization.*

1. ORTHODOX MAXISM AND NATURAL SCIENCE

[1] This particular comparison is made by Brehier, *Histoire de la philsosophie*, II, 908.

[2] Acton, *The Illusion of the Epoch*, p. 271.

[3] Cited in Plekhanov, *Izbrannye*, I, 490–491, and II, 303–304.

[4] Lenin, *Sochineniia*, I, 308. Cf. Rosa Luxemburg's comment: 'Marxism allied theoretical knowledge with the revolutionary *élan* of the proletariat; the one illuminated and fructified the other. Both aspects belong equally to the inner core of Marxism; if they are separated Marxism becomes a pitiful caricature of itself.' Quoted in Frölich, *Rosa Luxemburg*, p. 66.

[5] Cf. G. D. H. Cole, *Socialist Thought*, II, Chapter XI, for a critical account that recognizes the appeal of Marxism's claim to being scientific. By contrast H. Sée seems rather naïve in his remark: 'There is an obvious contradiction, a "dualism" [between the scientific and revolutionary aspects of Marxism], that a mind as sound as Marx would have repudiated, without any doubt, if the Hegelian dialectic had not accustomed him to endure similar contradictions. . . .' Sée, *Materialisme historique*, p. 48.

[6] Cf. Cole, *Socialist Thought*, and cf, also below, pp. 6 *et seq.*

[7] Cf. Professor Randall's description of the common characteristic of naturalistic philosophers: agreement 'that there was to be no sharp difference in intellectual methods in treating man and the other aspects of Nature of which he was taken to be a part'. See Krikorian, editor, *Naturalism*, p. 357. For Lenin's recognition that Marxist materialism is akin to naturalism, see Lenin, *Filosofskie tetradi*, p. 73. Cf. also H. Marcuse, *Reason and Revolution*, especially pp. 323 *et seq.*, for a sharp distinction between *dialectical* materialism and naturalism; Marcuse uses the latter term to signify a philosophy that aspires to be scientific but not revolutionary.

[8] See Marx and Engels, *The German Ideology, passim*, for the first major statement of this point of view. And cf. Marcuse, *Reason and Revolution.*

[9] Marx, *Capital*, I, 26.

[10] Marx and Engels, *Correspondence*, p. 210. For the German, see Marx and Engels, *Briefwechsel*, III, 413. Cf. also Marx and Engels, *Correspondence*,

p. 313; and Marx's ridicule of John Stuart Mill for the latter's lack of dialectics, in Marx, *Capital,* I. 654.

[11] Engels, *Anti-Dühring,* p. 31. It must be borne in mind that Engels read this work in manuscript to Marx. This fact, and the absence of any objections by Marx in the *Briefwechsel,* make it possible to regard *Anti-Dühring* as a statement of Marx's philosophy as well as Engels'.

[12] Cf. for example, Marx and Engels, *The German Ideology,* pp. 15 *et passim*; and Engels, *Ludwig Feuerbach,* p. 59.

[13] Engels, *Anti-Dühring,* p. 155.

[14] *Ibid.,* p. 15.

[15] See Riazanov, 'Marks i Engel's o "Dialektike prirody"', in Engels, *Dialektika prirody* (1930 edition), pp. xv–l.

[16] Marx and Engels, *Briefwechsel,* III, 218–219. Cf. also, IV, 476, 480 (and Marx and Engels, *Correspondence,* pp. 322–323), for Engels asking Marx's opinion on some ideas concerning the dialectics of nature. Marx replied that he must consult 'authorities', and turned the letter over to Schorlemmer, a chemist who shared the philosophy of Marx and Engels.

[17] Marx and Engels, *Briefwechsel,* III, 424–425.

[18] *Ibid.,* pp. 430–431 and 433–435.

[19] *Ibid.,* p. 432.

[20] Marx and Engels, *Sochineniia* (2nd ed.) II, 139.

[21] Marx, *Capital,* I 25. Cf. also Marx's scorn for the Hegelians to whom philosophy was a substitute for positive knowledge: 'they understood nothing but could write about everything'. Quoted in Marx-Engels-Lenin, *O biologii,* pp. 212–213.

[22] He referred to himself in a double capacity as 'a Party man' and 'a scientific man' in Marx and Engels, *Correspondence,* p. 313. Cf. also p. 105 for Marx ridiculing Lassalle for allegedly not being aware that 'to bring a science by criticism to the point where it can be dialectically presented is an altogether different thing from applying an abstract ready-made system of logic to mere inklings of such a system'.

[23] Marx, *Letters to Dr. Kugelmann,* p. 41.

[24] Marx and Engels, *Correspondence,* p. 102; *Briefwechsel,* II, 341.

[25] See, e.g., Marx, *Capital,* I, 337–338. Engels' addition to the footnote on p. 338 is justified by the exchange of letters between him and Marx on this matter. See Marx and Engels, *Correspondence,* pp. 220–223, and *Briefwechsel,* III, 470–473.

[26] Engels, *Dialectics of Nature,* pp. 152–153. The translation has been altered somewhat to accord more closely with the Russian version in Engels, *Dialektika prirody* (1952 edition), p. 159. Cf. pp. iii-v, for the dating of this passage.

[27] See Mehring, *Karl Marx,* pp. 531 *et seq.,* and Gustav Mayer, *Friedrich Engels,* pp. 236 *et seq.*

[28] Marx and Engels, *Correspondence,* p. 344.

[29] Engels, *Anti-Dühring,* p. 15.

[30] *Ibid.,* pp. 15–16.

[31] *Ibid.,* p. 17.

[32] *Ibid.*

[33] *Ibid.*, p. 19.

[34] See Mayer, *Friedrich Engels.*

[35] Engels, *Dialectics of Nature*, p. 155. The translation has been altered to accord with the Russian version in Engels, *Dialektika prirody* (1952 edition), p. 161. See pp. xiv–xviii, for reasons to believe that this version is the best so far.

[36] Engels, *Dialectics of Nature*, p. 183.

[37] *Ibid.*, p. 243, and p. 165 in the Russian edition.

[38] *Ibid.*

[39] Engels, *Dialectics of Nature*, p. 155.

[40] See the numerous references to Clausius in the Index of both English and Russian editions.

[41] Engels, *Dialectics of Nature*, pp. 211–213.

[42] *Ibid.*, pp. 24, and 211–213.

[43] See, e.g., Mises, *Positivism*, pp. 266–267.

[44] Engels, *Dialectics of Nature*, p. 213.

[45] Engels, *Dialektika prirody* (1952 edition), p. 26.

[46] Marx and Engels, *Briefwechsel*, III, 94–95.

[47] See Marx and Engels, *Correspondence*, pp. 125–126; and Engels, *Anti-Dühring*, pp. 77 *et passim*; and *Breifwechsel*, II, 547.

[48] See *Briefwechsel*, III, 547; and Engels, *Anti-Dühring*, *passim.*

[49] Marx-Engels, *Correspondence*, p. 126. Cf. on p. 125 Marx's statement that 'Darwin's book. . . . serves me as a basis in natural science for the class struggle in history'. This may be interpreted to imply that Marx was a 'social Darwinist' but, in the view of the present writer, such an interpretation cannot be sustained.

[50] Darwin's letter to Marx was first published in Russian translation, in *PZM*, 1931, No. 1–2, pp. 203–204. For an English version, see *The London Times*, May 18, 1931; and *Labour Monthly*, November, 1931, p. 702.

[51] See Engels, *Anti-Dühring*, pp. 79–83.

[52] The most well-known of such remarks occurred in the fragmentary essay, 'The Part Played by Labor in the Transition from Ape to Man'. This was originally published the year after Engels' death, in *Neue Zeit*, 1896, Jahrgang XIV, Bd. 2, pp. 545–554. For a convenient English version, see Engels, *Dialectics of Nature*, pp. 279–296. Cf. also *Anti-Dühring*, pp. 79–83.

[53] See Cole, *Socialist Thought*, II, Chapter 15, and III, *passim.*

[54] Reported by Engels in a letter to Lafargue. See Engels and Lafargue, *Correspondance*, II, 407.

[55] See Gay, *The Dilemma*, especially Chapter VI.

[56] See, e.g., Lenin, *Materialism* (1947 edition), pp. 10, 257.

[57] Bogdanov, for example, was a revisionist who tried to make a single universal science the basis of socialism.

[58] Lenin had considerable respect for Lafargue as a philosopher of orthodox Marxism. See, e.g., Lenin, *Materialism*, pp. 205–206. For an important case of Lafargue's influence on Soviet Marxism, see the account of Tseitlin below, Chapter 10, pp. 160 ff. For the Deborinite Maksimov

listing Lafargue ahead of Lenin as an orthodox philosopher of natural science, see Maksimov, 'K voprosu', *PZM*, 1924, No. 4–5, p. 138.

[59] Lafargue, *Social and Philosophical Studies*, pp. 56, 76.

[60] *Ibid.*, p. 74.

[61] For a French Marxist arguing that Marxism is the legitimate continuation of Comtian philosophy, see Semkovskii, ed., *Istoricheskii materializm* (1922 edition). pp. 244–261.

[62] Mehring. *Zur Geschichte der Philosophie*, p. 239. This part of the book is a reprint of an article published in 1909.

[63] Cf. Cole, *Socialist Thought*.

[64] *Gruppa 'Osvobozhdeniia truda'*, No. 5, p. 227.

[65] Kautsky, 'Ein Brief Über Marx und Mach', *Der Kampf*, 1909, No. 10, p. 452.

[66] *Ibid.*

[67] For evidence of Lenin's high regard for Mehring and Kautsky as philosophers, see, e.g., Lenin, *Materialism*, pp. 206–207, 255, 369. As a result of Lenin's high opinion, Kautsky's early philosophical works have been reprinted several times in the Soviet Union, in spite of Kautsky's hostile attitude towards Bolshevism. Mehring is still held in high regard by Soviet Marxists.

[68] Cole, *Socialist Thought*, III, Chapter 9. This interpretation is of course not original to Cole. Cf. Masaryk, *Spirit*, II, 220 *et passim*, for a similar analysis. Cf. also the subtle version of this interpretation in Marcuse, *Soviet Marxism*, pp. 147 ff.

[69] One of the earliest statements of revisionism—antedating Bernstein's—was Struve's *Kriticheskie zametki*, published in 1894. See Lenin's response in Lenin, *Sochineniia*, I, 315–484, and Plekhanov's somewhat belated response in Plekhanov, *Izbrannye filosofskie*, II, 504 *et seq.* For another, and more clearly philosophical, work of the Legal Marxists, see Berdiaev, *Sub'ektivizm*, published in 1901 with a long Preface by Struve. See especially pp. 115–116, for Berdiaev's rejection of dialectical materialism.

[70] See, e.g., Vandek and Timosko, 'Vstupitel'naia stat'ia', in Plekhanov, *Protiv filosofskogo revizionizma;* or Shchipanov, editor, *Iz istorii russkoi filosofii*, pp. 656 *et seq.*

[71] This was in the Introduction and notes that he wrote for the Russian translation of Engels' *Ludwig Feuerbach*. A convenient reprint that gives the variant forms of both editions is in Plekhanov, *Izbrannye filosofskie*, I, 451–503.

[72] See Bazarov, *et al.*, *Ocherki po filosofii marksizma*, pp. 67 *et passim.*

[73] Lenin, *Materialism*, p. 237 or *Sochineniia*, XIV, 219–220.

[74] See Lenin, *Sochineniia* (3rd ed.) XIII, 329–333, where Aksel'rod's review is reprinted. It was originally published in *Sovremennyi mir*, 1909, No. 7.

[75] Iushkevich, *Stolpy*, pp. 56–62.

[76] Bazarov, *Na dva fronta*, p. xxiii *et passim.* Cf. also Bogdanov, *Vera i nauka*, which was unavailable to the present writer. Secondary accounts of it show that Bogdanov made the same point. See Luppol, 'Materializm i

empiriokrititsizm', *PZM*, 1927, No. 1; and Sitkovskii, 'Lenin', in Kommunisticheskaia Akademiia, *Dvadtsat'piat' let*, pp. 117–118.

[77] See Plekhanov, *Izbrannye filosofskie*, I, 480–481. To be sure, this withdrawal did not change Plekhanov's essential view. Cf. also Deborin, 'Filosofiia', *Golos sotsial'demokrata*, asserting that the 'sensuous qualities of light, sound, taste, smell, etc., cannot be considered a reflection of things, as the naive realist thinks'. He concludes with a somewhat Lockian distinction: 'There exist perceptions [*predstavleniia*] that are reflections of objects themselves, and perceptions that reflect only relations between objects ' The reader will note that Deborin in the last sentence uses the term 'reflect' to describe the relation between perceptions and objects.

[78] Marx and Engels, *German Ideology*, p. 197.

[79] Engels' pointing to alizarin as the refutation of an unknowable thing-in-itself recurs time and again in the writings of Plekhanov and Lenin. See Engels, *Ludwig Feuerbach*, pp. 22–23.

[80] Lenin, *Sochineniia*, XXXIV, 338.

[81] Cf. Lunacharsky's comment that Plekhanov knew the German idealists very well, but 'I think that in general among the predecessors of Marxism, Plekhanov's everlasting friends, Holbach and Helvetius, were more akin to his somewhat Bazarovian mind than the great idealists'. Lunacharsky, *Revoliutsionnye siluety* (2nd ed,), pp. 41–42. 'Bazarovian' is a reference to Turgenev's famous character in *Fathers and Sons*.

[82] For the recurrence of his disagreement with Plekhanov, see Lenin, *Filosofskie tetradi*, especially pp. 173 and 327.

[83] *Ibid.*, p. 328.

[84] See, e.g., *ibid.*, pp. 193 and 202–205. It should be borne in mind that the concept of the priority of 'practice' has roots in Hegel as well as in the positivist elements of Marxism.

[85] Lenin, *Sochineniia*, XIV, 238, or *Materialism*, p. 257.

[86] See Lenin, *Materialism*, Chapter V.

[87] *Ibid.*, p. 267.

[88] Lenin, *Sochineniia*, XIV, p. 249, or *Materialism*, p. 269.

[89] See, e.g., Deborin, *Sovremennye*, p. 26.

[90] For one of many examples, see A. K. Timiriazev, 'Lenin', *PZM*, 1924, No. 2, pp. 224–225. For the basis of his argument, see Lenin, *Sochineniia*, XIV, 294, *et passim.*

[91] *Ibid.*, pp. 274–275, *et passim.*

[92] See Semkovskii, *Dialekticheskii materializm i printsip otnositel' nosti*, p. 70. For the contrary interpretaion of Lenin and of physics, see Mitkevich, *Osnovnye fizicheskie vozzreniia* (2nd ed.), p. 64.

[93] Plekhanov, 'Predislovie', in Deborin, *Vvedenie v filosofiiu*, pp. 32–34.

[94] Lenin, *Sochineniia*, XIV, 239.

[95] 'The struggle for existence will last as long as the world we live in, and nothing can do away with it; we can merely transform the struggle of all against all into a struggle of all for all.' Kautsky, *Der Einfluss der Volksvermehrung auf den Fortschritt der Gesellschaft*, p. 194. See also Chapter IV of Kautsky, *Ethics and the Materialist Conception of History*; and his *Vermehrung und Entwicklung in Natur und Gesellschaft*, for an extended

discussion. *Markizm i darvinizm* (Moscow, 1927) is a convenient collection of mainly pre-revolutionary writings on the subject.

[96] See Plekhanov, *Fundamental Problems of Marxism*, p. 29. This was originally published in 1908. See also Plekhanov, *God na rodine*, II, 86–91. See also Kautsky, *The Social Revolution*, pp. 12–17. This was originally a pair of lectures given in the Netherlands in 1902. Lafargue and Deborin used DeVries' theory in the same manner. But Stalin, apparently unaware of this line of argument, in 1906 used neo-Lamarckism to support 'orthodox' Marxism. See Stalin, *Sochineniia*, I, 301, 307–309.

[97] Lenin, *Sochineniia*, I, 398.

[98] See references in note 84 above to the assertions of the priority of 'practice' over theory. Cf. A. K. Timiriazev, 'Iz oblasti', *DVP*, 1928, No. 3, p. 43, for a case of Lenin's *Philosophical Notebooks* being used to support a positivistic argument.

2. LENIN AND THE PARTYNESS OF PHILOSOPHY

[1] N. Valentinov, *Vstrechi s Leninym*, p. 338.

[2] *Ibid.*, p. 357.

[3] M. B. Mitin, 'K voprosu', *RIK*, 1930, No. 19–20, pp. 36–37.

[4] See, e.g., K. Marx and F. Engels, *Sochineniia* (2nd ed.), II, 89, for an early statement of this theory, in *The Holy Family*. The most famous statement is of course in *The Communist Manifesto*. For the passage cited by Lenin in coining the term 'partyness', see K. Marx, *The Eighteenth Brumaire of Louis Bonaparte* (London, 1926), pp. 17–18.

[5] V. I. Lenin, *Sochineniia*, I, 169–170 *et passim*.

[6] *Ibid.*, pp. 380–381. That the term was a coinage may be seen by reference to V. Dal', *Tolkovyi slovar' zhivogo velikorusskago iazyka*, various editions.

[7] See Lenin, *What Is To Be Done?*, various editions, for the original statement of this theory.

[8] Lenin, *Sochineniia*, I, 398.

[9] *Ibid.*, IV, 66.

[10] *Ibid.*, XXXIV, 21.

[11] *Ibid.*, pp. 5–6, 21.

[12] *Ibid.*, XXXVII, 195, and XXXIV, 20–21, and XIII, 411.

[13] *Ibid.*, XIII, 412.

[14] For the editors' correspondence concerning Aksel'rod's article, see *Leninskii sbornik*, III, 261, 269, 271, 273.

[15] See Lenin, *Sochineniia*, XIII, 411. Bogdanov's three books were: *Osnovnye elementy istoricheskogo vzgliada na prirodu* (St. Petersburg, 1899), *Poznanie s istoricheskoi tochki zreniia* (St. Petersburg, 1901), and *Empiriomonizm* (Moscow, 1904), I.

[16] Lenin, *Sochineniia*, XIII, 412.

[17] L. I. Aksel'rod, 'Novaia raznovidnost' revizionizma', *Iskra*, Nov. 5 (18), 1904, p. 2. Actually, the correspondence cited in note 14 suggests that Plekhanov was the first to ask her for the article. Cf. also L. I. Aksel'rod, 'Iz proshlogo', *Zapiski Instituta Lenina*, I, 104, for her reminiscence of Lenin's request.

[18] G. V. Plekhanov, *Sochineniia*, XIII, 203–211.

[19] Lenin, *Sochineniia*, VIII, 357.

[20] *Ibid.*

[21] Plekhanov, *Sochineniia*, XIII, 274.

[22] *Ibid.*, pp. 274–279

[23] Lenin, *Sochineniia*, VIII, 438. Cf. also *Leninskii sbornik*, V, 361, 366.

[24] Lenin, *Sochineniia*, VIII, 315.

[25] Lenin, 'Partiinaia organizatsiia i partiinaia literatura', *Sochineniia*, X, 26–31.

[26] *Ibid.*, p. 29.

[27] *Ibid.*, p. 28.

[28] Lenin, *Sochineniia*, XIII, 412.

[29] *Ibid.* The visitor to the *dacha* was N. A. Rozhkov. See the reminiscence of Ia. Berzin in *Pravda*, Jan. 21, 1928.

[30] The secondhand report is by Valentinov, *Vstrechi s Leninym*, pp. 323–324. Valentinov writes that he heard this from Bogdanov in 1927. Lenin, *Sochineniia*, XIII, 412, is circumstantial evidence in support of this version, for Lenin, remembering the 'explanation in love' in 1908, assumed that it was among his effects in St. Petersburg, which suggests either that he had never sent it to Bogdanov, or that Bogdanov had returned it.

[31] *Ibid.*

[32] *Ibid.*, p. 416.

[33] A. Bogdanov, 'Otkrytoe pis'mo Plekhanovu', *Vestnik zhizni*, 1907, No. 7. The present writer was unable to consult this item.

[34] V. I. Lenin and A. M. Gorky, *Pis'ma* (Moscow, 1958), pp. 205–207. Unfortunately, only excerpts are given from these letters, which are published here for the first time.

[35] *Ibid.*, p. 206.

[36] Lenin. *Sochineniia*, XXXIV, 332, 335–336.

[37] V. Bazarov and others, *Ocherki po filosofii marksizma*. For Lenin's 'raging with indignation', see Lenin, *Sochineniia*, XIII, 415.

[38] *Ibid.*

[39] A. Bogdanov, 'Ernst Mach und die Revolution', *Neue Zeit*, February 14, 1908, pp. 695–700.

[40] *Ibid.*, pp. 695–696.

[41] For the quotation from Gorky, see Lenin and Gorky, *Pis'ma*, p. 43. For the argument among the editors concerning Gorky's article, see Lenin, *Sochineniia*, XIII, 411–417.

[42] *Ibid.*, p. 415.

[43] See *Protokoly*, pp. 120–122, *et passim*, for references to the agreement of February, 1908 as a *postanovlenie* or decree. Apparently it has not survived, or not been published.

[44] *Ibid.*

[45] See *Neue Zeit*, March 20, 1908, p. 898. For the announcement in *The Proletarian*, see Lenin, *Sochineniia*, XIII, 410.

[46] *Ibid.*, pp. 411–417.

[47] For his feverish study, see Lenin, *Sochineniia*, XXXIV, 337. On

October 27 Lenin wrote his sister that the book was finished. *Ibid.*, XXXVII, 315.

[48] See, e.g., P. Iudin, 'Lenin', *PZM*, 1931, No. 9–10; D. Rudnik, 'Materializm', in Ral'tsevich and Iankovskii, eds., '*Materializm i empiriokrititsizm*' *V. I. Lenina;* and Shcheglov, *Bor'ba Lenina protiv bogdanovskoi revizii marksizma.*

[49] B. D. Wolfe, *Three Who Made a Revolution*, pp. 496–517.

[50] Lenin, *Sochineniia*, XXXIV, 339.

[51] A. Deborin, 'Filosofiia Makha', *Golos sotsialdemokrata*, 1908, No. 4–5, pp. 11–12.

[52] Lenin, *Sochineniia*, XV, 19–20.

[53] *Ibid.*

[54] See Lenin, *Sochineniia* (3rd ed.; Moscow, 1935), XIII, 334.

[55] *Ibid.*

[56] L. I. Aksel'rod-Ortodoks, 'Dve techeniia', *Na rubezhe*, p. 259.

[57] Some of the contemporary reviewers were reminded of the polemical style of Russia's extreme rightist groups. See, e.g., S. L. Frank, 'Filosofskie otkliki', *Russkaia mysl'*, 1910, No. 4, p. 142; and P. Iushkevich, *Stolpy filosofskoi ortodoksii*, pp. 42–43.

[58] See, e.g., Mitin, 'K voprosu', *RIK*, 1930, No. 19–20, p. 37. Citing Lenin's book as a model of partyness, he notes with satisfaction the 'coarseness' of Lenin's polemical style.

[59] Lenin, *Sochineniia*, XIV, 70–72, *et pass.*

[60] *Ibid.*, p. 340.

[61] See especially Lenin's letter to Vorovskii of July 1, 1908. *Ibid.*, XXXIV, 345.

[62] See G. V. Plekhanov, *Sochineniia*, XIX, *passim.*

[63] Lenin, *Sochineniia*, XXXVII, 323–324.

[64] Shantser, a supporter of Bogdanov on political issues, wrote a criticism of Lunacharsky in January, 1909. When Bogdanov protested that the resolution of February, 1908 prohibited such an article, Shantser withdrew it. Thereupon Lenin intervened, and Kamenev wrote a substitute article. See *Protokoly*, pp. 43, 271, *et pass.*

[65] L. Kamenev, 'Ne po doroge', *Proletarii*, Feb. 12 (25), 1909. Cf. Lenin, *Sochineniia* (3rd ed.), XIII, 342–343.

[66] Lenin, *Sochineniia*, XXXVII, 339. Actually Kamenev sent out the call for the conference, presumably acting for Lenin.

[67] *Protokoly*, pp. 145–148, 25, 27, 51, 66.

[68] *Ibid.* Note especially Lenin's statement on p. 27: 'When he [Plekhanov] is ready to give up his [political] error, the question is this: do we attract him by the article against Lunacharsky, or do we repulse the minority of Party-Mensheviks and of orthodox Marxist Mensheviks for the sake of the Bogdanov-Lunacharsky anti-Marxist propaganda? We have not made a deal with Plekkanov against Lunacharsky, but we will say who is making advances to whom. When Plekhanov kicks out Potresov [the chief liquidator], I am ready to extend my hand to him.'

[69] See *KPSS*, I, pp, 212–230, for the published resolutions. See *Protokoly*, p. 47 for Skrypnik's motion.

[70] See Lenin, *Sochineniia*, XIV, 363, for the date.

[71] *Ibid.*, XV, 371–381.

[72] *Protokoly*, pp. 155–156.

[73] *Ibid.*, pp. 42–43

[74] *Ibid.*

[75] *Ibid.*, pp. 120–123.

[76] *Ibid.*

[77] Lenin, *Sochineniia*, XVI, 242, 357.

[78] See *Ibid.*, XXXV, 42–44, for a letter to Gorky explaning the basis of the re-established unity.

[79] *Ibid.*, XVII, 53.

[80] *Ibid.*, pp. 53–55.

[81] *Ibid.*, XX, 456–457.

[82] *Ibid.* Cf. the letter that Lenin, Zinoviev, and Kamenev published in 1914, explaining the new break with Bogdanov. Printed in *ibid.*, pp. 76–77, without credit to Zinoviev and Kamenev.

[83] The phrase is Plekhanov's, cited by Iushkevich, *Stoply*, p. 37. For Plekhanov grumbling at Lenin, see Plekhanov's *Sochineniia*, XIX, 61. Cf. Aksel'rod's review of Lenin's book; reprinted in Lenin, *Sochineniia* (3rd ed.), XIII, 329–333.

[84] Plekhanov, *Sochineniia*, XIX, 22.

[85] Lenin, *Sochineniia*, XXI, 69–70. For the reference to Plekhanov's political 'idiocy', see *Ibid.*, XXXVI, 237.

[86] To be sure, their explicitly Bolshevik line might have been more tolerant of divergent opinions than Lenin's supra-Bolshevik orthodoxy was; but one cannot be sure of this, for the 'Machists' were always a dissident minority within Bolshevism.

3. INTRA-PARTY POLITICS AND PHILOSOPHY

[1] See, e.g. Mitin, *Dialekticheskii i istoricheskii materializm*, Part I, pp. 236, 237. Cf. also the relevant articles in *BSE*, and in the various editions of Rozental' and Iudin, eds., *Kratkii filosofskii slovar'*.

[2] See Mitin, 'Na filosofskom fronte', *Kommunisticheskaia revoliutsiia*, 1929, No. 8; and his 'O pravoi', *Ibid.*, No. 10–11.

[3] See P. Vostokov, 'La philosophie russe durant la période post-revolutionnaire', *Le monde slave*, 1932, No. 11, pp. 286–305, No. 12, pp. 432–457; V. Sesemann, 'Die bolschewistische Philosophie in Sowjet-Russland', *Festschrift N. O. Losskij zum* 60 *Geburtstage*, pp. 163–171; J. Danzas, 'La pensée philosophique en U.R.S.S.,' *La vie intellectuelle*, 1936, No. 41, pp. 422–445; N. Berdiaev, 'The "General Line" of Soviet Philosophy', *American Review*, October, 1933, pp. 536–560; N. O. Losskii, *Dialekticheskii materializm v SSSR* (Paris, 1934); P. Prokof'ev, 'Krizis sovetskoi filosofii', *Sovremennye zapiski*, 1930, Vol. 43, pp. 471–488.

[4] See G. A. Wetter, *Der dialektische Materialismus*, Chapters VI and VII. I. M. Bochenski, *Der sowjetrussische dialektische Materialismus*, and Henri Chambre, *Le Marxisme en Union Sovietique*, are largely derivative from Wetter in their accounts of the matter in question.

[5] See Somerville, *Soviet Philosophy*, especially p. 228.

[6] See Bauer, *The New Man*, Chapter II. Bauer's account of the philosophical controversies is largely derived from J. F. Hecker, *Moscow Dialogues*.

[7] That is the chief method by which Bauer arrived at the conclusion that Stalin sponsored the Deborinite victory over the mechanists in order to facilitate his struggle with Bukharin and the right deviation. See Bauer, *The New Man*, Chapter II, especially pp. 31–33.

[8] See Bukharin, 'Po skuchnoi', *KN*, 1923, No. 1, p. 228.

[9] Sarab'ianov, *PZM*, 1922, No. 3, p. 69.

[10] Another example of a leading mechanist—Stepanov—defining the relationship of freedom and necessity in a strikingly different manner from the stereotype of the mechanist, is perhaps in order: '. . . objective causes place a certain limit on possible achievements. But still [the Party] has always recognized that this boundary is not absolute but relative, supple, pliable, and the more organized, the more conscious, the more planned that human activity is, the greater is this pliability.' Stepanov, *Skvortsov-Stepanov politicheskii boets*, p. 21.

Actually, as the reader will see in the following Chapters, the problem of determinism and free will was not central to the controversy between the mechanists and the Deborinites.

[11] For this sequence of events, see Stalin, *Sochineniia*, XI, 222-238, 318–325, and XII, 1–107. A shrewd observer might have detected hints of schism within the Party in the summer of 1928, but hardly earlier.

[12] Bauer, *The New Man*, p. 33. The Conference of Marxist-Leninist Institutions met from the 8th to the 13th of April. The plenary meeting of the Central Committee that heard Stalin denounce Bukharin met from the 16th to the 23rd of April; the Sixteenth Party Conference, at which the criticism of Bukharin was first made public, met from the 23rd to the 29th of April.

[13] For Deborin's main speech, see Deborin, *Sovremennye*, pp. 1–40. For Shmidt's main speech, see Shmidt, *Zadachi* pp. 7–26.

[14] *VKA*, 1929, Kn. 32/2/, p. 221.

[15] *VKA*, 1929, Kn. 33/3/, p. 282.

[16] The full text of Stalin's speech is in his *Sochineniia*, XII, 1–107.

[17] Lenin, *Sochineniia*, XXXII, 62–78.

[18] Stalin, *Sochineniia*, XII, 69, *et seq.*

[19] Cf. also Stalin, *Sochineniia*, XII, 148, for Stalin, in December, 1929, citing Lenin's criticism of Bukharin's *Ekonomika perekhodnogo perioda.* Lenin made much of Bukharin's infatuation with the philosophy of Bogdanov. Bogdanov was a leading mechanist in the Soviet period, and the Deborinites, who had been criticizing him throughout the controversy, had begun to show the connection between Bukharin and Bogdanov by December, 1929, Stalin, however, did not mention Bogdanov or philosophy, but confined himself to Lenin's criticism of Bukharin's economic theory.

[20] Stalin, *Sochineniia*, XII, 72.

[21] See Deborin, 'Oktiabr' i marksistsko-leninskaia dialektika', *Pravda*,

November 10, 1929, for Deborin's first considerable attempt to show a connection between the mechanists in philosophy and the right deviation in politics.

[22] The stenographic record of the philosophers' conference is in *VKA*, 1929, Kn. 35–36, pp. 227–279. This conference was held in a series of meetings on November 14 and 26, December 3, 12, and 19. The Conference of Marxist Agrarian Economists assembled on December 20, 1929, and heard Stalin speak on December 27. See Stalin, *Sochineniia*, XII, 378, n. 16, and pp. 141–172 for his speech.

[23] For Sten's admission of membership in the Left opposition, see *Raznoglasiia*, pp. 121–122. For Karev's see *ibid.*, p. 157.

[24] Stepanov, *Dialekticheskii materializm i deborinskaia shkola*, p. 48.

[25] See, e.g., *VKA*, 1926, Kn. 16, pp. 186–187.

[26] The letter was published in the course of a later controversy. See Deborin's letter in *Pravda*, 24 August 1930.

[27] *Ibid.*

[28] See *PZM*, 1927, No. 10–11, pp, 243, 85, 120. For a charge that the deviationist Preobrazhenskii was a follower of the mechanist philosopher Bogdanov, see *Bol'shevik*, 1927, No. 14, pp. 89–96. But cf. Preobrazhenskii's criticism of Stepanov's mechanism in *VKA*, 1925, *kn.* 11, pp. 3–5. Except for this passing criticism of Stepanov, Preobrazhenskii played no part in the philosophical discussions.

[29] Between 1930, when the Deborinites were first charged with support for Trotskyism, and 1936, no other names were mentioned. Charges made after 1936 cannot be accepted as evidence of what happened in the 'twenties.

[30] *VKA*, 1928, *Kn.* 30/6/, p. 15.

[31] For a report of the explusions and admissions, see *Nauchnyi rabotnik*, 1928, No. 5–6, p. 112.

[32] Stepanov, *Dialekticheskii materializm i deborinskaia shkola.* The first 68 pages of this collection of articles constitute Stepanov's last philosophical polemic.

[33] For hostile comments see Sarab'ianov 'Dialektika', *PZM*, 1922, No. 3; Gonikman, 'Dialektika t. Bukharina', *PZM*, 1922, No. 3, and cf. also *PZM*, 1924, No. 3, pp. 31 *et passim*, for one of the last criticisms of Bukharin's philosophy before 1929. Sarab'ianov became a leading mechanist, Gonikman a leading Deborinite, in the philosophical controversy. No doubt Lenin's criticism of Bukharin as one who did not understand dialectics had something to do with the initial attack of Gonikman and Sarab'ianov on Bukharin's book. For Lenin's criticism see *Sochineniia*, XXXII, 62–78. For a defence of Bukharin's book against Gonikman and Sarab'ianov, see A. Kon, 'Kritika, "kritikov"', *PZM*, 1922, No. 5–6. For Bukharin's reply to his critics, see his 'K postanovke', *VSA*, 1923, Kn. 3; and 'Po', *KN*, 1923, No. 1.

[34] For an early example of the uniformly favourable references to Bukharin by philosophers of both blocs, see the Milonov-Varjas debate. The Deborinite Milonov cites Bukharin in his own behalf in *VM*, 1924, Kn. 1, pp. 120–121, note 1. The mechanist Varjas quotes Bukharin in *his*

behalf in *VKA*, 1924, Kn. 9, p. 285. In neither case, however, was the reference to Bukharin central to the argument.

One of many striking instances of the surprising silences occurs in the Deborinite Stoliarov's polemic against the mechanist Sarab'ianov's interpretation of dialectical contradiction as the opposition between forces moving in different directions. Stoliarov shows that the view is the same as Bogdanov's but fails to note that Bukharin too argued similarly. See *PZM*, 1925, No. 10–11, pp. 73–76, *et passim.*

Another striking indication of the Deborinites' non-committal attitude toward Bukharin in the period 1924–1929 is the carefully neutral account that Luppol gave in 1928 of the controversy occasioned by Bukharin's *Theory of Historical Materialism.* See Gordon, Luppol, and Volgin, *Obshchestvennye nauki SSSR, 1917–1927* (M., 1928), pp. 20–21.

35 *VKA*, 1929, Kn. 34/4/, pp. 20–21. The author was Bobrovnikov.

36 For an example of Karev writing plainly as an opponent of Stalin's theory of socialism in one country, see *PZM*, 1926, No. 7–8, pp. 225–227, and *PZM*, 1924, No. 10–11, p. 231. These Trotskyite comments occurred in reviews of books on Lenin and Leninism as a social theory. In Karev's writings on the philosophy of natural science there was no reference to the question of Trotskyism or to Bukharin's understanding of Marxist philosophy.

37 The man in question was Innokentii Nikolaevich Stukov. For his participation in the Left opposition, see Carr, *Interregnum*, pp. 315, 371. For a brief account of his participation in a public debate as a mechanist against Riazanov, who spoke as a Deborinite, see the latter's Preface to Engels, *Dialektika prirody* (Moscow, 1930), p. ix.

38 See references in note 33 above.

39 See Bukharin, 'Pis'mo v redaktsiiu Pravdy', *Pravda*, April 19, 1923. The article that inspired Bukharin's anger was 'Zametki chitatelia o knige tov. Bukharina', *PZM*, 1922, No. 11–12. It was signed 'Plekhanovets', and when Bukharin demanded that the editors reveal the author's identity, in view of the rumour that he was an enemy of the Party, the editors refused. Bukharin's letter to *Pravda* followed.

40 In 1929, just before the taboo on criticism of Bukharin was removed, one party to a dispute over historical materialism urged the opposing party to learn a certain elementary point by consulting 'N. I. Bukharin's *Theory of Historical Materialism* and other works where historical materialism is treated popularly'. *VKA*, 1929, Kn. 32/2/, p. 261. It should be borne in mind that the sub-title of Bukharin's book was *A Popular Textbook of Marxist Sociology*. Note also *VKA*, 1928, Kn. 26/2/, pp. 12–14, where Pokrovsky praises Bukharin's contribution to the development of Soviet Marxist social thought but ignores the *Theory of Historical Materialism.*

41 Zinov'iev, 'Leninizm i dialektita', *Bol'shevik*, 1925, No. 16, pp. 3–16.

42 For an analysis of this speech, see below, pp. 97ff.

43 Stalin, *Sochineniia*, XI, 221. (The obituary appeared originally in *Pravda*, October 9, 1928). For further evidence of Stalin's aloofness from the philosophical controversy, see XI, 74–77, where Stalin stressed the need for 'red specialists' in a speech of May, 1928, but said nothing about

the connection between the training of 'red specialists' and the penetration of dialectical materialism in natural science. These two issues were intimately connected with the philosophical controversy.

[44] See Stalin, *Sochineniia*, VI, 69 *et seq.*

[45] *Sochineniia*, VI, 90. The official translation in *Problems of Leninism* (Moscow, 1947), p. 27, improves Stalin's style.

[46] Bammel'. 'Literatura o leninizme', *PIR*, 1924, No. 4, pp. 52–53.

[47] The second installment of Bammel''s review is in *PIR*, 1924, No. 6. It should be noted that the Trotskyite Karev, whose criticism of Stalin on a question of political theory has already been noted (note 36 above), showed a more respectful attitude toward Stalin's book than Bammel' displayed. Perhaps Bammel''s political orthodoxy emboldened him.

[48] The only other instance discovered by the present writer of a Deborinite citing Stalin in a philosophical discussion before the second half of 1929 is in *PZM*, 1926, No. 7–8, pp. 33–34. The Deborinite is Stoliarov, and he contrasts a quotation from Stalin with a quotation from Trotsky with the purpose of denigrating the latter. The connection with Stoliarov's subject is rather remote.

[49] Bammel' was probably condemned as an 'enemy of the people' some time after 1935, when his *Demokrit v ego fragmentakh i svidetel'stvakh drevnosti* was published; for citations of the book in Soviet writings since 1936 uniformly omit mention of Bammel''s name.

[50] See Stalin, *Sochineniia*, XII, 141–172.

4. THE CULTURAL REVOLUTION AND 'BOURGEOIS' SCIENTISTS

[1] Ushakov, *Tol'kovyi slovar' russkogo iazyka*, I, 1546.

[2] In a letter to Molotov, for example, Lenin wrote: 'The chief thing that we lack is culturedness [*kul'turnost'*], the ability to govern. . . . Economically and politically NEP fully guarantees us the possibility of building the foundation of a socialist economy. It is "only" a matter of the cultural forces of the proletariat and its vanguard.' Lenin. *Sochineniia*, XXXIII, 223–224.

[3] Cited in Iaroslavskii, *O roli intelligentsii*, p. 17. O. Iu. Shmidt described the majority of teachers in higher education as 'reactionary [*chernosotennye*] rather than liberal, and in general character bureaucratic rather than socially conscious'. In either case, they were certainly not, in their majority, sympathetic to the Bolsheviks. Cf. the picture given by Gorky, who intervened with the Bolshevik authorities on behalf of persecuted scholars, in his *Culture and the People*. pp. 13, 35–36, *et pass.* For the quotation from Shmidt, see *BSE* (1st ed.), XIV, 32.

[4] Quoted in Bukharin, *Ataka*, pp. 206–207.

[5] *Narodnoe obrazovanie*, 1927, No. 11–12, pp. 22–23.

[6] For a convenient review of the history of Soviet higher education, see Shmidt, 'Vuzy', *BSE* (1st ed.), XIV.

[7] See Mamai, *Kommunisticheskaia partiia*, pp. 66 *et pass.* The government appears to have feared that the sons and daughters of the intelligentsia

were going through the *rabfaki* as a method of avoiding the discrimination practised against them in admissions to the higher schools.

[8] See Shmidt, 'Vuzy', *BSE* (1st ed.), XIV, 35; and Shmidt and Shmulevich, editors, *Nauchnye kadry, passim.*

[9] See Butiagin and Saltanov, *Universitetskoe obrazovanie*, p. 49. For a detailed account of the early Bolshevik struggles against the hostility of teachers, see Korolev, *Ocherki po istorii sovetskoi shkoly, passim.* Unfortunately Korolev does not deal with higher education.

[10] See again Mamai, *Kommunisticheskaia partiia.* The basic decree was 'O vysshikh uchebnykh zavedeniiakh R.S.F.S.R. (Polozhenie)', *Sobranie uzakonenii*, 1921, No. 65, item 486. For the rules concerning appointment of staff, see 'Polozhenie o nauchnykh rabotnikakh Vysshikh Uchebnykh Zavedenii', *ibid.*, 1921, No. 80. item 695.

[11] Only in February, 1929 was the Congress of the Union of Scientific Workers presented with a resolution that made ideological agreement with the régime mandatory for membership. See Luppol's account in *Antireligioznik*, 1929, No. 6, pp. 13–14.

[12] See Luppol, ed., *Kul'turnaia revoliutsiia*, pp. 28–29.

[13] *Izvestiia*, 1925, No. 287.

[14] Luppol, ed., *Kul'turnaia revoliutsia*, pp. 28–29.

[15] *KPSS*, I, 674.

[16] For an anonymous author demanding the closing of the Academy, see *PZM*, 1923, No. 1, pp. 190–191. For the exemption of the Academy's press from the censorship, see the decree of June 6, 1922 in *Sobranie uzakonenii*, 1922, No. 40.

[17] See Lenin, *Sochineniia*, XXVII, 368–369, for Lenin's project of the decree of The Council of People's Commissars that established the Academy. For the decree of 25 June 1918 see *Sobranie uzakonenii*, 1918, No. 49, *stat'ia* 573.

[18] See 'Ob uchrezhdenii institutov po podgotovke krasnoi professury', *Sobranie uzakonenii*, 1921, No. 12, item 79.

[19] See *KPSS*, I, 877, Cf. also Mamai, *Kommunisticheskaia partiia*, p. 36.

[20] See M. N. Laidov (Liadov in rigorous transliteration), 'Functions of a Communist University', *Labour Monthly*, July, 1926, pp. 435–440. He was the Rector of Sverdlov Communist University, the first and most important of the network.

[21] See below, pp. 81 ff.

[22] *KPSS*, I, 878. The resolution does not actually speak of the non-Communist schools as such. It refers to 'the school of *Glavprofobr*', a compound abbreviation for the Chief Administration of Professional and Polytechnical Schools and Higher Educational Institutions, one of the two agencies within the Commissariat of Education (the Academic Centre was the other) that had charge of the non-Communist institutions of higher learning. Cf. also *ibid.*, pp. 732–733 for a similar definition of the function of the *Sovpartshkol* (Soviet Party School) by the 12th Congress.

[23] See *VKA*, 1928, Kn. 26/2, pp. 249 *et seq.*

[24] See Pokrovsky's account of the Commission's work in *Trudy Instituta Krasnoi Professury*, I, 5–6.

[25] For a report of the dropping of courses in philosophy, see Borichevskii, 'Neskol'ko slov o tak nazyvaemoi "russkoi filosofii",' *Kniga i revoliutsiia*, 1922, No. 3 (15), pp. 31 *et seq*. Cf. the decree of March 4, 1921 (*Sobranie uzakonenii*, 1921, No. 19, item 119), which lists the obligatory subjects in higher education. 'Historical materialism', 'the development of social forms', and other courses in social philosophy were included, but the closest thing to general philosophy or the philosophy of natural science was something called 'cosmic physics'.

[26] *VKA*, 1929, Kn. 33/3, pp. 284–285. The situation is hardly surprising in view of a report made a year earlier that up to 1928 only four graduates of the Institute of Red Professorship had been sent to teach in Bielorussian institutions of higher learning (three of them were sent to teach in Bielorussia because they were 'oppositionists', i.e., opposed to the Party line). See *VKA*, 1928, Kn. 27/3, p. 295, In general one gets the feeling that the repeated claims made by Communist officials during the 'twenties, to the effect that the teaching of social sciences (and philosophy was included in this term) was in Marxist hands, must be taken with a grain of salt. See, for example, the *Sbornik obshchestva istoricheskikh, filosofskikh i sotsial'nykh nauk pri Permskom Universitete* (Perm, 1927), vypusk II, which is utterly non-Marxist, and even rather cool toward the Soviet régime. Vypusk III, which appeared in 1929, is significantly different in this respect.

[27] For brief accounts of the Association, and lists of the Institutes within it, see the articles in *BSE* (1st ed.), III, 638, and *BSE* (2nd ed.), III 271–272. Some of the Institutes appear to have grown out of the decree of March 20, 1921, 'O plane organizatsii fakul'tetov obshchestvennykh nauk', *Sobranie uzakonenii*, 1921, No. 19. Note especially point six of that decree. RANION itself was formed only in 1923; it was dissolved in 1930.

[28] *VKA*, 1928, Kn. 27/3/, p. 297.

[29] *Ibid*., p. 298.

[30] For reports on the origin of the Institute, see *VKA*, 1924, Kn. 8, p. 374; *PIR*, 1924, No, 2, p. 304; *VKA*, 1923, Kn. 6, pp. 420–421; and *PZM*, 1923, No. 1, pp. 189–190. Cf. also M. Smit and A. K. Timiriazev, eds., *Statisticheskii metod.*

[31] Cf. accounts cited in note 30.

[32] For Friche's report on the Institute as part of RANION still in 1928, see *VKA*, 1928, Kn. 27/3/, p. 297.

[33] For an example of the work of the largely non-Marxist Institute of History within RANION, see its *Sbornik statei* (Moscow, 1926), vypusk 1. Note especially (pp. 523, *et seq.*) the account of the Institute's origin and activity.

[34] In March, 1928 Pokrovsky said that Marxists would have monopolized social science from 1917, if the Soviet government had not given support to 'bourgeois' social science. And such a monopoly, he commented, would have been bad, 'because social science, which is itself a reflection of the class struggle, grows strong precisely in struggle, and if it should be entirely released from this struggle and that order of things were established of which they tell abroad, that among us only Communists

may carry on social research, then this would be bad'. *VKA*, 1928, Kn. 26/2/, p. 15. Cf. also p. 25, where he boasted that the State Press itself published 'bourgeois' historical works. But note his concluding remarks on p. 30, that it will be a fine thing—say, five years ahead—to survey Soviet social science and see no vestiges of 'bourgeois' scholarship.

[35] *VKA*, 1929, Kn. 32/2/, p. 213.

[36] For the original decree see *Sobranie uzakonenii*, 1921, No. 80, item 685. For the establishment of *Glavlit*, see *Ibid.*, 1922, No. 40. See Mamai, *Kommunisticheskaia partiia*, p. 116 for statistics concerning the relative proportions of private and state publishing to 1927. The private portion showed a steady decline. The last privately published work in the field of philosophy, to the knowledge of the present author, was V. E. Barykina, *Dialekticheskii metod Gegelia i Marksa* (Orel, 1930).

[37] For his work as an official in *Giz*, see his biography in *BSE* (1st ed,), XXX, 514. It should be noted that he came to *Giz* from a private press. Cf. his 'Aksioma', *BSE* (1st ed.) II, 34–41, for evidence of his non-Marxist philosophy of science. In the 'thirties and 'forties Kagan became a full-fledged Marxist.

[38] *Vnutreniaia forma slova* and *Vvedenie v etnicheskuiu psikhologiiu*. Both were published by the State Academy of Fine Arts, a member of RANION. Cf. reviews in *VKA*, 1927, Kn. 23, pp. 250–265, and 1928, 29/5/, pp, 280–286. It is perhaps symptomatic of the changing times in 1928 that the latter review concludes: 'It is necessary to check the rebirth of mysticism in the U.S.S.R. while it is still in embryo.' The earlier review was quite respectful of Shpet and his book.

An extreme case of the publication of non-Marxist works by a Communist-controlled press is reported in *PZM*, 1923, No. 4–5, pp. 284–285. The Proletarian press published a translation of a German professor's polemic against Marxism, explaining in the editorial introduction that the class-conscious proletarian should be aware of his adversary's arguments. The reviewer in *PZM* urged proletarians to boycott the book.

[39] *Prostranstvo, vremia, dvizhenie* (Petrograd, 1923).

[40] The English original is in *Nature*, Nov. 3, 1929, pp, 689–700. The Russian translation is the lead article in *Priroda*, 1929, No. 1.

[41] Timiriazev announced his support of the Bolsheviks in a new edition of *Nauka i demokratiia* that was published in 1920. Lenin's letter of 'enraptured' thanks is in Lenin, *Sochineniia*, XXXV, 380.

[42] See P. A. Florenskii, *Mnimosti v geometrii.*

[43] Cf. O. D. Chwolson, *Hegel, Haeckel, Kossuth und das Zwölfte Gebot* (Brunswick, 1906). His international reputation as a physicist was gained largely by his six-volume textbook of physics, which was translated into German, French, and Spanish. See his biography in *BSE* (1st ed.), LIX, 477–478.

[44] Quoted in *Mysl' i slovo*, 1917, p. 379, from Khvol'son's pamphlet, *Znanie i vera v fizike*, which was a reprint of his introductory lecture.

[45] See Lenin, *Sochineniia*, XIV, 334. Khvol'son's memoir concerning his relations with the Soviet government is in *Nauchnyi rabotnik*, 1927, No. 12, pp. 21–27.

[46] Quoted in *PZM*, 1922, No. 11–12, p. 132, from Khvol'son's *Teoriia otnositel'nosti* (1922 ed.), p. 72.

[47] The later edition appeared in 1925. Already in 1924 his *Kharakteristika razvitiia fiziki* induced a Soviet Marxist reviewer to note that Khvol'son had abandoned his fideism. See VM, 1924, No. 1, p. 206. And indeed Khvol'son argued in this book that the main *trouble* with the new physics was its inability to explain more and more of its most fruitful concepts.

[48] *PZM*, 1926, No. 12, p. 226.

[49] Semkovskii's proud citation of the review is in Shmidt, *Zadachi*, p. 42.

[50] For a Soviet reprint of the essay, see Vernadskii, *Ocherki i rechi*, II, 5–40. For Lenin's praise, see *Sochineniia*, XIV, 287. Cf. XXXVII, 348. For a Bolshevik citing the article with approval, see *Molodaia gvardiia*, 1926, No. 5, p. 160. It was used in 1927–28 as required reading in a course on the history and philosophy of science at Moscow University. See *PZM*, 1927, No. 9, p. 243. Cf. Novinskii, 'Za liniiu', *RIK*, 1930, No. 19–20, pp. 106 *et pass.*, for an effort to disillusion those comrades who thought of Vernadskii as an unwitting dialectical materialist.

[51] See, e.g., V. I Nevskii's comments in *PZM*, 1922, No, 7–8, pp. 119–123. Cf. also references in note 50.

[52] See his biographies in *BSE* (1st ed.), X, 306–7; and *BSE* (2nd ed.), VII, 499–502. Cf. also *Priroda*, 1921, No. 4–6, pp. 86–87, for news of his activity in the Crimea.

[53] Vernadskii, *Ocherki i rechi*, I, 157. This was written before the Bolsheviks came to power, but it was printed in 1922.

[54] Steklov, *Galileo Galilei*, p. 99.

[55] *Ibid.*, pp. 99–100. For the information that Steklov was Vice President of the Academy of Sciences, see P. P. Lazarev, *Ocherki istorii russkoi nauki*, p. 243.

[56] See *PZM*, 1924, No. 4–5, pp. 140–141. The Soviet Marxist was A. A. Maksimov.

[57] Karev used the expression, 'the dictatorship of Marxism', persistently. See, e.g., Deborin, *Sovremennye*, p. 107. Note also his statement in *Bol'shevik*, 1924, No. 10, p. 53, that 'it is not possible to come out against Marxism with an undisguised attack', and his pugnacious acceptance of the label 'dogmatic'. The present author found two other Deborinites using the phrase, 'dictatorship of Marxism', or similar expressions. See *PZM*, 1927, No. 10–11, p. 46 for Bammel' using it, and Luppol, *et al.*, *Obshchestvennye nauki SSSR*. pp. 6–7, for Luppol speaking more realistically: the 'new class that has arrived at a socio-political dictatorship . . . *has begun to realize* also its ideological dictatorship'. (Italics added.) For a subsequent critic of the Deborinites sneering at '"the dictatorship of using the phrase, 'dictatorship of Marxism', or similar expressions. See *Bol'shevik*, 1931, No. 5, p. 90. The Stalinist critic, Kammari, preferred the term 'leadership' (*rukovodstvo*) or 'hegemony' in describing the position of Marxism within the Soviet Union. It is possible to see in this an analogy with the argument in political theory between the Stalinists and the Trotskyites over the question whether the Party's position in the Soviet

government was to be described as 'dictatorship' (the Trotskyite position) or 'leadership' (the Stalinist position). Karev, it should be noted, was a Trotskyite for a time during the 'twenties.

5. THE CULTURAL REVOLUTION AND MARXIST PHILOSOPHERS

[1] For a summation of such complaints in a resolution of the Twelfth Party Congress, see *KPSS*, I, 744–745. For a critique of the journal, *Ateist*, see *Antireligioznik*, 1927, No. 6, pp. 84–90. Cf. also *Antireligioznik*, 1926, No. 1, p. 5. For a vivid description of public debates between antireligionists and priests in the villages, see *Antireligioznik*, 1926, No. 7, pp. 66–67. Cf. also *Kommunisticheskaia revoliutsiia*, 1928, No. 15, p. 72 *et pass.*

[2] Lenin, *Sochineniia*, XIX, 491.

[3] See references in note 1 above.

[4] See Curtiss, *Russian Church*, p. 207, *et pass.*

[5] *Antireligioznik*, 1927, No. 10, p. 35.

[6] Curtiss, *Russian Church*, p. 201.

[7] *KPSS*, I, 744–745.

[8] See the quotations from Kalinin's and Lunacharsky's speeches at a conference of anti-religionists, in Martel, *Le mouvement antireligieux*, pp. 141–142.

[9] Even such an unusual Soviet Marxist as Aksel'rod the Orthodox, who had a doctorate in philosophy from a West European university, was not immune to the influences of the anti-religious movement. In debates with fellow Marxists she might grant that there was 'something more' to the human brain than a collection of electrons in motion, but when, in a public debate, the Metropolitan Vvedenskii used the 'something more' to prove the existence of the supernatural, Aksel'rod dispensed with dialectical subtleties in her reply, and implicitly accepted the reduction of thought to the motion of electrons. See *KN*, 1927, No. 3, pp. 179–180, and No. 5, p. 162.

[10] The words are Gorky's from an essay written in 1929. See Gorky, *Culture and the People*, p. 70.

[11] *PZM*, 1922, No. 1–2, pp. 3–7.

[12] *Ibid.*, pp. 70–73.

[13] Lenin, *Sochineniia*, XXXIII, 208. Shchedrin was a famous 19th-century author.

[14] See Krupskaia, 'Obstanovka', *PZM*, 1933, No. 1, pp. 147–149, for a reminiscence that attributes Lenin's article to the demands of the anti-religious campaign.

[15] Lenin, *Sochineniia*, XXXIII, 206–207.

[16] *Ibid.*

[17] *Ibid.*, pp. 207–208.

[18] See *PZM*, 1928, No. 12, pp. 216 *et seq.*, for news of its fusion with the Society of Militant Materialists to form the Society of Militant Materialist Dialecticians.

[19] For a clear illustration of the impulses pushing Soviet Marxists in these two directions, even without Lenin's double programme, see Ter-

Oganesian's comment on a speech that Sarab'ianov gave to the Party's Moscow Committee on the subject of Marxism and natural science. Ter-Oganesian, 'O nazrevshem voprose', *PZM*, 1923, No. 1, pp. 189–190.

[20] See, e.g., Kanatchikov, 'Mysli o partiino-vospitatel'noi rabote', *Pravda*, 1922, August 16. For a report on the subject at the end of the period under review, see Pokrovsky, 'O podgotovke nauchnoi smeny', *Kommunisticheskaia revoliutsiia*, 1929, No. 13.

[21] *KPSS*, I, 673.

[22] *Ibid.*, pp. 735–736.

[23] The Statute was enacted in November, 1926. It set three main tasks for the Academy, including 'struggle for the rigorous application of the point of view of dialectical materialism, both in the social and in the natural sciences. . . .' 'Polozhenie (ustav) o Kommunisticheskoi Akademii pri. Ts. I. K. Soiuza SSR', Article 1, in *Sobranie zakonov*, otdel I, 1926, No. 3, *stat'ia* 34.

[24] See above, pp. 68–69.

[25] See *VKA*, 1925, No. 12, pp. 390–392.

[26] Kommunisticheskaia akademiia, *Za povorot*, pp. 5–6.

[27] *Ibid.*, pp. 6, 18–19.

[28] See *VKA*, 1929, Kn. 31/1/, pp. 242–246, for a detailed report of the activities in the Section. For reports of the activities of the various Societies organized by the Section, see *EIM*, 1929, No. 1, pp. 178–181.

[29] See *PIR*, 1926, No. 2, for a composite review by S. Vasil'ev of a batch of such pamphlets.

[30] *VKA*, 1928, Kn. 27/3/, pp. 311–316. For an earlier report that tells how the Department grew out of a seminar that sprang up spontaneously outside of the Academy of Sciences in 1923–1924, see *PZM*, 1927, No. 5, pp. 186–189.

[31] See *VKA*, 1929, Kn. 32/2/, p. 234, for a report that sets 'the subjection of Marxist natural scientists to its influence' as one of the Department's goals.

[32] See Lenin, *Sochineniia*, XIX, 61, and cf. Lunacharsky, *Revoliutsionnye siluety* (2nd ed,), p. 25.

[33] See above, pp. 18 ff. They were, during the 'twenties, in charge of training philosophers at the Institute of Red Professorship.

[34] The change of names was effected at a meeting of April 17, 1924. See *VKA*, 1924, Kn. 8, pp. 391–393 for a transcript of the discussion that attended the decision. It should be noted that until April, 1919 the Socialist Academy was open not only to non-Party Marxists but even to actual members of the Menshevik and Social-Revolutionary Parties.

[35] For the figures cited, see *VKA*, 1928, Kn. 26/2/, p. 246, Cf. *Trudy Instituta Krasnoi Professury*, pp. 9–10, for Pokrovsky reporting in 1923 that half of the teachers in the Institute of Red Professorship were not Party members.

[36] *VKA*, 1924, Kn. 8, pp. 391–392.

[37] *Ibid.* Riazanov had a considerable history of deviation from Bolshevism. See, e.g., Popov, *Outline History*, II, 454, 19, *et pass*, See *VKA*,

1924, Kn. 8, pp. 392–393, for an interesting exchange between two non-Communist members of the Academy, Bogdanov and Gorev, on the change of name in relation to the non-Party Marxists within the Academy.

38 Lenin, *Sochineniia*, XXXIII, 20.

39 Quoted in Mamai, *Kommunisticheskaia partiia*, p. 44.

40 Nikolai Andreevich Gredeskul was the former Cadet who became a prominent Soviet Marxist philosopher. Cf. his contribution to *Intelligentsiia* (S.P., 1910), and cf. also Stepanov, *Dialekticheskii materializm*, p. 48. N. A. Karev, another prominent Marxist philosopher in the Soviet period, was a Social-Revolutionary before the Revolution. See *Bol'shevik*, 1931, No. 5, p. 82.

41 Quoted in *Raznoglasiia*, p. 39. Though the note is not printed in Lenin's *Sochineniia*, it appears to be genuine.

42 Lenin, *Sochineniia*, XXXII, 73.

43 See, e.g., Lunarcharsky, *Etiudy* (1922 ed.), pp. 5–27; and his *Ot Spinozy do Marksa* (1925); and his *Protiv idealizma* (1924).

44 Lenin, *Sochineniia*, XXVIII, 76. Cf. the account of the Congress in Lenin, *Sochineniia* (3rd ed.), XXIII, 563.

45 See *Proletarskaia kul'tura*, 1920, No. 17–19, p. 74.

46 See Lenin, *Sochineniia*, XXIX, 308, for Lenin joking about *Proletcult* at the opening of the conference. Closing the conference ten days later, his attitude was quite different. See *ibid.*, p. 343.

47 See Lenin, *Sochineniia*, XXXI, 532 for the date. For Nevskii's essay, see Lenin, *Sochineniia* (3rd ed.), XIII.

48 See Lenin, *Sochineniia* (3rd ed.), XXV, 636–637, for a long footnote telling the story.

49 *Ibid.*

50 *Proletarskaia kul'tura*, 1920, No. 17–19, p. 78.

51 'O proletkul'takh; pis'mo Ts. K. R. K. P.', *Pravda*, 1 Dec. 1920.

52 See the Statute cited in note 23.

53 See *VKA*, 1924, Kn. 8, p. 374, for a record of attendance at the meetings of the Academy's Presidium during 1923–1924. The reports of meetings in later years do not give a formal record of attendance, but one may infer that the Party chiefs did not attend from the fact that they did not speak. The only exception to this rule was the Trotskyite Preobrazhenskii, who attended infrequently.

54 *VSA*, 1923, Kn. 6, pp. 420–421.

55 The election is reported in *PIR*, 1927, No. 2, p. 225. Cf. *VKA*, 1927, Kn. 20, p. 295, for Pokrovsky congratulating the meeting of the Academy for not taking the position that the Academy was above politics.

56 See *VKA*, 1927, Kn. 22, pp. 7–18.

57 See above, p. 55. The cryptic quotation is from the announcement of the expulsion in *Nauchnyi rabotnik*, 1928, No. 5–6, p. 112. Those expelled were Trotsky, Preobrazhenskii, Radek, Rakovskii, and V. M. Smirnov.

58 See below, pp. 203 ff.

59 The thinker was I. P. Razumovskii. See *VSA*, 1923, Kn. 4, p. 265. The label of 'Talmudist' was pinned on Razumovskii by V. Rumii,

'Otvet odnomu iz talmudistov', *PZM*, 1923, No. 8–9. The argument concerned the nature of ideology.

6. MECHANISM AS A TENDENCY

[1] Semkovskii coined the phrase 'Soviet Americanism'. See Deborin, ed., *Sovremennye*, p. 70. Complaints of 'revolutionary pragmatism' were chronic throughout the 'twenties. See e.g., *PZM*, 1922, No. 7–8, pp. 165–166; 1926, No. 7–8, p. 131; *Kommunisticheskaia revoliutsiia*, 1928, No. 17–18, p. 23. Cf. also Gorky, *Culture and the People*, pp. 139–144 *et pass.* for a colourful discussion of this attitude. The reader will note that Gorky himself was not entirely immune to it.

[2] Enchmen, *Vosemnadtsadt' tezisov.* Neither this nor Enchmen's other pamphlet was available to the present author. It was necessary to piece together Enchmen's views from the polemics against him.

[3] Bukharin attributed Enchmen's views not only to a vulgar notion of Pavlov's work but also to 'the school of Lavrov and Mikhailovskii, which traces its genealogy from Spencer and the "organic" (biological) school in sociology'. Bukharin, *Ataka*, p. 150. Bukharin repeatedly alluded to Enchmen's former membership in the Social-Revolutionary Party (*Ibid.*, pp. 132, 137, 150), ostensibly to prove Enchmen's derivation from Lavrov and Mikhailovskii.

[4] Some critics pointed out, for example, that Pavlov's faith in neurophysiology as alone capable of delivering man from the 'dark powers' that lead him 'to wars and revolutions and their horrors', was not essentially different from Enchmen's biological approach to social science. The quotation from Pavlov, originally published in 1923, may be found in English in Pavlov, *Lectures*, p. 41.

[5] Quoted in *PZM*, 1923, No. 4–5, p. 290. Also in *PZM*, 1927, No. 10–11, p. 82.

[6] Quoted in Samoilov, *Detskaia bolezn'*, pp. 219 *et pass.* Also in *PZM*, 1923, No. 4–5, p. 288, and *PZM*, 1927, No. 10–11, p. 82. Enchmen, it should be noted, conceded that there were some truly revolutionary elements in Marxism. They were to be preserved in the Theory of the New Biology.

[7] Quoted in *VKA*, 1927, Kn. 24, p. 17. Also in *PZM*, 1923, No. 4–5, pp. 285–286.

[8] Enchmen, *Teoriia novoi biologii.* The source of the bibliographical information concerning Enchmen is Rozanov, *Filosofsko-sotsiologicheskaia literatura.* Bukharin quoted from a typescript of Enchmen's ('Psychology before the Bar of Renascent Positivism'), which was apparently never printed. See Bukharin, *Ataka*, pp. 129 *et seq.*

[9] Enchmen's critics were unanimous in describing his followers as Communist student youth. For insight into the unusual type of students that they were, note that seventy per cent of the first class to graduate from the three-year programme of Sverdlov Communist University had come to the University in 1921 directly from factories. *Izvestiia*, 18 June, 1924, p. 5.

[10] Reported by Deborin, *VKA*, 1927, Kn. 24, p. 17.

[11] 'Stariki i molodezh' v nashei partii', *Pravda*, 14 Dec. 1923. A collection of anti-Enchmenist articles was published in 1924: Girinis, ed., *Ocherednoe izvrashchenie*. For Bukharin's polemics, see his collected essays, *Ataka*.

[12] See, e.g., Bauer, *The New Man in Soviet Psychology*, p. 25.

[13] See *Zapiski kom. universiteta im. Sverdlova*, 1923, No. 1, p. 277. Note also p. 249.

[14] S. Minin, 'Filosofiiu za bort'! *Armiia i revoliutsiia*, Kharkov, 1922, May, No. 5. The journal was not available to the present writer.

[15] 'Filosofiiu za bort'! *PZM*, 1922, No. 5–6.

[16] For the editorial note see *ibid.*, p. 122. For the long criticism (by V. Rumii) see pp. 127–130.

[17] S. Minn, 'Kommunizm i filosofiia', *PZM*, 1922, N. 11–12.

[18] The periodicals in which these articles appeared were not available to the present author. For reports on them see *PZM*, 1923, No. 1, pp. 204 *et pass.*, and *PZM*, 1924, No. 1, pp. 253–254.

[19] According to a polemicist in *VM*, 1924, Kn. 1, pp. 363–365, a book by D. M. Maksimov, *Vvedenie v izuchenie marksizma*, argued a case very similar to Minin's. The book was not available to the present author.

[20] *PZM*, 1922, No. 5–6, pp. 123, 127.

[21] *PZM*, 1922, No. 11–12, p. 192. Plekhanov's comments on Saint-Simon and Comte may be found in his *Sochineniia*, VIII, 91.

[22] *PZM*, 1922, No. 11–12, pp. 193, 195.

[23] See Bukharin, *Ataka*, p. 150. Note also that Sarab'ianov, a future leader of the mechanist faction, argued against Enchmen that the reduction of social science to biology was possible, but only for 'some very, very distant time'. *PZM*, 1924, No, 8–9, pp. 298–299.

[24] Kanatchikov, 'Mysli o partiino-vospitatel'noi rabote', *Pravda*, 1922, August 16, p. 2. Cf. also Samoilov, *Detskaia bolezn'*. This entire book is a polemic against Enchmen, Minin, and a few others, but Samoilov himself displays many characteristically mechanist ideas.

[25] See below, pp. 108 ff.

[26] See above, pp. 48–49.

[27] Nevertheless, one can perceive the union of political 'voluntarism' and philosophical mechanism in his political writings. See, e.g., Trotsky, *The Third International After Lenin*, p. 84, where he wrote: 'The role of the subjective factor in a period of slow, organic development can remain quite a subordinate one. . . . But as soon as the objective prerequisites have matured, the key to the whole historical process passes into the hands of the subjective factor, that is, the party'.

[28] For other writings by Trotsky that touch upon the philosophy of natural science, see 'Radio, nauka, tekhnika i obshchestvo', *KN*, 1927, No. 2; 'Darvinizm i marksizm', *Pravda*, 1923, Nos. 139, 140; 'Pis'mo vserossiiskomu s'ezdu nauchnykh rabotnikov', *Pravda*, 1923, No. 267.

[29] Trotsky, 'D. I. Mendeleev i Marksizm', *Sochineniia*, XXI, 281.

[30] *Ibid.*, p. 283.

[31] *Ibid.*, p. 271.

[32] All phenomena, Trotsky says in one place, are aggregates of physical particles, whether society or the earth's core or an amoeba. 'But this, to be sure, does not mean that each phenomenon of chemistry can be directly reduced to mechanics, still less that each social phenomenon can be directly reduced to physiological and further to chemical and mechanical laws. Such, it may be said, is the ultimate goal of science'. *Ibid.*, pp. 275–6. For another expression of Trotsky's reductionism, see *KN*, 1927, No. 2. pp. 135–6.

[33] Trotsky, *Sochineniia*, XXI, 275–6.

[34] See above, note 4.

[35] The neuro-physiologist Bekhterev, if we may believe his many critics, felt that sunspots might be the ultimate determinant of human behaviour and hence of social development. The notion seems to have been fairly popular at the time. See, e.g., A. L. Chizhevskii, *Fizicheskie faktory istoricheskogo protsessa*.

[36] See the quotations from Trotsky in Babakhan, 'V zashchitu', *PZM*, 1923, No. 4–5. Cf. also *PZM*, 1924, No. 2, p. 250.

[37] *VKA*, 1924, Kn. 7; note Bukharin's complaint about the banner at the Institute of Red Professorship inscribed 'Marxism in Science [*nauka*] and Leninism in Tactics'.

[38] See above, pp. 59–60.

[39] See Bukharin's autobiography in *Entsiklopedicheskii slovar' Granat*, XLI, *chast'* I, pp. 54–55, where he ironically confesses a 'heretical inclination toward the empirio-critics' before the Revolution, without making clear whether he still suffers from the heresy. Cf. also Bukharin *Ataka*, pp. 135–136, where he declares himself opposed to empiriocriticism, 'especially in its Machist formulation'. Apparently he did not consider Bogdanov such a Machist. See, e.g., his defence of Bogdanov's 'organizational science' in an exchange of notes with Lenin; in Lenin, *Filosofskie tetradi* (1933 ed.), pp. 431–432. Cf. also the uniformly favourable references to Bogdanov in Bukharin, *Historical Materialism*, especially pp. 83, 171. But cf. Bukharin's obituary of Bogdanov in *Pravda*, 8 April 1928, where he writes that Bogdanov departed from Marxist theory. See also, Bukharin, *Ekonomika perekhodnogo perioda*, *passim*, for his extensive use of Bogdanov's terminology. Cf. *Pravda*, 22 Nov. 1921.

[40] *VSA*, 1923, Kn. 3, p. 9. For Bogdanov's disagreement, see *VKA*, 1925, Kn. 10, pp. 96 *et pass.*

[41] See Bukharin, 'K postanovke', *VSA*, 1923, Kn. 3, and cf. his *Historical Materialism* (N.Y., 1928), pp. 91, 148–149 and Chapter V in general.

[42] See, e.g., Philipp Frank, *Le principe de causalité et ses limites*, pp. 121–122 *et pass.*

[43] See VSA, 1923, No. 3, pp. 3–6. It will be noted that Bukharin argued that Marx and Engels were right in rejecting mechanical materialism because the science of mechanics was not dialectical in their day. The development of physics since their time had allegedly made mechanics as dialectical as any other science.

[44] See Bukharin, *Ataka*, pp. 78 *et seq.*, for Bukharin in 1913 answering

Struve's charge of scholasticism in Marx by distinguishing between Marx's 'figurative expressions' and his actual meaning. For Bukharin's sensitivity to criticisms of Marxism see his *Ataka*, a collection mostly of polemical articles, and note also, in his *Historical Materialism*, the virtually continuous dialogue between Bukharin and an impressive array of critics.

[45] For Bukharin's exposition of the theory of equilibrium see *Historical Materialism, passim*. For his defence of the principle against his critics see his article in *VSA*, 1923, Kn. 3.

[46] *Ibid.*, p. 6.

[47] In writing his textbook of Marxist sociology Bukharin wavered between the terms *dialectical* and *dynamic* when characterizing Marxist philosophy. See *Historical Materialism*, Chapter III, Section c. (pp. 63 *et seq.* of the English translation).

[48] Bukharin, *Ataka*, pp. 150–151.

[49] See above, pp. 56 ff. for the attitude toward Bukharin on the part of established theorists. For his popularity among the Party youth see, *e.g.*, Sarab'ianov's polemic against Bukharin, in the course of which he alludes to Bukharin's 'tremendous authority among our youth'. *PZM*, 1922, No. 3, p. 62. For another of Bukharin's critics (Gorev) making a similar allusion, see *PZM*, 1923, No. 10, p. 242.

[50] *VKA*, 1929, Kn. 34/4/, pp. 20–21. The author was Bobrovnikov.

[51] E. Kol'man, ed., *Na bor'bu za materialisticheskuiu dialektiku*, p. 5.

[52] Bukharin, *Ataka*, p. 171. The quotation from Pavlov may be found in translation on p. 49 of Pavlov, *Lectures on Conditioned Reflexes*.

[53] One of the many indications is the surprising frequency of mechanistic sentiments in the philosophical parts of articles on biology and physics in the volumes of the *Great Soviet Encyclopedia* that were published during the 'twenties. Deborin was the philosophical editor of the *Encyclopedia*, but he was apparently obliged to approve mechanistic articles written by non-Marxist specialists, who alone were qualified to write on a number of scientific topics that involved philosophical aspects of biology and physics. See, *e.g.*, the articles cited by Stepanov, *Dialekticheskii materializm i deborinskaia shkola*, pp. 10 *et seq.*

[54] The only total exception to that generalization known to the present author occurs in *PZM*, 1924, No. 1, p. 267. There A. Maksimov estimates the majority to be religious in outlook. But see *PZM*, 1926, No. 1–2, p. 198, for the same author estimating the predominant mood of natural scientists to be mechanistic. The Communist mathematician O. Iu. Shmidt on one occasion estimated that the majority of Russia's natural scientists were simply followers of the latest philosophical vogues among west European scientists (Shmidt, *Zadachi*, p. 10), and on another occasion felt that the predominant mood was a naïve materialism (*Raznoglasiia*, p. 103). In 1929 a young Deborinite who had spent the past three years proselytizing among the natural scientists of Leningrad reported that their philosophical outlook was predominantly positivist (*Raznoglasiia*, pp. 110–111).

[55] *PZM*, 1928, No. 1, pp. 174–175. The Deborinite was Gredeskul.

[56] *PZM*, 1926, No. 4–5, pp. 62–63.

[57] *KN*, 1924, Kn. 2, p. 183. M. M. Zavadovskii should not be confused with B. M. Zavadovskii, another biologist who became a Deborinite.

7. THE FIRST CHALLENGE TO MECHANISM, 1922–1924

[1] *PZM*, 1923, No. 4–5, p. 115.

[2] See his autobiography in *Entsiklopedicheskii slovar' granat*, v. 41, *chast'* I, columns 346–350. Note his deviation from Leninism in the period 1909–1911.

[3] M. N. Liadov, 'Razvernutaia "rabochaia demokratiia" i akademism', *Pravda*, 1923, No. 287 (18 December); 'O studenchestve, akademizme i demokratii', *Pravda*, 1924, No. 6 (8 January).

[4] *Ibid.*

[5] See above, p. 85.

[6] For biographical data concerning Gonikman, see Lenin, *Sochineniia*, (3rd ed.), XXVII, 568. The slight blemish in his Party record was a brief membership in a semi-Menshevik group in 1917 and 1918. The polemic against Liadov is 'Protiv filosofskogo likvidatorstva', *Pravda*, 1924, No. 8 (10 January).

[7] *PZM*, 1924, No. 3, p. 36. Gonikman's paper was published in three parts, the first coinciding with his article in *Pravda* against Liadov. For the information that the paper originated in Deborin's seminar at the Institutue of Red Professorship, see *PZM*, 1924, No. 4–5, p. 243, note 1.

[8] Liadov's reply to Gonikman is 'Spasaet li filosofiia ot opportunizma?' *Pravda*, 1924, No. 11 (13 January). This article was subjected to an extended critique in *Under the Banner of Marxism* by V. Rumii (*PZM*, 1924, No. 1, pp. 240–249), but Liadov published no rejoinder.

[9] Karev, 'O deistvitel'nom', *PZM*, 1924, No. 4–5. Gonikman replied in 'Revizor', *PZM*, 1924, No. 10–11, and Karev replied to the reply in *ibid.*, pp. 222–229.

[10] Karev's discovery of his Hegelianism occurred in his dispute with Varjas. See below, pp. 121–122.

[11] See, e.g., *KN*, 1924, No. 2 (March), pp. 318–323, for another student of Deborin's, I. K. Luppol, defending the journal, *Under the Banner of Marxism*, against such complaints.

[12] Sapir, 'Protiv', *SK*. 1923, Nos. 24, 25. Struminskii, *Psikhologiia* (Orenburg, 1923).

[13] A. Troitskii was a graduate student of Deborin's. See *PZM*, 1923, No. 11–12, pp. 299–304 for his criticism of Struminskii. See *PIR*, 1924, No. 5, for Bammel''s criticism of Struminskii. For another rare instance of a Deborinite venture into the discussions of the Marxist psychologists, see N. Karev, 'Marksistskaia kritika i kritika marksizma', *Bol'shevik*, 1924, No. 10. Note too that criticism of Freudianism was a fairly regular feature of Deborinite writings.

[14] *SK*, 1923, No. 24, p. 335.

[15] Sapir, indeed, turned up at the Conference of April, 1929, which terminated the controversy between the Deborinites and the mechanists,

and spoke as a Deborinite. See Shmidt, *Zadachi*, pp. 65–70. Note also that Sarab'ianov dropped some incidental criticism of A. K. Timiriazev's understanding of Marxist philosophy. See *SK*, 1923, No. 24, p. 327. A. K. Timiriazev was, like Sarab'ianov, a future leader of the mechanist faction.

[16] Adoratskii made Lenin's acquaintance in 1904. See Adoratskii, *K voprosu* (Moscow, 1933), pp. 61 *et pass.* Cf. Lenin, *Sochineniia*, XXXV, 379, for a solicitous letter of 1920, inquiring after Adoratskii's health. For Lenin's advice to compile the *Correspondence*, and for the comment on 'scholarly fools', see Adoratskii, *K voprosu*, pp. 71–73.

[17] *Pis'ma Marksa i Engel'sa* (Moscow, 1922), pp. xvii-xlvi. The citations below are from a second, enlarged edition of 1923.

[18] Lenin's remark is in his *Sochineniia*, XIX, 503. Though written in 1913 or 1914 the article in which the remark appears was not published until 1920, and was therefore fresh in Adoratskii's mind when he wrote his Preface.

[19] It was during the first part of the First World War, i.e., just after Lenin made the remark about dialectics being the focus of the correspondence, that he studied Hegel intensively. See his *Filosofskie tetradi.*

[20] Marx expressed the desire in a letter to Engels. See *Selected Correspondence* (N.Y., 1936), p. 102.

[21] *Pis'ma Marksa i Engel'sa* (M., 1923), p. xxx.

[22] *Ibid.*, p. xxiv.

[23] See, e.g., *Ibid.*, p. xxiii, for Adoratskii's distinction between science and ideology. In Adoratskii's view, it would appear that Marxism is not an ideology. In the middle 'twenties, when he was not participating in the philosophical discussions, Adoratskii wrote articles for the *Entsiklopediia gosudarstva i prava*, in which he showed his continued belief in such a conception of Marxism and ideology. See, e.g., II, 35, where Adoratskii says: 'To the degree that [the proletariat smashes illusions], it severs itself in this manner from legal ideology.'

[24] The articles of Reisner that are particularly apropos are: 'Problemy psikhologii v teorii istoricheskogo materializma', *VSA*, 1923, Kn. 3; 'Refleksy Pavlova i ideologiia Marksa', *Oktiabr' mysli*, 1924, No. 5–6; 'Freid i ego shkola o religii', *PIR*, 1924, Nos. 1, 3. Cf. also his posthumous 'Ideologiia i politika', *VKA*, 1929, Kn. 33/3/.

[25] Adoratskii, 'Ob ideologii', *PZM*, 1922, No. 11–12.

[26] Razumovskii, 'Sushchnost'', *VSA*, 1923, Kn. 4; cf. also 'Nashi "zamvriplekhanovtsy"', *PZM*, 1923, No. 12. By the end of the 'twenties Razumovskii had left Saratov University and was an important legal theorist in Moscow.

[27] See Luppol, 'O novom', *PZM*, 1924, No. 12, pp. 109–110. For Razumovskii's complaint that he had been misunderstood, see *PZM*, 1925, No. 4, pp. 215–219. For Razumovskii's further shift, see below, pp. 201–202.

[28] See Borichevskii, 'Dogmaticheskoe', *Kniga i revoliutsiia*, 1922, No. 3 (15).

[29] L. Rudas (V. Rudash in the Russian alphabet), 'Ortodoksal'nyi

marksism', *VKA*, 1924, Kn. 8, 9; completed in *VKA*, 1925, Kn. 10. For the editorial note expressing explicit agreement, see Kn. 8, p. 281.

30 Bammel''s review is in *PZM*, 1923, No. 6–7.

31 Ivanovskii's reply to Bammel' is in *PZM*, 1923, No. 10, immediately followed by Bammel''s rejoinder. It should be noted that as late as 1928 a Deborinite quoted Ivanovskii with approval. See *PZM*, 1926, No. 6, p. 109, for a case in 1926, and *VKA*, 1928, Kn. 26/2/, p. 94, for the case in 1928.

32 The German original is *Marxismus und Philosophie* (Berlin, 1923). The Russian translation, *Marksizm i filosofiia* (Moscow, 1924), was unavailable to the present author.

33 Karev was probably the reviewer in *Bol'shevik*, 1924, No. 7–8. Troitskii was probably the reviewer in *PZM*, 1924, No. 4–5.

34 Fogarasi reviewed the German original in *VSA*, 1923, No. 6.

35 Borichevskii, 'Idealisticheskaia legenda o Kante', *VSA*, 1923, Kn. 4. The note promising a reply from Deborin is on p. 285. Deborin's criticism is 'Legkomyslenny kritik', *VKA*, 1924, K,. 7.

36 See Orlov, 'Zdravyi smysl i ego ideolog', *PZM*, 1925, No. 1–2.

37 Orlov, 'Chto takoe materiia?' *KN*, 1924, July.

38 For Tseitlin's debate with Orlov concerning the theory of relativity, see *PZM*, 1924, No. 3. For his criticism of Orlov's views on Descartes, see 'Ratsional'nyi i formal'nyi dialekticheskii materialism', *VM*, 1924, No. 1.

39 See Luppol, 'Neskol'ko', *VM*, 1925, No. 3.

40 For evidence of Luppol's neutrality in 1927, see Luppol, Volgin, and Gordon, *Obshchestvennye nauki*, pp. 13–14.

41 See below, Chapters 18 and 19.

42 *PZM*, 1924, No. 1, p. 117.

43 *PZM*, 1924, No. 1, p. 135.

44 See Thalheimer's article in *PZM*, 1925, No. 1–2. A German translation appeared in *Unter dem Banner des Materialismus*, 1925, Heft 2 (July).

45 See Shmidt, *Zadachi*, pp. 40–45.

46 *PZM*, 1924, No. 3, p. 105.

47 See Timiriazev, 'Teorii', *PZM*, 1924, No. 7–8.

8. THE FORMATION OF FACTIONS, 1924–1926

1 Quoted in Luppol, *Na dva fronta*, p. 174. The italics are added.

2 The chief evidence that the Deborinites did not overlook Semkovskii's subtle criticism of March, 1924, is the frequency with which they cited it, once the controversy came into the open.

3 Semkovskii, *Etiudy po filosofii marksizma*, pp. 163–164.

4 *Gruppa osvobozhdeniia truda*, Kn. 2, p. 5.

5 *PIR*, 1924, No. 3, pp. 213–214. Bammel' was the Deborinite reviewer.

6 Sten was a publicist, but by the end of the 'twenties he was also a professor of philosophy at the Institute of Red Professorship. See *NR* (1930), p. 364. The present author infers from the latter fact, and from

occasional references to Sten as 'Deborin's student', that he must have been among the first class at the Institute, 1921–1924.

[7] Rudnianski, *Besedy po filosofii materializma.*

[8] *PZM*, 1924, No. 8–9, p. 289. (Cf. also *Bil'shovik*, 1925, No. 2–3, pp. 107–114.) Another Deborinite (probably Luppol) saddled Aksel'rod with alleged Enchmenism in D. Rakhman's study of Locke, because she had written an Introduction to the book. See *PZM*, 1924, No. 10–11, pp. 247–251.

[9] In 1925 Aksel'rod prefaced an essay on Spinoza with a criticism of the Deborinite view, without mentioning the names of Deborin and his students. For an English translation, see Kline, *Spinoza in Soviet Philosophy*, pp. 61–62.

[10] See below, pp. 127 ff.

[11] For Varjas' original sketch of his book in a paper read at the Communist Academy, see 'Istoriia filosofii', *VKA*, 1924, Kn. 9. Bogdanov's criticism is on pp. 318 *et seq.* The first Deborinite criticism of Varjas' paper (precipitated by his presentation of it to the Institute of Red Professorship), is K. Milonov, 'Ob odnom', *VM*, 1924, Kn. 1. Varjas' reply is in *PZM*, 1924, No. 12, pp. 283–291. In that same issue, *PZM*, 1924, No. 12, pp. 50–85, appears Karev's criticism of Varjas. Varjas' reply, containing the charge of Hegelianism, appears in *PZM*, 1925, No. 5–6, pp. 215–237. It is noteworthy that Bogdanov did not persist in his criticism of Varjas. In the debates within the Institute of Scientific Philosophy in April-May, 1926, both spoke as critics of the Deborinites. By the time Varjas' *Istoriia novoi filosofii* appeared as a book (in 1926), the factional conflict was fully under way. For Deborinite polemics against it see *PZM*, 1926, No. 7–8, pp. 206–225 (by Asmus), and *PZM*, 1926, No. 9–10, pp. 214–220 (by Luppol), and *VKA*, 1926, Kn. 18, pp, 273–281 (by Dmitriev), and *VKA*, 1926, Kn, 19, pp. 146–160 (by Milonov).

[12] *PZM*, 1925, No, 5–6, p. 261. For Karev's quarrel with Gonikman over Hegelianism, see above, pp. 109–110.

[13] The phrase quoted is from Bammel', 'O nashem filosofskom', *PZM*, 1927, No. 12, pp. 34–35. Note that Deborin referred to the complaints about 'scholasticism' on his entry into the controversy. See below, p. 124.

[14] Stepanov's booklet originally appeared in the spring of 1924 as an appendix to the second edition of his translation of Gorter, *Istoricheskii materializm.* A separate printing, *Istoricheskii materializm i sovremennoe estestvoznanie*, appeared about the same time that Sten's criticism was published in *Bol'shevik*, 1924, No. 11, pp. 82–89.

[15] The mere fact that *Bol'shevik* published Sten's polemic did not imply full support by some higher authority. As late as March, 1930, when the direct authority of the Central Committee was being established in the field of philosophy, *Bol'shevik* carried a philosophical article that reflected neither the Deborinite viewpoint, which was then under attack, nor the viewpoint that was in process of replacing it. (Furshchik, 'O liberal'nom i marksistskom ponimanii etiki', *Bol'shevik*, 1930, No. 6). The unrepentant mechanist, Varjas, evidently considering Furshchik a mechanist,

described the article as 'official', since it appeared in *Bol'shevik*, Deborin dismissed Varjas' notion with a sneer. See *PZM*, 1930, No. 6, p. 18.

[16] The possibility is judged remote because, neither in his original criticism of Stepanov, nor in the polemics following did Sten allude to Stepanov's position as a leading Stalinist. See again, Chapter 3.

[17] Stepanov, *Istoricheskii materializm i sovremennoe estestvoznanie* (1925 ed.), pp. 56–57. An entire section of the book is entitled 'The Identity of Contemporary Natural Science and Materialist Philosophy', pp. 83 *et seq.* For a more detailed examination of Stepanov's philosophy, see Chapter 9.

[18] *Bol'shevik*, 1924, No. 15–16, p. 118.

[19] The only exception to this rule was Deborin's defence of Kant against Borichevskii. See above, p. 114.

[20] See *Bol'shevik*, 1924, No. 15–16, p. 115: 'In view of the fact that the dispute between Comrades Sten and Stepanov has taken on a special character, the editors intend to transfer further discussion to the journal, *Under the Banner of Marxism*.'

[21] See *Bol'shevik*, 1924, No. 14, p. 85. The quotation was from Deborin's *Vvedenie*.

[22] *PZM*, 1925, No. 1–2, p. 5.

[23] *Ibid.*

[24] When Stepanov informed the readers of *Under the Banner of Marxism* that he had got support from the Timiriazev Institute, he took care to point out that it 'has on its staff many Communists'. *PZM*, 1925, No. 3, p. 219.

[25] Quoted by Stepanov, *ibid.*, p. 238. Cf. also p. 267 for a letter from G. G. Bosse, the botanist who was then assistant director of the Institute, notifying the public of the discussion and the resolution. A stenographic report was later printed: Bosse, ed., *Mekhanisticheskoe estestvoznanie i dialekticheskii materializm.* Unfortunately, the present writer could not obtain this book.

[26] *PZM*, 1924, No. 6–7, p. 69.

[27] Kol'man, 'K voprosu o sluchainosti', *PZM*, 1926, No. 6. Orlov's reply is in *PZM*, 1926, No. 9–10: 'Teoriia sluchainosti i dialektika'.

[28] Deborin, *Lenin i krizis noveishei fiziki*, p. 27.

[29] Deborin's concluding speech was printed in *LM*, 1926, No. 2. Aksel'rod printed two of the three speeches she gave in the course of the debates, in *KN*, 1927, No. 5. A. K. Timiriazev printed one of his speeches in *DVP*, 1928, Sb. 3. Both Aksel'rod and Timiriazev appended considerable reminiscences of the debates to their speeches.

[30] Bogdanov was so quoted by Aksel'rod in *KN*, 1927, No. 5, p. 146.

[31] See *Pravda*, 1926, March 21. The Deborinite was Sapozhnikov.

[32] For Deborin telling how incensed the mechanists were at Sapozhnikov's review, see *LM*, 1926, No. 2, p. 3. See *ibid.*, *passim*, for Deborin's repeatedly calling his opponents revisionists, and rebuffing the efforts of some (e.g., Aksel'rod) to represent the conflict as a difference of opinions among orthodox Marxists.

[33] *KN*, 1927, No. 5, p. 152.

[34] The words were Deborin's in his summing up. *LM*, 1926, No. 2, pp. 25–26.

[35] The present writer has chosen the terms 'Deborinite' and 'mechanist' to describe the opposing factions, as the least of many evils. The terms are pejoratives, but at least they are *both* pejoratives. To some extent, moreover, they were accepted by those they were supposed to slander. For Stepanov's acceptance of the term 'mechanist' to describe his position, see below, Chapter 9. For a Deborinite casually referring to himself as a Deborinite, see, Shmidt, *Zadachi*, p. 90. (The Deborinite was Iurinets.)

[36] This was, for example, the theme of Timiriazev's speech of 27 April 1926. See *DVP*, 1928, No. 3, pp. 25 *et seq*. For Aksel'rod to the same effect, see *KN*, 1927, No. 5, pp. 141–142.

[37] See Lenin, *Sochineniia*, (3rd ed.), XXVII, 591, for a brief biographical sketch. At the time that Sorin wrote his article on the philosophical controversy, he was Assistant Director of the Institute of Lenin; his article was presumably a report on Lenin's philosophical notes that were still unpublished.

[38] *Pravda*, August 1, 1926.

[39] For the Resolution concerning literature, see Brown, *The Proletarian Episode*, pp. 235–240. It should be noted that this Resolution was a refusal to shut off controversy in literature.

[40] 'True, a portion of the patented Marxists and especially of the Marxist natural scientists have turned tail at a crucial moment, but the social movement around dialectics continues to develop in breadth and depth'. *PZM*, 1926, No. 12, p. 13.

[41] For the sources and component elements of the anti-religious movement, see above, Chapter 6. The leading mechanists contributed by the anti-religious movement were Stepanov and Sarab'ianov.

[42] The League grew out of a society called 'Friends of the Newspaper *Bezbozhnik*', which was formed in Moscow in the fall of 1924. See Curtiss, *The Russian Church and the Soviet State*, p. 206.

[43] *Antireligioznik*, 1926, No. 1, p. 3, for the quotation from the editorial.

[44] *Antireligioznik*, 1926, No. 1, p. 72. The reviewer was I. Omskii. Note also Sarab'ianov's article in the same number of the journal; while it contains the philosophical views that made him a leading member of the mechanist faction, it does not openly polemicize against the Deborinites.

[45] *Antireligioznik*, 1927, No. 2, p. 41.

[46] For one of many admissions of this by the Deborinites, see the quotation from Deborin in note 40 above. See also the quotation from Samoilov on p. 105 above, and the accompanying admission by a Deborinite that Samoilov's views represented '*communis doctorum opinio*'.

9. THE MECHANIST FACTION: PROPAGANDISTS AND PHILOSOPHERS

[1] *VKA*, 1928, Kn. 30/6/, p. 10.

[2] Gorky, *Autobiography*, p. 568.

[3] See the definitions of *propagandist* and *agitator* in Ushakov, *Tolkovyi slovar'*. The distinction derives from Lenin's *What Is to Be Done*?

[4] See Lenin, *Sochineniia*, XXXVII, 339.

[5] *Ibid.*, XXXV, 46–48.

[6] For a convenient reprint of Lenin's enthusiastic preface, see Lenin, *Sochineniia*, XXXIII, 217–218. Stepanov later referred to this preface as an endorsement of his views on the philosophy of natural science. See, e.g., Stepanov, *Dialekticheskii materializm*, p. 42.

[7] See Stepanov. *Elcktrifikatsiia RSFSR*, pp. 4–31. Cf. also Gorter, *Istoricheskii materializm* (1919 ed.), for Stepanov's Introduction, in which Stepanov disagrees with some of Gorter's views, but not with Gorter's belief that Marxism is exclusively a social theory with no implications for the philosophy of natural science. Stepanov's handbook, *Istoricheskii materializm i sovremennoe estestvoznanie*, was originally an appendix to the second edition of Gorter's book, published in 1924. Between 1919 and 1924 Stepanov's experience in the anti-religious movement convinced him that Gorter's renunciation of a Marxist interest in natural science could not be accepted. (Gorter was a Dutch Communist).

[8] *PZM*, 1925, No. 3, p. 206.

[9] See Stepanov, *Istoricheskii materializm i sovremennoe estestvoznanie*; Cf. also his *Dialekticheskii materializm i deborinskaia shkola*, a convenient collection of his polemics against the Deborinites.

[10] Stepanov, *Dialekticheskii materializm i deborinskaia shkola*, p. 39.

[11] *Bol'shevik*, 1924, No. 14, p. 85.

[12] Stepanov, *Istoricheskii materializm i sovremennoe estestvoznanie*, pp. 56–57. One of the chapter headings in this book reads: 'The Identity of Contemporary Natural Science and Materialist Philosophy.'

[13] Stepanov, *Dialekticheskii materializm*, pp. 26–27.

[14] The characterization in Pokrovsky's. See his obituary of Bogdanov in *VKA*, 1928, Kn. 26/2/, pp. viii–ix. Cf. also Krivtsov, 'A. A. Bogdanov', *PZM*, 1928, No. 4, p. 185, for Bogdanov's refusal to rejoin the Party when invited to do so in 1917.

[15] Bogdanov, 'Avtobiografiia', *Entsiklopedicheskii slovar' Granat*, XLI, *chast'* 1, p. 31.

[16] See N. Semashko, 'Smert' A. A. Bogdanova', *Pravda*, April 8, 1928, for an account of the fatal experiment.

[17] See above, p. 86, for Lenin's commissioning of an essay against Bogdanov for the second edition of *Materialism and Empirio-criticism* in 1920. See A. Udal'tsov, 'K kritike Bogdanova', *PZM*, 1922, No. 7–8, especially pp. 82–83, for an account of Bogdanov's meteoric popularity before Lenin's action.

[18] *VKA*, 1927, Kn. 21, p. 281.

[19] Bogdanov, 'Avtobiografiia', p. 30.

[20] *VKA*, 1927, Kn. 21, p. 263. It should be noted that Bogdanov himself, on occasion, appealed to Marx as an authority. See, e.g., *VKA*, 1924, Kn. 7, pp. 359–360, and Kn. 9, p. 318.

[21] *VKA*, 1927, Kn. 21, p. 289.

[22] See A. Bogdanov, *Filosofiia zhivogo opyta*; also his *Tektologiia*. Cf. also Bogdanov, 'Uchenie ob analogiiakh', *VSA*, 1923, Kn. 2, pp. 78–79; 'Uchenie o refleksakh i zagadki pervobytnogo myshleniia', *VKA*, 1925,

Kn. 10, pp. 67–96; 'Predely nauchnosti rassuzhdeniia', *VKA*, 1927, Kn. 21, pp. 244–290.

[23] *KN*, 1927, No. 5, p. 146.

[24] Timiriazev's speech in the Institute of Scientific Philosophy may be found in *DVP*, 1928, No. 3, pp. 25–39, with comments preceding and following.

[25] *KN*, 1927, No. 5, p. 162.

[26] For biographical data concerning Aksel'rod, see the biographies in *BSE*, 1st and 2nd editions, volumes 2 and 1, respectively. Cf. also Aksel'rod, *Etiudy i vospominaniia*, especially pp, 17–36, for her reminiscences of Plekhanov.

[27] See *LM*, 1926, No. 2, p. 29. She is reported to have defended her former Menshevik views also in 'Moi otvet', *DVP*, 1927, No. 2; unfortunately, the present writer did not have access to this number of the journal. In 1929 Deborin claimed that her 'attitude toward October' (i.e., toward the Bolshevik Revolution) was still unknown. *PZM*, 1929, No. 10–11, pp. 9–10.

[28] *KN*, 1927, No. 5, pp. 155–156.

[29] For evidence in 1936 of Aksel'rod's continued refusal to recant, see *PZM*, 1936, No. 1, p. 28.

[30] See above, pp. 127–128.

[31] *KN*, 1927, No. 3, p. 171.

[32] *PZM*, 1928, No. 1, p. 210.

[33] Aksel'rod (Esther Luba Axelrod), *Tolstois Weltanschauung und ihre Entwicklung*. See *Leninskii sbornik*, II, 326–327, for correspondence between Lenin and Aksel'rod on the possibility of a Russian edition. (It did not appear until 1922.) For her writings on esthetics during the Soviet period, see 'Metodologicheskie voprosy iskusstva', *KN*, 1926, Nos. 6, 7, 12; 'Estetika N. G. Chernyshevskogo', *VKA*, 1929, *kn.* 34/4/. In 1934, after the Deborinites had been condemned, she restated her basic philosophy in a pamphlet, *Idealisticheskaia dialektika Gegelia i materialisticheskaia dialektika Marksa.*

[34] Askel'rod, *Filosofskie ocherki*, and *Protiv idealizma*. The last edition of the latter came out in 1933. Her 1934 pamphlet, cited in note 33, was her last publication. She died in 1946.

[35] Aksel'rod, 'Spinoza i materializm', *KN*, 1925, No. 7, for her first extended polemic against the Deborinites. Cf. pp. 120 ff. above, for her brief attack on them in 1924. Her polemics against the Deborinites are collected in *V zashchitu dialekticheskogo materializma*, which was not available to the present writer.

[36] *KN*. 1927, No. 5, p. 158.

[37] *Ibid.*, p. 162.

[38] *Ibid.*, pp. 144 and 149. Cf. also Aksel'rod, *Marks kak filosof*, especially pp. 33–34.

[39] *KN*, 1927, No. 5, p. 160.

[40] *Ibid.*, p. 162. For Aksel'rod stressing activism as the soul of dialectics in a paper written before the controversy with the Deborinites began, see 'Deistvennost' i dialektika v filosofii K. Marksa', *VSA*, 1923, Kn. 4.

[41] See above, pp. 18 ff.

[42] *KN*, 1927, No. 5, p. 151, Cf. *KN*, 1927, No. 6, pp. 195, and 202–203, *et pass.*, for the Deborinite, Karev, on the same subject. Cf. Aksel'rod, 'Spinoza i materializm', *KN*, 1925, No. 7, pp. 144–168. There is a translation of the article just cited in Kline, *Spinoza in Soviet Philosophy*. Cf. also Kline, pp. 14–47, for a thorough account of the disputes among Soviet Marxists concerning Spinoza.

[43] Aksel'rod, 'Nadoelo!' *KN*, 1927, No. 3, p. 180.

[44] In the heat of the debate within the Institute of Scientific Philosophy Aksel'rod said that universals 'reflect' objective relationships among particulars. See *KN*, 1927, No. 5, pp. 138–140. Karev seized upon this mementary slip in his polemic against Aksel'rod: *PZM*, 1928, No. 9–10, pp. 27–35.

[45] For Semkovskii's views on Darwinism and Marxism, see his pamphlet, *Chto takoe marksizm*? It was not available to the present writer, who relied for his knowledge of it on a favourable review in *PZM*, 1922, No. 7–8, p. 188. For his lectures to Ukrainian scientists, see Semkovskii, *Dialekticheskii materializm i printsip otnositel'nosti.*

[46] See, e.g., A. K. Timiriazev, 'Otvet tov. Semkovskomu', *PZM*, 1925, No. 8–9, pp. 170–190. And see Semkovskii, *Dialekticheskii materializm i printsip otnositel'nosti, passim.*

[47] Reported by Rozanov, 'Kievskaia', *PZM*, 1927, No. 5, p. 189. Cf. also *PZM*, 1929, No. 5, pp. 136–138, 149–151, 159–162.

[48] *VKA*, 1928, Kn. 26/2, p. 281.

[49] See Deborin, *Sovremennye*, pp. 64–71; and Shmidt, *Zadachi*, pp. 40–45.

[50] In Deborin, *Sovremennye*, p. 67.

[51] *Ibid.*, p. 70.

[52] At a meeting of the Society of Materialist Dialecticians just after the Conference, a number of miltant young Deborinites charged Semkovskii with a lack of sincerity and thoroughness in his shift to the Deborinites. See *PZM*, 1929, No. 5, pp. 136–138, 149–151, 159–162, *et pass*. Note that Semkovskii, defending himself, repeated the metaphor given in the text above.

[53] See his biography in *BSE* (1st ed,), IX, 45, and *NR* (1930), p. 42, He was a teacher at Moscow State University, and a leading figure at RANION and the Timiriazev Institute.

[54] Varjas, 'O tom', *PZM*, 1925, No. 5–6, p. 218.

[55] Varjas, 'Istoriia filosofii', *VKA*, 1924, Kn. 9, p. 307.

[56] *Ibid.*, pp. 299–300; see also *VKA*, 1928, Kn. 26/2/, p. 252.

[57] Varjas, 'Ob obshchikh zakonakh', *DVP*, 1928, No. 3, pp. 119–120.

[58] *VKA*, 1927, Kn. 9, p. 261.

[59] *Ibid.*, p. 315. For Varjas accusing the Deborinites of a tendency toward 'Hegelian panlogism', see *PZM*, 1925, No. 5–6, p. 217.

[60] For a Deborinite accusing Varjas of hypostasizing logic, see G. Dmitriev's review of Varjas' *Istoriia novoi filosofii*, in *VKA*, 1926, Kn. 18, pp. 276–279.

[61] See, e.g., Varjas, 'Formal'naia', *PZM*, 1923, No. 6–7, expecially pp.

210–213. Unfortunately, the present writer was unable to consult Varjas, *Logika i dialektika.*

[62] See again *VKA*, 1924, Kn. 9, pp. 260 *et pass.* It must be borne in mind that Varjas had a *tendency* in this direction, and not as strong a tendency as can be found in Lafargue or Tseitlin.

[63] *PIR*, 1926, No. 1, p. 204. This is a review of Sarab'ianov, *Osnovnoe v edinom.*

[64] See above, p. 49.

[65] For Sarab'ianov defending philosophy against the attack of Enchmen, see Sarab'ianov, 'Enchmenizm', *SK*, 1924, No. 27.

[66] See. e.g., Sarab'ianov, *Istoricheskii materializm* (1925 ed.), p. 152.

[67] See, e.g., Sarab'ianov, 'O nekotorykh', *PZM*, 1925, No. 12, pp. 194–196, where Sarab'ianov explicitly disagrees with Stepanov's identification of mechanistic and dialectical materialism, but agrees with the reduction of mental processes to mechanical motion in the brain.

[68] See. e.g., Sarab'ianov, *Istoricheskii materializm* (1922 ed.), p. 92. Cf. also *PZM*, 1925, No. 12, p. 193, for Sarab'ianov explicitly agreeing with Bogdanov's definition of dialectical contradiction as the opposition of forces moving in different directions.

[69] See. e.g., Engels, *Dialectics of Nature*, pp. 174–175.

[70] Sarab'ianov, *Istoricheskii materializm* (1925 ed.), p. 159.

[71] Sarab'ianov, *Istoricheskii materializm*, (1922 ed.), p. 17. It should be noted that Sarab'ianov expressed disagreement with Aksel'rod's alleged view that secondary qualities are *entirely* subjective.

[72] Sarab'ianov, 'Kak inye tovarishchi', *PZM*, 1926, No. 6, pp. 62–65.

[73] Sarab'ianov, 'Nazrevshii vopros', *SK*, 1923, No. 20, pp. 231–232. For a Deborinite commenting on this example of dialectics, see Bammel', 'O nashem', *PZM*, 1927, No. 12, pp. 54–55.

[74] See above, Chapter 2.

[75] See *PZM*, 1926, No. 6, pp. 73–75.

[76] *Ibid.*, pp. 68–73.

[77] *PZM*, 1925, No. 12, pp. 80–81. For Sarab'ianov using his peculiar epistemology in his anti-religious writing, see his 'Nauka', *Antireligioznik*, 1926, No. 2, p. 10.

[78] *PZM*, 1926, No. 6, pp. 65–70; also *PZM*, 1925, No. 12, p. 181.

[79] Bosse, ed., *Mekhanisticheskoe estestvoznanie*, pp. 52–53. This collection of the speeches at the Timiriazev Institute was not available to the present writer, who obtained the above citation from Stoliarov, *Dialekticheskii materializm* (1930 ed.), pp. 57–58.

[80] Quoted by Bammel' in *PZM*, 1927, No. 12, p. 47, from Bosse, ed., *Mekhanisticheskoe estestvoznanie*, p. 52.

[81] Quoted in a review of Bosse in *Molodaia gvardiia*, 1926, No. 3, p. 217.

[82] Note especially, Borichevskii, *Vvedenie v filosofiiu nauki* (1922) and *Drevniaia i sovremennaia filosofiia v ee predel'nykh poniatiiakh* (1925). The latter was marked 'Part I', but Part II never appeared. In the period before 1926 Borichevskii also had frequent articles on philosophical subjects in the journals *Zapiski nauchnogo obshchestva marksistov, Kniga i revoliutsiia,* and *Krasnyi student.* His first publication after 1926 was *Kratkii ocherk*

istorii drevnego materializma (1930), which was published 'under the editorship' of two members of the new, postmechanist opposition to the Deborinites.

[83] It is significant that the mechanist journal *DVP* carried no articles by Borichevskii. Cf. also Orlov, 'Zdravyi', *PZM*, 1925, No. 1–2, for a leading mechanist's criticism of Borichevskii's philosophy of mathematics. Borichevskii did not speak at the Conference of April, 1929, but a Deborinite speaker said that Borichevskii had recently abandoned the mechanist faction. See Deborin, *Sovremennye*, p. 164.

10. THE MECHANIST FACTION: NATURAL SCIENTISTS

[1] See G. G. Bosse, ed., *Mekhanisticheskoe estestvoznanie i dialekticheskii materializm.* Unfortunately, this collection of the speeches at the Timiriazev Institute was not available to the present writer, who obtained his knowledge of it from scattered references in various works. The biologists among those who spoke for the mechanists were M. M. Mestergazi, S. S. Perov, N. S. Poniatskii, and G. G. Bosse.

[2] For autobiographical and bibliographical data on Bosse, see Lipshits, *Russkie botaniki*, I, 259–263. It should be noted that Bosse omitted some of his philosophical and popular works of the 'twenties. See, e.g., his *Opyt posobiia po biologii*, reviewed in *VKA*, 1926, Kn. 18, pp. 173–174.

[3] Bosse, ed., *Mekhanisticheskoe estestvoznanie*, pp. 63–64, as cited in *VKA*, 1926, Kn. 16, p. 176. Cf. also *PZM*, 1926, No. 1–2, d. 198; and the review in *Molodaia gvardiia*, 1926, No. 3, pp. 216–217; and *VKA*, 1927, Kn. 20, pp. 257–259.

[4] Timiriazev, Varjas, Perov, 'Zaiavlenie', *DVP*, 1928, No. 3, p. 333.

[5] See again Lipshits, *Russkie botaniki*, I, 259–263.

[6] That M. M. Mestergazi did quit the mechanist faction is indicated not merely by his absence from such crucial meetings as the Conference of April, 1929, but also by his silence concerning the philosophical controversy when he participated in biological discussions at the Communist Academy. See. e.g., his 'Epigenezis', *VKA*, 1927, Kn. 19, pp. 187–233. His basic philosophy, it will be noted, was still mechanistic, but he no longer defended the mechanist faction or attacked the Deborinites. Cf. also, *EIM*, 1929, No. 3, pp. 127–129.

N. S. Poniatskii not only quit the mechanist faction but seems to have made an effort to assimilate the Deborinite philosophy. See *Anti-religioznik*, 1928, No. 9, p. 95.

[7] See *DVP*, 1926, No. 2, pp. 28–29, 42, 218, for some of the contributors to this mechanist collection making an effort to transcend the conflict of mechanism and vitalism as biological theories.

[8] For biographical information see 'Samoilov, A. F.', *BSE*, both editions and Koshtoiants, 'Aleksandr Filippovich Samoilov', in Samoilov, *Izbrannye stat'i*. Cf. also Pavlov's memoir, and others, in *Kazanskii meditsinskii zhurnal*, 1931, No. 4–5.

The article in question was: Samoilov, 'Dialektika', *PZM*, 1926, No. 4–5.

[9] *Ibid.*, pp. 80–81.

[10] *Ibid.*, pp. 65 *et pass.* Samoilov felt that Engels had anticipated Boltzmann and Smoluchowski in his criticism of Clausius' prediction of the universe's 'heat death'. What was more impressive—for Samoilov was not a physicist but a physiologist—was his praise of Engels' criticism of the Faulhorn experiment performed by Fick and Wislicenus.

[11] *Ibid.*, pp. 78–79. For a Deborinite claiming that by this and similar concessions Samoilov had actually abandoned mechanistic in favour of dialectical materialism, see Gredeskul, 'Byt'', *PZM*, 1928, No. 1, pp. 190 *et pass.*

[12] Samoilov, 'Dialektika', *PZM*, 1926, No. 4–5, pp. 80–81.

[13] Samoilov described the controversy simply as a 'polemic between representatives of contemporary natural science, the defenders of the mechanist view on nature, on the one hand, and the defenders of dialectical materialism, on the other hand'. *Ibid.*, p. 62. For the most part the mechanists ignored Samoilov's article, while the Deborinites constantly cited it to prove the essential anti-Marxism of their opponents. For an unusual case of a mechanist arguing that *Deborin* had the same understanding of contemporary natural science as Samoilov, see Timiriazev, 'Voskresaet', *VKA*, 1926, Kn. 17, pp. 116–119.

[14] Another prominent biologist who supported the mechanist faction for a time but finally desisted was B. M. Kozo-Polianskii. See his biography and bibliography in Lipshits, *Russkie botaniki*, IV, 249–258. See *DVP*, 1926, No. 2, for his last known defence of the mechanist faction.

[15] See Duchinskii, 'Darvinizm', *PZM*, 1926, No. 7–8. For a few biographical data concerning Duchinskii, see *NR* (1930).

[16] Duchinskii, 'Darvinizm', *PZM*, 1926, No. 7–8, pp. 98 *et pass.*

[17] See *PZM*, 1927, No. 2–3, pp. 256–258.

[18] *PZM*, 1927, No. 7–8, pp. 278–279.

[19] Cf. the similar case of Vasili Slepkov, also claimed as a mechanist by Stepanov. See Slepkov's denial in a letter to *PZM*, 1927, No. 4, pp. 253–254. For Slepkov using 'mechanism' and 'materialiam' as synonyms, and in general revealing his essentially mechanistic philosophy, before he wrote the letter, see *VKA*, 1927, Kn. 19, pp. 223–224. The article that prompted Stepanov's claim was Slepkov, 'Vitalizm', *PZM*, 1926, No. 9–10.

[20] Duchinskii, 'Osnovnye', *DVP*, 1929, No. 4. This number of the mechanist journal was not available to the present writer, who relied for his knowledge of Duchinskii's article on Grebenev, 'Kriticheskaia', *VKA*, 1929, Kn. 34/4/, p. 251.

[21] See below, Chapter 19.

[22] See Deborin, ed., *Sovremennye*, pp. 55–56, 84–85, and 99–100. The three mechanist leaders were Timiriazev, Perov, and Varjas. For Duchinskii's attack on the Deborinites, see note 20.

[23] See again, Chapter 19.

[24] See Shmidt, *Zadachi*, pp. 45–48, 76–80, for the speeches of E. S. Smirnov and Iu. M. Vermel'. The doubtful case is G. E. Koritskii. See *EIM*, 1929, No. 3, pp. 136–138.

25 Duchinskii, 'Neodarvinizm', *PZM*, 1930, No. 2–3.

26 For biographical data see *NR* (1930) and *NR* (1928). For his part in founding the mechanist faction, see note 1 above. For a report of his participation in the debates within the Institute of Scientific Philosophy in the spring of 1926, see Deborin, 'Nashe', *LM*, 1926, No. 2, p. 3. For a report of his participation in a major debate between the mechanists and the Deborinites (staged in the Meierhold Theatre in December, 1927) see 'Korennye', *PZM*, 1928, No. 1. For his speech at the Conference of April, 1929, see Deborin, *Sovremennye*, pp. 78–85.

27 For a report of his unrepentant mechanism in 1933, see Institut Marksa-Engel'sa-Lenina, *Marksizm i estestvoznanie*, p. 68. For a report of his still unrepentant mechanism in 1936, see *PZM*, 1936, No. 1, p. 54. For his early support of Lysenko, see *Spornye voprosy genetiki i selektsii*, the record of the meeting of the Academy of Agricultural Sciences of December, 1936, pp. 318–330. Cf. also *The Situation in Biological Science*, pp. 139–149, for his part in the same Academy's climactic meeting of August, 1948.

28 The evidence is not merely the absence of the problem from his publications that were available to this writer, but also his absence from the discussions of heredity that were held in the Communist Academy and regularly reported in the *Vestnik* (Bulletin) of that body.

29 See Perov, *Iavlenie tozhdestva v belkakh*. Unfortunately, this pamphlet was unavailable to the present writer, who relied for his knowledge of it on the review in *EIM*, 1929, No. 2, pp. 193–195.

30 Perov, 'Dialektika', *DVP*, 1929, No. 4, p. 131. This number of the mechanists' journal was not available to the present writer, who relied for his knowledge of it on Grebenev, 'Kriticheskaia', *VKA*, 1929, Kn. 34/4/, p. 252.

31 Perov, 'Dialektika', *DVP*, 1928, No. 3, pp. 55–75, and especially p. 57.

32 See note 27.

33 See the last two items cited in note 27 above.

34 See *EIM*, No. 1, pp. 178–181, for a report on these societies.

35 *Ibid.*

36 For representative discussions of physical and biological topics, see Chapters 18, 19. The professional qualifications of the participants were found in *NR*, 4 volumes.

37 For biographical data, see Lenin, *Sochineniia* (3rd ed.), XXVII, 593.

38 For K. A. Timiriazev's radicalism, see his *Nauka i demokratiia* (1920 ed.). Note especially the dedication to his parents for inspiring in him 'a seething hatred for any injustice, especially social'. The attribution of his father's radicalism to the Decembrist rising is on pp. 412–413.

39 For A. K. Timiriazev's resignation, see Lenin, *Sochineniia* (3rd ed.), XXVII, 593. There is an extensive biographical literature concerning K. A. Timiriazev. See Ivanov and Figurovski, eds., *Istoriia estestvoznaniia.*

40 See Maynard, *Russia in Flux*, especially p. vii for Pares' agreement.

41 A. K. Timiriazev, *Filosofiia nauki*, I, 3.

42 *Ibid.*, p. 6.

43 *Ibid.*, pp. 6–7.

[44] See A. K. Timiriazev, 'Pokhod', *SK*, 1922, No. 18; 'Dialektika', *PZM*, 1923, No. 4–5; 'Ob ideologicheskoi baze', *Nauchnyi rabotnik*, 1925, No. 1.

[45] A. K. Timiriazev, 'Dialektika', *PZM*, 1923, No. 4–5, p. 115.

[46] See Timiriazev's speech in Bosse, ed., *Mekhanisticheskii*, p. 24. This collection was unavailable to the present writer, who relied on the quotations in Samoilov, 'Dialektika', *PZM*, 1926, No. 4–5, p. 74, and Stoliarov, 'Filosofiia', *PZM*, 1926, No. 7–8, p. 35.

[47] See, e.g., Timiriazev, 'Lenin', *PZM*, 1924, No. 2, pp. 224–225.

[48] See, e.g., the quotation in Samoilov, 'Dialektika', *PZM*, 1926, No. 4–5, p. 74.

[49] A. K. Timiriazev, 'Voskresaet', *VKA*, 1926, Kn. 17, pp. 134–135. It should be borne in mind that both Timiriazev and the Deborinites recognized, however ill-naturedly, that their positions on reduction had come practically to coincide. See e.g., *Ibid.*, p. 129. Cf. Deborin, 'Mekhanisty', *VKA*, 1927, Kn. 19.

[50] A. K. Timiriazev, 'Iz oblasti', *DVP*, 1928, No. 3, pp. 44.

[51] For Timiriazev 'willingly accepting' the accustion of 'tailism', only to redefine the term, see Deborin, *Sovremennye*, pp. 51–52.

[52] For the charge of 'nihilism', see *VKA*, 1929, Kn. 34/4/, p. 254. For the charge of 'On-Guardism', see *PZM*, 1927, No. 2–3, pp. 188 *et seq.* For background concerning the meaning of 'On-Guardism', see Brown, *The Proletarian Episode, passim.*

[53] A. K. Timiriazev, 'Voskresaet', *VKA*, 1926, Kn. 17, p. 136.

[54] See Chapter 18.

[55] For biographical data, see *NR*, 1930.

[56] See above, p. 115.

[57] See especially Tseitlin, 'Problema', *DVP*, 1928, No. 3, p. 146.

[58] See Tseitlin's review of Semkovskii, *Dialekticheskii*, in *PZM*, 1926, No. 4–5, pp. 221–223.

[59] For the original of these ideas in Lafargue, see above, p. 15. For Tseitlin to the same effect, see especially his participation in the discussion of Asmus, 'Dialektika', *VKA*, 1928, Kn. 25/1/, pp. 165–169. The actual example given in the text above, of a civilized child learning arithmetic easily, was taken from Engels.

[60] Tseitlin, *Nauka i gipoteza*; 'Metod', *PZM*, 1924, No. 6–7; 'Ratsional'nyi', *VM*, 1924, No. 1; 'O matematicheskom'. *VKA*, 1927, Kn. 23.

[61] Tseitlin, *Nauka i gipoteza*, pp. 174–176. Tseitlin's first announcement of such ideas was an article entitled 'The Method of Proof of the Law of Interaction of Gravitational and Electrical Masses of Newton-Cavendish-Maxwell in Comparison with the Method of Research of K. Marx and F. Engels'. See Tseitlin, 'Metod', *PZM*, 1924, No. 6–7.

[62] See A. K. Timiriazev, 'Teoriia', *PZM*, 1924, No. 7–8, No. 10–11, and 'Otvet na vozrazheniia tov. Tseitlina', *PZM*, 1924, No. 12.

For another mechanist disagreeing with Tseitlin, see Orlov, 'O printsipe', *VM*, 1926, No. 3.

[63] Tseitlin, 'O matematicheskom', *VKA*, 1927, Kn. 23, p. 164.

[64] *Ibid.* Note that there is an ambiguity in Tseitlin's words: it is not

clear whether he meant that the physical object called *War and Peace* or the aggregate of ideas and emotions expressed by that physical object could be expressed in mathematical formulas.

[65] Deborin, *Sovremennye*, pp. 159–160. For the chemist, F. M. Perel'-man, expressing a similar view, see p. 134.

[66] *Ibid.*, p. 182.

[67] Biographical data may be found in *NR*, 1930, p. 357. The present writer assumes that he was a chemist from the fact that he held the position of 'Senior Chemist' at the State Scientific Chemico-Pharmaceutical Institute, even though Orlov listed his specialty as 'methodology of the natural and exact sciences'. The articles published in 1916 were: Orlov, 'Realism', *Voprosy filosofii i psikhologii*, 1916, Kn. 131 (1), pp. 1–35, and 'Ob induktivnom', 1916, Kn. 135 (v), pp. 356–388.

[68] See A. P. Primakovskii, *Bibliografiia*, and Ivanov and Figurovskii, eds., *Istoriia estestvoznaniia*. The last publications of Orlov were 'Ischislenie' *Matematicheskii sbornik*, 1928, v. 35, pp. 263–286, and some contributions to the last two numbers of *Dialektika v prirode*, the mechanist publication that expired in 1929.

[69] Orlov, 'Dialektika', *VSA*, 1923, Kn. 6, p. 107.

[70] See Orlov, 'Ob induktivnom', *Voprosy filosofii i psikhologii*, 1916, Kn. 135 (v), pp. 356–388.

[71] See Orlov, 'Chto', *KN*, 1924, Kn. 4; and 'Klassicheskaia', *PZM*, 1924, No. 3; and 'Novye', *VM*, 1924, No. 2.

[72] Orlov, 'Zadachi', in *Teoriia otnositel'nosti i materializm*, p. 5.

[73] *Ibid.*

[74] *PZM*, 1923, No. 10, p. 263.

[75] Orlov, 'O zakonakh sluchainykh iavlenii', *PZM*, 1924, No. 7–8; Orlov, *Logika estestvoznaniia*, Chapter VI, pp. 134–168; Orlov, 'Teoriia', *PZM*, 1926, No. 9–10.

[76] Orlov, 'Chto', *KN*, 4, p. 224. Kol'man's attack is 'K voprosu', *PZM*, 1926, No. 6.

[77] *Ibid.*, pp. 153, *et seq.*

[78] See Orlov, 'O ratsionalizatsii umstvennogo truda', *PZM*, 1926, No. 12. His first contribution to the mechanists was 'Mekhanika', *DVP*, 1926, No. 2.

[79] The present writer had access to two polemics in which Orlov clearly stated his reasons for siding with the mechanists. See the account of his speech at a crucial meeting in March, 1927, in *PZM*, 1927, No. 4, pp. 220–221; and see especially Orlov, 'O dialekticheskoi', *DVP*, 1928, No. 3.

[80] Orlov, 'Realizm', *Voprosy filosofii i psikhologii*, 1916, Kn. 131 (1), pp. 1–35.

[81] Orlov, 'O dialekticheskoi', *DVP*, 1928, No. 3, p. 159.

[82] *Ibid.*, p. 161.

[83] For the clearest evidence of the close similarity between the mechanist and Deborinite positions on the nature and necessity of reduction, see A. K. Timiriazev, 'Voskresaet', *VKA*, 1926, Kn. 17, p. 129, where Timiriazev quotes Karev and then exclaims: 'If that is so, then the dispute is ended.'

[84] *PZM*, 1927, No. 4, p. 220.

[85] Orlov, 'O dialekticheskoi', *DVP*, 1928, No. 3.

[86] From the unsigned editorial, 'Marksistskoe mirovozzrenie', *DVP*, 1928, No. 3, pp. 15–16.

[87] The term, 'mechanist', was a pejorative epithet; though some mechanists willingly described themselves by it, most preferred to call themselves 'Marxist natural scientists', or 'so-called "mechanists"', or simply 'dialectical materialists'. The Deborinites, in the course of nearly five years of controversy, attempted to pin many labels on the faction, ranging from 'Freudian' (Varjas was enamoured of Freud) to 'crawling empiricist'. Several times the Deborinites described their opponents as positivists, but the term did not stick, probably because mechanists like A. K. Timiriazev were ardent opponents of 'Machism' or 'Humism'.

II. DEBORIN AND HIS STUDENTS

[1] The principal leaders of the Deborinite faction, after Deborin, were I. K. Luppol, Ia. E. Sten, and N. A. Karev. All three were students in Deborin's seminar early in the 'twenties. Other prominent Deborinites who issued from the seminar were V. A. Iurinets, S. L. Gonikman, A. K. Stoliarov, Konstantin Milonov, and M. B. Mitin.

[2] Luppol, *et al.*, *Obshchestvennye nauki*, pp. 15–16.

[3] Milonov, 'Gegel', *VKA*, 1928, Kn. 25/1/, p. 20. Cf. also Karev's encomium of Deborin at the end of the Conference of April, 1929. *PZM*, 1929, No. 5, pp, 165–169.

[4] For biographical data on Deborin, see *BSE* (1st ed.), XX, 758. For Deborin's pre-revolutionary criticism of Machism and Bolshevism, see above p. 34.

[5] See above, p. 85.

[6] Deborin, 'Oktiabr'', *Pravda*, November 10, 1929.

[7] For Deborin against Spengler, see Deborin, 'Gibel', *PZM*, 1922, No. 1–2; against Freud, see Deborin, 'Freidizm', *VM*, 1925, No. 4; against Christian socialism, see Deborin, 'Poslednee', *VM*, 1924, No. 1.

[8] See Deborin, *Vvedenie*, for Deborin's one full-length book, which was originally published before the Revolution. During the 'twenties he published a large number of articles, many of which were collected in Deborin, *Filosofiia i marksizm*; Deborin, *Ocherki po istorii materializma*; Deborin, *Dialektika i estestvoznanie*; and Deborin, *Lenin kak myslitel'*. (The 1929 edition of the last title is the most complete.)

[9] See Karev, 'Na putiakh', *PZM*, 1923, No. 2–3.

[10] *Ibid.*, p. 226.

[11] Note that Deborin was not made a member of the Institute of Scientific Philosophy when it was reorganized in 1923 in order to extend Marxist-Leninist thought 'beyond the boundaries of social science'. For the staff in 1923, and probably still in 1924, see *VKA*, 1924, Kn. 8, p. 374.

Not also the fact that nearly all of Deborin's writings on the philosophy of natural science were a product of his controversy with the mechanists. See Deborin, *Dialektika i estestvoznanie*.

[12] See *Trudy Instituta Krasnoi Professury*, I, 13. Note that Aksel'rod the Orthodox was the joint author of the quoted statement, as also of the programme of training in philosophy at the Institute of Red Professorship. Apparently she broke with Deborin because of his stress on Hegel, which appeared inordinate to her.

[13] For Lenin's instructions to Soviet philosophers, see above, pp. 79 ff.

[14] For Deborin's admission, at the end of 1926, that very little had been done by *Under the Banner of Marxism* in response to Lenin's demand for anti-religious work, see *PZM*, 1926, No. 12, p. 9. Note also the almost complete absence of contributions to *Antireligioznik*, the theoretical journal of the anti-religious movement, by Deborin or his students.

[15] See Deborin, 'Oktiabr'', *Pravda*, Nov. 10, 1929.

[16] *PZM*, 1925, No. 1–2, p. 5.

[17] Deborin, *Lenin i krizis noveishei fiziki*, p. 27.

[18] Deborin, 'Engel's', *PZM*, 1925, No. 10–11, p. 18. For mechanists expressing horror at this assertion, see, e.g., Timiriazev, 'Voskresaet', *VKA*, 1926, Kn. 17, pp. 128–129, Cf. also *VKA*, 1927, Kn. 20, pp. 263–264, for the Deborinite Podvolotskii defending Deborin's assertion against three mechanists.

[19] Deborin, *Lenin i krizis noveishei fiziki*, p. 5.

[20] For the Marxist tradition in this matter, see above, p. 19. For Deborin unequivocally asserting that 'theory must above all withstand the test of practice', see *VKA*, 1927, Kn. 20, pp. 263–264. For Deborin giving such reassurances in his speech at the Conference of April, 1929, see Deborin, *Sovremennye*, pp. 38–39.

[21] Deborin, *Lenin kak myslitel'* (1929 ed,), p. 59.

[22] Deborin, 'Lenin', in Lenin, *Materializm i empiriokrititsizm* (1925 ed.), pp. xxii–xxv.

[23] *Ibid.* Cf. also Deborin, *Lenin, kak myslitel'* (1924 ed.), pp. 35–36 and 38.

[24] Probably the most colourful of Deborin's criticisms of the mechanists' tendency to make an ontology of current physical theories was the following:

'Of course everything consists equally of electrons, a head of cabbage and the head of a mechanist. Nevertheless, whatever our attitude toward the mechanists may be, we must say that the complete indentification of a mechanist's head with a head of cabbage will advance us very little toward an understanding of the mechanist's thought. Assertions of the kind, "thought is a crowd of electrons", come cheap because there is no content in them.' Deborin, 'Nashi', *LM*, 1926, No. 2, p. 19.

[25] Deborin, 'Mekhanisty', *VKA*, 1927, Kn. 19, p. 45.

[26] See e.g., *ibid.*, pp. 49–50.

[27] *LM*, 1926, No. 2, p. 18.

[28] *VKA*, 1927, Kn. 19, p. 61.

[29] *Ibid.*

[30] Deborin, 'Engel's', *PZM*, 1926, No. 1–2, No. 3, No. 9–10. Reprinted in Deborin, *Dialektika i estestvoznanie* (4th ed.), pp. 102–188.

[31] Aksel'rod, 'Otvet', *KN*, 1927, No. 5, p. 137.

[32] Deborin, *Dialektika i estestvoznanie* (4th ed.), p. 127.

[33] Shmidt, *Zadachi,* p. 118.

[34] See below, Chapter 19.

[35] Deborin, 'Lenin', *PZM,* 1924, No. 2, pp. 14–15; reprinted in Introduction to Lenin, *Materializm i empiriokrititsizm* (1925 ed.), pp. xxi-xxii.

[36] Deborin, 'Nashi', *LM,* 1926, No. 2, p. 12.

[37] Deborin, *Lenin i krizis noveishei fiziki,* pp. 24–27.

[38] See Deborin, *Sovremennye,* pp. 28–32, 57.

[39] See. e.g., Deborin, 'Nashi', *LM,* 1926, No. 2, pp. 25–26. And cf. Deborin, 'Engel's', *PZM,* 1925, No. 10–11, p. 10.

[40] Deborin, *Sovremennye,* pp. 28–37.

[41] Deborin, 'K piatiletiiu', *PZM,* 1926, No. 12, p. 13. And cf. p. 15, for further stress on the 'youth' as the main source of support for the Deborinites.

[42] See above, p. 173.

[43] Deborin, 'K piatiletiiu', *PZM,* 1926, No. 12. pp. 14–15.

[44] The only exception to Deborin's failure to offer empirical proof for this conviction was his citation of Engels' ideas that had been allegedly confirmed by the development of natural science after Engels. The most notable example that Deborin offered was Engels' criticism of Clausius' version of the second law of thermodynamics. See Engels, *Dialectics of Nature,* pp. 201–202, and 216. Cf. J. B. S. Haldane's footnote on p. 24, in which he too claims that Engels anticipated future developments in thermodynamics. Haldane points to the work of Milne; Soviet Marxists of the 'twenties adduced the work of Smoluchowski and Boltzmann.

[45] The present writer believes that the work of G. K. Bammel' and I. K. Luppol was probably the most noteworthy. One should also add the work of V. F. Asmus, who was a leading Deborinite though not actually a student of Deborin's in the literal sense. Bammel' did his most important work on the ancient materialists. See, e.g., his 'Istochniki', *PZM,* 1923, Nos. 6–7, 8–9, 12; and 'Materialisticheskaia', *VKA,* 1926, Kn. 14. Asmus did his notable work on the development of dialectics in modern philosophy. See, e.g., his *Ocherki.* Luppol concentrated on the French materialists of the eighteenth century. His book on Diderot was translated into French: *Diderot, ses idées philosophiques.* Cf. also the collection: Luppol, *Istoriko-filosofskie etiudy.*

Cf. also Kline, *Spinoza,* for more samples of Deborinite work in the history of philosophy. The Deborinite writings that treat of Hegel impressed the present writer by their dull obscurity.

[46] Egorshin, 'K voprosu', *PZM,* 1926, No. 7–8, pp. 124, 128, *et pass.* Cf. also Egorshin, 'Estestvoznanie', *PZM,* 1926, No. 6, pp. 112–113, *et pass.*

[47] Vyropaev, 'Eshche ob empirizme', *PZM,* 1926, No. 11, pp. 140–141.

[48] The Deborinites repeatedly countered mechanist complaints of scholasticism by calling for the training of philosophers in one or another of the natural sciences. See, e.g., *PZM,* 1925, No. 8–9, pp. 286–287. But little was actually done to implement this programme, and, in the resolutions of the Conference of April, 1929, it was still projected as a plan for

future fulfilment. See 'Rezoliutsiia', *VKA*, 1929, Kn. 32/2/, p. 243, and 246.

[49] See, e.g., *VKA*, 1927, Kn. 21, p. 296.

[50] See *Arkhiv Marksa-Engel'sa*, 1925, No. 2, pp. 190–191. In the English, and the more recent Russian, editions the metaphor is somewhat softened. See Engels, *Dialectics of Nature*, p. 243, and Engels, *Dialektika prirody* (1952 ed.), p. 165.

[51] N. M(atusov), 'Zadachi', *Iunyi kommunist*, 1930, No. 7, pp. 53-54.

[52] Sten's speech at the Conference of April, 1929, in Deborin, *Sovremennye*, p. 175. Cf. *Molodaia gvardiia*, 1929, No. 1, p. 77.

The quotation from Marx may be found in Marx and Engels, *German Ideology*, p. 197. The newest Soviet Russian translation may be found in Marx-Engels, *Sochineniia* (2nd ed.), III, 1. For the German original see Marx-Engels, *Gesamtausgabe*, Section 1, V, 533.

[53] Stoliarov, *Dialekticheskii materializm* (1930 ed.), pp. 41–42.

[54] Vishnevskii, in a review of Stepanov's book, *PZM*, 1924, No. 12, p. 310.

[55] See Dynnik, 'Uchenie', *VKA*, 1929, Kn. 34/4/, pp. 163–194. For similar Deborinite studies, see V. Brushlinskii, 'O kategorii mery u Geglia', *VKA*, 1929, Kn. 35–36; G. Dmitriev, 'Logika Gegelia i logika marksizma', *VKA*, 1929, Kn. 35–36. Cf. S. Ia. Ianovskaia, 'Kategoriia kolichestva u Gegelia i sushchnost' matematiki', *PZM*, 1928, No. 3, for one of the isolated Deborinite efforts to apply Hegelian dialectics to actual problems of the philosophy of natural science, in this case, defining the subject of mathematics. To be sure the Deborinites touched on such problems in the course of their polemics, and occasionally became rather heavily involved. (See e.g., Grib, 'Dialektika i logika kak nauchnaia metodologiia', *PZM*, 1928, No. 6). But extended studies of a non-polemical nature, such as Ianovskaia's, were rare.

[56] See Dynnik, 'Uchenie', *VKA*, 1929, Kn, 34/4/, pp. 182–192, for the discussion. Cf. *PZM*, 1925, No. 7, pp. 222–226, for one of the few instances of a Deborinite warning against an excessive admiration for Hegel.

[57] Karev, 'Neskol'ko', *PZM*, 1926, No. 4–5, p. 84. For Timiriazev's expression of agreement, see his 'Voskresaet', *VKA*, 1926, Kn. 17, p. 129.

12. DEBORINITE NATURAL SCIENTISTS

[1] One of the latest examples is Luppol, 'Filosofiia', in Luppol *et al.*, *Obshchestvennye nauki*, p. 16, *et passim*, where Luppol describes the mechanists as 'materialist empiricists' and 'materialist natural scientists'.

[2] See, e.g., *Ibid.*, p. 17.

[3] Ioffe, 'O polozhenii', *PZM*, 1937, No. 11–12, p. 138.

[4] See Maksimov's biography in *BSE* (2nd ed.), LI, 188. For his work with Timiriazev, see Gol'dshtein, 'Kruzhok', *PZM*, 1923, No. 4–5.

[5] For Stepanov calling him a physicist, see Stepanov, *Dialekticheskii materializm i deborinskaia shkola*, pp. 43–48. For Maksimov's enthusiastic report of the Department of the History and Philosophy of Natural

Science, see Maksimov, 'Programmy', *PZM*, 1928, No. 12. For a less enthusiastic report, see Agol's comment in Shmidt, *Zadachi*, p. 104.

[6] Maksimov, 'Ob istochnikakh', *PZM*, 1926, No. 1–2.

[7] Cf. Maksimov, 'O zadachakh', *Molodaia gvardiia*, 1926, No. 9; and Maksimov, 'Sovremennoe estestvoznanie', *PZM*, 1927, No. 10–11.

[8] See below, Chapter 18.

[9] See Hessen, 'The Social and Economic Roots of Newton's Principia', in *Science at the Crossroads* (London, 1931).

[10] For biographical data see *NR* (1930).

[11] See Gessen and Egorshin, 'V s'ezd russkikh fizikov', *PZM*, 1927, No. 1, p. 141; and Timiriazev, 'Po povodu', *PZM*, 1927, No. 2–3.

[12] Gessen and Egorshin, 'Ob otnoshenii', *PZM*, 1927, No. 2–3, pp. 194–195.

[13] *Ibid.*, p. 199.

[14] See Gessen, *Osnovnye idei teorii otnositel'nosti*, pp. 66, 83, 163, 165, *et passim*. Unfortunately this book was not accessible to the present writer, who obtained the citations from several sources. See, e.g., Maksimov, 'Filosofiia', *PZM*, 1936, No. 1, pp. 48–49, Cf. also A. K. Timiriazev, 'Teoriia', *Tekhnicheskaia entsiklopediia*, XV, 367.

[15] See Deborin, *Sovremennye*, p. 132, for a brief reference to the incident. It occurred at the founding of the Society of Militant Materialist Dialecticians. Cf. *PZM*, 1928, No. 12, pp. 216–219, for the list of charter members, which does not include Maksimov. Apparently the incident was precipitated by Gessen's criticism of a paper that Maksimov gave at this meeting. See Maksimov, 'Metodologiia', *PZM*, 1929, No. 7–8, pp. 147–179.

[16] Maksimov, 'N'iuton', *PZM*, 1927, No. 4, pp. 36. Italics added.

[17] Stalin, *Sochineniia*, XI, 226.

[18] See *VKA*, 1928, Kn, 30/6/, p. 264. Both Maksimov and Gessen were members of the bureau in charge of the Institute's Section of the Dialectics of Natural Science, but Gessen was also a member of the board (*pravlenie*) in charge of the entire Institute.

[19] See *EIM*, 1929, No. 1, for a list of the editors, Gessen included.

[20] For Maksimov's statement that illness kept him from the Conference, see *PZM*, 1929, No. 9, p. 225. For a Deborinite at the Conference disowning Maksimov's interpretation of relativity, see Deborin, *Sovremennye*, p. 147. For the announcement of Gessen's (or Semkovskii's) view as the Deborinite view, see speeches by Gessen, Semkovskii, *et al.*, in *ibid.*

[21] See below, Chapter 17.

[22] See Chapter 19.

[23] See Agol's speeches in Deborin, *Sovremennye*, pp. 90–91, and Shmidt, *Zadachi*, pp. 103–104.

[24] For Agol's conversion to militant 'Morganism', see below, Chapter 19. For his tribute to Kammerer, see Agol, 'Pamiati', *Antireligioznik*, 1926, No. 11.

[25] See references in note 23.

[26] See, for example, the speeches of B. M. Zavadovskii and M. L.

Levin in Shmidt, *Zadachi*, pp. 95–103, and 80–90, and Deborin, *Sovremennye*, pp. 147–156. M. L. Levin as late as 1928 did not connect the biological and philosophical disputes. See his comments on a paper given by Smirnov, 'Novye', *VKA*, 1928, Kn. 25/1/, pp. 199–205.

[27] For biographical data see *NR* (1925) and *NR* (1930). Cf. *VKA*, 1927, Kn. 21, pp. 298–303, for Agol criticizing Levit's 'Lamarckism' in 1927.

[28] See Levit's speech in Shmidt, *Zadachi*, pp. 58–63.

[29] It is significant that Bauer, *The New Man*, suggests ideological affinities between the Deborinite philosophy and some schools of Soviet Marxist psychology, but gives no evidence of an explicitly Deborinite position in the psychological controversies of the late 'twenties.

For efforts by psychologists to gain an advantage over opponents by suggesting a connection between the opponents and the mechanist faction, see Frankfurt, 'Mekhanisticheskii', *VKA*, 1928, Kn. 26/2/; Kornilov, 'Mekhanicheskii', *PZM*, 1926, No. 4–5, and 'Vozzreniia', *Psikhologiia*, 1929, v. 2, *vypusk* 1, pp. 3–15. As far as the present writer knows, none of the Deborinite leaders endorsed the psychological theories of Frankfurt and Kornilov, the latter did not participate in the campaign against the mechanist faction, and the opponents of Kornilov and Frankfurt did not defend the mechanist faction.

[30] See the speeches of A. N. Zalmanzon, I. D. Sapir, and A. B. Zalkind, in Shmidt, *Zadachi*, pp. 54–58, 65–70, and 112–118, respectively.

[31] See Shmidt, *Zadachi*, p. 56. Presumably he was referring to Deborin's silence in the face of the tactics of Kornilov and Frankfurt described in note 29.

[32] For biographical information on Shmidt see *BSE* (1st ed.), LXII, 556–557.

[33] See Shmidt's part in the discussion of Timiriazev, 'Teoriia', *VKA*, 1924, Kn. 7, pp. 365–366.

[34] *Ibid.*, p. 368.

[35] See above, pp. 82–83.

[36] See Kommunisticheskaia akademiia, *Za povorot*, pp. 18–20.

[37] See 'Pervaia', *VKA*, 1928, Kn. 26/2/, pp. 262–265. The 'goading' was inadvertently administered by Riazanov. *Ibid.*, p. 251.

[38] *Ibid.*, pp. 262–265.

[39] See 'Nashi zadachi', *EIM*, 1929, No. 1.

[40] Shmidt, *Zadachi*, p. 8.

[41] *Ibid.*, p. 19.

[42] *Ibid.*

[43] *Ibid.*, pp. 21–22, for Shmidt's original theses. For the resolution as amended by the Deborinite editorial committee, see *ibid.*, pp. 127–130. The theses in question were numbers seven and eight in Shmidt's draft, and number eight in the final version.

[44] *Ibid.*

[45] *Ibid.* The only trace of this thesis in the final Resolution was a fleeting demand for experimentation in the elaboration of the philosophy of natural science.

[46] *Ibid.*, pp. 13–14, for Shmidt's review of the subject in April, 1929.

[47] See Shmidt, *Zadachi*, pp. 121–122.

[48] Shmidt, *Zadachi*, p. 24.

[49] *Ibid.*, p. 25. Cf. also the editorial, 'Nekotorye itogi i perspektivy', *EIM*, 1929, No. 3. Shmidt was very likely the author of the editorial in question.

13. SOCIAL THEORISTS IN THE DEBORINITE FACTION

[1] See above, Chapter 3.

[2] For Riazanov's flat announcement in April, 1924 that he was not a Leninist, see above, p. 84. As far as the present writer knows, this was the only one of Riazanov's protests against the cult of Lenin that found its way into print. For a report of Riazanov's verbal protests, including a threat to give a formal speech (*doklad*) on the subject, see *Za marksistsko-leninskoe uchenie o pechati*, pp. 14–15.

[3] For his fate in 1931, see below, pp. 263–267.

[4] See *Arkhiv Marksa-Engel'sa*, 1925, No. 2.

[5] See Riazanov, 'Novye dannye', *VSA*, 1923, Kn. 6, p. 364.

[6] In November, 1925, Riazanov presented this argument in a public debate with the mechanist Stukov. See Riazanov, 'Predislovie', in Engels, *Dialektika prirody* (1930 ed.).

[7] See 'Pervaia', *VKA*, 1928, Kn. 26/2/, pp. 249–252, 281, 284–286, 290–291, and Kn. 27/3/, p. 311–312.

[8] See above, p. 67. For an earlier report of Riazanov's derision, see *VKA*, 1926, Kn. 16, pp. 285–286.

[9] *VKA*, 1928, Kn. 26/2/, p. 252.

[10] *Ibid.*

[11] *Ibid.*, p. 281.

[12] It is noteworthy that Riazanov did not participate in the Conference of April, 1929.

[13] See above, pp. 111 ff.

[14] See his contributions to the *Entsiklopediia gosudarstva i prava.* Note especially his article, 'Ideologiia (pravovaia)', *ibid.*, II, 13–16.

[15] The institutions where they could hardly have avoided each other were the Communist Academy and Sverdlov Communist University. In the latter each was an important official in the Department of Marxist Philosophy. See *Zapiski Kom. Univ. im. Sverdlova,* 1923, No. 1, pp. 249–251.

[16] See above, p. 124.

[17] For a report of Stepanov's becoming head of the Lenin Institute in 1926, see Stepanov's obituary in *Antireligioznik*, 1928, No. 10. p. 5.

[18] See Luppol's complaint in his *Lenin i filosofiia*, repeated and stressed by the reviewer in *VKA*, 1927, Kn. 19, p. 244. See Bammel', 'O nashem', *PZM*, 1927, No. 10–11, pp. 72–73, for a recollection of Adoratskii's 'Mininism'.

[19] For the first volume of Lenin's notes with Deborin's Preface, see *Leninskii sbornik*, IX. For Adoratskii's report of unspecified alterations in the Preface, see *Raznoglasiia*, pp. 185–186.

[20] Adoratskii, 'O rabotakh', *VKA*, 1929, Kn. 32/2/, pp. 198–210.

[21] *Ibid.*, p. 210.

[22] See Razumovskii, 'Filosofskaia reviziia', *PZM*, 1926, No. 7–8.

[23] See *ibid.*, and also Razumovskii, 'V debriakh', *VKA*, 1929, Kn. 31/1/, pp. 104–112.

[24] See, e.g., Razumovskii, 'Filosofiia prava', *Entsiklopediia gosudarstva i prava*, III, 1436–1440.

[25] Razumovskii altered his definition of 'ideology' somewhat in the first, 1924, and especially the second, 1927, editions of his *Kurs istoricheskogo materializma*, but one can still perceive his old attitude in his new formulations. See *PZM*, 1924, No. 12, pp. 109–112, for a Deborinite noting the shift in Razumovskii's view of 'ideology' in 1924, but still expressing dissatisfaction with the result. See the further exchange in *PZM*, 1925, No. 4, pp. 215–225. Cf. finally, *PZM*, 1927, No. 7–8, p. 261, for a Deborinite making peace with Razumovskii's formulation of 1927. The present writer, however, notes that in 1927 Razumovskii still avoided the assignment of 'theoretico-cognitive' meaning to ideology, and said only that '*real economic content*' manifested itself in ideologies.

[26] The quoted phrase is from Pashukanis; see his 'General Theory of Law and Marxism', in Babb and Hazard, *Soviet Legal Philosophy*, p. 118. Cf. Podvolotskii, *Marksistskaia teoriia prava*, which was published in two editions, 1923 and 1925, with a laudatory Preface by Bukharin. Podvolotskii's booklet was originally a paper in Bukharin's seminar at Sverdlov Communist University.

[27] Podvolotskii, 'Dialektika', *VKA*, 1927, Kn. 20, p. 260.

[28] See Polvolotskii's speech in Deborin, *Sovremennye*, pp. 124–130.

[29] Podvolotskii and Gessen, 'Filosofskie korni', *PZM*, 1929, No. 9.

[30] See 'Kritika', *VKA*, 1929, Kn, 35–36, pp. 227–296. The main speakers were Krivtsov, Razumovskii, and Podvolotskii.

[31] See above, Chapter 3.

[32] *VKA*, 1929, Kn. 35–36, p. 280. Italics in the original. Cf. p. 296, for Podvolotskii's repetition of this at the close of his speech.

[33] For Podvolotskii's repetition of this argument, see his 'Dialektika', *VKA*, 1927, Kn. 20, p. 264.

[34] *VKA*, 1929, Kn. 35–36, pp. 269–270.

[35] *Ibid.*, pp. 271–272. Italics added.

14. CLOSING THE CONTROVERSY, 1926–1929

[1] The Deborinite was Agol. See Deborin, *Sovremennye*, pp. 90–91.

[2] This Deborinite was Gonikman. See *ibid.*, p. 168.

[3] See above, p. 124.

[4] See, e.g., Sapozhnikov, 'Novaia', *Bol'shevik*, 1925, No. 16; and Troitskii, 'Zazorno', *Bol'shevik*, 1925, Nos. 17–18 and 19–20. The Ukrainian *Bil'shovik* did the same; see Gonikman, 'Voinstvuiushchii', *Bil'shovik*, 1925, No. 2–3.

[5] See, e.g., A. Tr(oitskii)'s review of Engels' *Dialectics in Nature*, in *Pravda*, 1925, No. 203; Lepin's review of Semkovskii's *Dialekticheskii materializm* in *P avda*, May 23, 1926.

Re The Young Guard, see, e.g., Egorshin, 'Sovremennoe', *Molodaia gvardiia*, 1926, No. 5; and K. Milonov, 'Dialektika', *op. cit.*, 1926, No. 6.

Re Chronicles of Marxism, see Deborin, 'Nashi', *LM*, 1926, No. 2.

Other important journals that followed a Deborinite editorial policy were *PIR* (*The Press and the Revolution*), *RIK* (*The Revolution and Culture*), and *Kommunisticheskaia revoliutsiia*.

[6] Perhaps the earliest sign of the Deborinite policy of the journal was the hostile review of Stepanov's *Istoricheskii materializm* in *PZM*, 1924, No. 12, pp. 307–314. In the next issue, *PZM*, 1925, No. 1–2, Deborin's proclamation of a crusade against the mechanists was the first item. Until the middle of 1924, Vaganian-Ter, who sympathized with the Deborinites, was the journal's 'responsible editor', and there was a large list of Soviet Marxists who 'take part'. In *PZM*, 1924, No. 4–5, p. 316, there was a notice that Vaganian-Ter had left the editorial board, and an 'editorial college' had been established, consisting of Pokrovsky, Stepanov, Deborin, and Karev. For the change at the end of 1926, see the back cover of *PZM*, 1926, No. 12.

[7] See, e.g., Bammel', 'Dialektika', *VKA*, 1926, Kn. 15; and cf. the deprecatory editorial note on the article by A. K. Timiriazev in *VKA*, 1926, Kn. 17, p. 116. Note that Deborin's reply to Timiriazev in *VKA*, 1926, Kn. 19, is printed without editorial footnote. Cf. *DVP*, 1928, No. 3, pp. 114 and 117, for Varjas referring to Deborin as the editor of *VKA*.

For the Deborinite policy of *VM*, see *VM*, 1925, *passim*. On January 7, 1927, the 'general meeting' of the Society of Militant Materialists adopted a Deborinite resolution with one abstention and no opposing votes. See 'Rezoliutsiia', *PZM*, 1926, No. 12, p. 236. Cf. Stepanov's letter of protest in *PZM*, 1927, No. 2–3, pp. 256–258. At the end of 1928 the Society became even more strongly Deborinite, renaming itself the Society of Militant Materialist Dialecticians. See *PZM*, 1928, No. 12, pp. 216, *et seq.*

[8] See Karev's remarks in Deborin, *Sovremennye*, pp. 106–107.

[9] See above, pp. 128–129.

[10] See Aksel'rod, 'Nadoelo', *KN*, 1927, No. 3, for a polemic printed without an editorial disclaimer of responsibility. Aksel'rod, 'Otvet', *KN*, 1927, No. 5, had an editorial note disclaiming responsibility, but so did the reply of Karev, 'Spinoza', *KN*, 1927, No. 6.

[11] See Stepanov, 'Retsenziia', *Izvestiia*, Aug. 23, 1925.

[12] Sarab'ianov, 'O filosofskikh', *Iunyi kommunist*, 1928, No. 2.

[13] See, e.g., G. A. Gurev's review of V. G. Fridman's *Vozmozhno li dvizhenie?* in *Antireligioznik*, 1927, No. 11, pp. 87–89. Gurev himself was subsequently criticized for declaring, in his *Mirovedenie bezbozhnika*, that 'there is absolutely no difference in principle between so-called living and dead matter'. See *Antireligioznik*, 1928, No. 9, p. 95, for N. S. Poniatskii's objection.

[14] *Antireligioznik*, 1928, No. 10, pp. 5–6.

[15] See N. Gubanov, 'Obshchestvo', *Antireligioznik*, 1929, No. 6, pp. 18–21.

[16] See below, pp. 252–253.

[17] *VKA*, 1928, Kn. 30/6/, p. 14.

[18] 'Pervaia', *VKA*, 1928, Kn. 26/2/, p. 275.

[19] *Ibid.*

[20] *Ibid.*, p. 286.

[21] *Ibid.*

[22] *Ibid.*

[23] *Ibid.*

[24] *Ibid.*

[25] Pokrovsky, 'Obshchestvennye', *VKA*, 1928, No. 26/2/.

[26] The proposal was originally made by Miliutin at a meeting of the Communist Academy's Plenum in June, 1926, See *VKA*, 1926, Kn. 16, pp. 292 and 303. For Miliutin's apparently neutral defence of the proposal, see the report of the meeting of the Plenum in January, 1927, in *VKA*, 1927, Kn. 20, pp. 300–301. For Pokrovsky's similarly 'neutral' endorsement of the proposal, see pp. 292–294.

[27] In the discussion of January, 1927 Miliutin casually mentioned that Deborin was being made the head of the new Section of Philosophy. *Ibid.*, pp. 300–301.

[28] See above, p. 192.

[29] See the report in *VKA*, 1928, Kn. 30/6/, p. 248. Cf. also p. 264 for further detail, including the list of key personnel, who were solidly Deborinite except, significantly, for those in charge of the Section of Historical Materialism. A. A. Maksimov, who was one of the three placed in charge of the Section of the Dialectics of Natural Science, was another special case.

[30] See Deborin, *Sovremennye*, p. 194.

[31] Stepanov, *Dialekticheskii materializm*, p. 43.

[32] See Deborin, *Sovremennye*, pp. 90–91. Cf. Tseitlin's remarks, pp. 156–158.

[33] For Shmidt, see above, p. 195. For the Resolution, see Shmidt, *Zadachi*, p. 129.

[34] For the bureaucratic history of the fusion, see 'Deiatel'nost'', *VKA*, 1929, Kn. 33/3/, pp. 275–276, and *VKA*, 1929, Kn. 34/4/, p. 259.

[35] See *VKA*, 1929, Kn. 34/4/, p. 259.

[36] See Deborin, *Sovremennye*, and Shmidt, *Zadachi*, for the stenographic record of the two sessions, with the respective Resolutions appended. Other sessions of the Conference heard a variety of reports, largely concerning the work of various Marxist-Leninist institutions. Besides the philosophical speeches of Deborin, Shmidt, and Adoratskii, Kritsman gave the keynote for a discussion of economics. For a description of the Conference, see *VKA*, 1929, Kn. 32/2/, pp. 195–198.

[37] See above, p. 194.

[38] For the quotation from Shmidt, see above, p. 196. For other Deborinites echoing Shmidt's sentiment, see Shmidt, *Zadachi*, pp. 32–37, 94, and Deborin, *Sovremennye*, p. 112.

[39] Shmidt, *Zadachi*, p. 129.

15. 'CLASSICAL' AUTHORITY AND THE CULTURAL REVOLUTION

[1] The time of its appearance is shown by the reviews in the newspapers. See *Izvestiia*, 1925, August 23, and *Pravda*, 1925, September 6. On the book itself, see above, pp. 8 ff.

[2] Luppol, *Na dva fronta*, pp. 171–172. This is a later reprint of an article published in *PIR*, 1925, No. 8.

[3] Sapozhnikov, 'Novaia', *Bol'shevik*, 1925, No. 16, p. 87.

[4] Stepanov, 'Engel's', *PZM*, 1925, No. 8–9, pp. 54–55.

[5] Deborin, 'Engel's', *PZM*, 1925, No. 10–11, especially pp. 21–23.

[6] See, e.g., Stepanov, *Dialekticheskii materializm*, p. 62. For the mechanist Sarab'ianov explicitly agreeing with Deborin that Stepanov was wrong in arguing the existence of two periods in Engels' work, see *PZM*, 1925, No. 12, p. 181.

[7] See, e.g., Timiriazev, 'Voskresaet', *VKA*, 1926, Kn. 17, pp. 130–133. Cf. also Klemm, ed., *Osnovy*, especially the Preface by Varjas, as reported in *VKA*, 1927, Kn. 24, pp. 294–301.

[8] *Arkhiv K. Marksa i F. Engel'sa*, 1925, No. 2, p. 130. Cf. Engel's, *Dialektika prirody* (1952 ed.), p. 26.

[9] Before Lenin's death in January, 1924, the most common attitude toward Lenin as a philosopher was probably summed up in the unsigned review of Lenin's *Works* that was published in *PZM*, 1922, No. 4, p. 129: 'N. Lenin did not add a single word to Marx, but he taught how to understand him and transformed his theory into the mighty weapon of proletarian revolution.' For Stalin in April, 1924 saying that Lenin added nothing to the Marxist world view, see above, pp. 59–60. About the same time Bukharin objected to the spirit epitomized in a banner that he observed at the Institute of Red Professorship: '"Marxism in science, Leninism in tactics."' Bukharin, 'Lenin', *VKA*, 1924, Kn. 7, p. 32. From 1924 Lenin's authority as a philosopher increased steadily, but until 1930 there was no objection to Deborin's remark that 'Plekhanov was above all a theorist . . . , Lenin . . . a practical man, a politician, a chief'. Deborin, 'Lenin', in Lenin, *Materializm i empiriokrititsizm*, 1925 edition, p. viii.

[10] See above, p. 200.

[11] See, e.g., Deborin, *Sovremennye*, pp. 92–93, for an angry exchange between Varjas and Deborin on this subject. Cf. also A. K. Timiriazev, 'Iz oblasti', *DVP*, 1928, No. 3, p. 43; and cf. Egorshin's rebuttal in *VKA*, 1928, Kn. 27/3/, p. 272.

[12] Chichikalov, 'Itogi', *Bol'shevik*, 1929, No. 19, p. 87.

[13] B. M. Zavadovskii, 'Darvinizm', *VKA*, 1926, Kn. 14, p. 273.

[14] B. M. Zavadovskii was charged with 'revisionism' and other sins in the genetics controversy of 1948, but the particular quotation given above was not exhumed. See his defence in Lysenko *et al.*, *The Situation*, pp. 334–360.

[15] See Levit's speech in Shmidt, *Zadachi*, pp. 58–63.

[16] See Vermel' in Shmidt, *Zadachi*, pp. 76–79.

[17] *VKA*, 1927, Kn. 21, p. 266.

[18] Luppol, *Na dva fronta*, p. 178.

[19] Deborin, 'Nashi', *LM*, 1926, No. 2, pp. 34–38.

[20] See *VKA*, 1924, Kn. 7, pp. 359–360, and *VKA*, 1924, Kn. 9, p. 318.

[21] Aksel'rod, 'Nadoelo', *KN*, 1927, No. 3, pp, 171, 175.

[22] Karev, 'Spinoza', *KN*, 1927, No. 6, p. 192.

[23] Sarab'ianov, 'O nekotorykh', *PZM*, 1925, No. 12, p. 181.

[24] Deborin, *Sovremennye*, p. 189. Italics added.

[25] See Maksimov's review of Timiriazev's *Estestvoznanie i dialekticheskii materializm* in *PZM*, 1925, No. 8–9, pp. 302–309. At the time Maksimov wrote this review he had not yet joined the Deborinite faction, and his disagreement with Timiriazev on the subject of Einstein's theory seems to have been free of rancour.

[26] See, e.g., Shmidt, *Zadachi*, pp. 57–58; Deborin, 'Lenin', in Lenin, *Materializm i empiriokrititsizm* (1925 ed.), p. xxxv; Deborin, 'Engel's', *PZM*, 1925, No. 10–11, pp. 17–19; Agol, 'Metafizika', *PZM*, 1926, No. 3, pp. 149–150.

[27] For Stepanov and Sarab'ianov, see above, Chapter 9, and cf. references in note 7 above.

[28] See above, pp. 65, 97–98.

[29] See Luppol, ed., *Kul'turnaia revoliutsiia*, p. 64. For absolute figures see Veger, 'Desiatiletie', *FNIT*, 1933, No. 11–12, p. 116, Note that Veger inflated his figures by 30 per cent. Cf. *Nauchnyi rabotnik*, 1929, No. 1, p. 4.

[30] See Shmidt, *Zadachi*, p. 75.

[31] See *VKA*, 1929, No. 32/2/, p. 212. Note that this figure is for 1929.

[32] See the resolutions of the Central Committee and of the Fifteenth Party Congress, in *KPSS*, II, pp. 398–404, 346, and in *VKA*, 1929, Kn. 34/4/, p. 261.

[33] See Shmidt, 'Vysshye uchebnye zavedeniia v SSSR', *BSE* (1st ed,). XIX, 32–36, for many figures on this subject.

[34] See Pokrovsky, 'O podgotovke', *Kommunisticheskaia revoliutsiia*, 1929, No. 13, pp. 62–64.

[35] See Luppol and Gessen, 'O kruzhkakh', *Kommunisticheskaia revoliutsiia*, 1928, No. 14, pp. 81–82 *et pass.* Cf. Kol'man, 'Ob obostrenii', 1929, No. 1, p. 19: 'Who [will prevail over] whom? That question also applies to the preparation of new scientific cadres, and with ever greater persistence demands a solution. Whom are we training in our universities, in our scientific research institutes; for whom are we preparing a change-over [*smena*], for ourselves or for the bourgeois professors?'

[36] Reported in Pokrovskii, 'O podgotovke', *Kommunisticheskaia revoliutsiia*, 1929, No. 13, p. 64. Cf. Luppol, 'Perspektivy', *Nauchnyi rabotnik* 1927, No. 5–6.

[37] The quotation is from Ostrovitianov, 'O perspektivakh', *Nauchny rabotnik*, 1929, No. 7–8, p. 4.

[38] See, e.g., *VKA*, 1929, No. 35-36, pp. 381–385.

[39] Vyshinsky, 'O podgotovke', *Kommunisticheskaia revoliutsiia*, 1928, No 23–24, p. 109.

[40] *KPSS*, II, 404.

[41] *Sobranie zakonov*, 1927, *otdel* 1, No. 3, *stat'ia* 367, Article 22. For a convenient review of the new provisions, see Luppol, 'K vyboram', *Nauchnyi rabotnik*, 1928, No. 11. As the time approached for the election in 1928, an editorial in *Izvestiia* reminded the existing members of the provisions of Article 22. See 'Pered', *Izvestiia*, 1928, July 21.

[42] The quotation is from Il'in, 'K voprosu', *Izvestiia*, 1928, July 22. For the announcement of the elections, see 'Vybory', *Izvestiia*, 1928, April 14.

[43] See *Nauchnyi rabotnik*, 1929, No. 7–8, p. 86.

[44] Reported by Luppol, 'Antireligioznaia', *Antireligioznik*, 1929, No. 6, pp. 13–14.

[45] The quotation is from Ostrovitianov, 'O perspektivakh', *Nauchnyi rabotnik*, 1929, No. 7–8, p. 7.

[46] For Kalinin's vision, see above, p. 65.

[47] These words of Pokrovsky's were reported in indirect discourse by Torbek, 'Deiatel'nost'', *VKA*, 1929, Kn. 33/3/, p. 270.

[48] *Ibid.*

[49] 'Marksistskoe', *DVP*, 1928, No. 3, pp. 16–17. This was an unsigned editorial.

[50] See Deborin, *Sovremennye*, p. 165.

[51] *Ibid.*, p. 168.

[52] See 'Pervoe', *PZM*, 1929, No 5, p. 165. Karev was the speaker who wound up this meeting of the Society of Militant Materialist Dialecticians.

[53] See *VKA*, 1929, Kn. 35–36, p. 390.

[54] See Brown, *Proletarian Episode*, pp. 11, 266–267.

[55] Miliutin, 'O direktivakh', *VKA*, 1929, Kn. 32/2/, pp. 220–221.

[56] *Ibid.*, pp. 222–223.

[57] The quoted words are those of the young biologist, Iu. M. Vermel'. See Shmidt, *Zadachi*, p. 80. For Tseitlin to the same effect, see Deborin, *Sovremennye*, p. 158.

[58] Deborin, *Sovremennye*, p. 39.

[59] See 'Novyi etap', *PZM*, 1929, No. 5, p. 2.

[60] See 'Pervoe', *PZM*, 1929, No. 5, p. 135.

[61] For Tseitlin's recalcitrance, see note 57. See above, p. 212, for the Timiriazev Institute continuing its mechanist publications after the Conference of April, 1929. As late as 1936 there were still complaints that Timiriazev, Varjas, Aksel'rod, and Perov had refused to recant.

[62] For Deborin worrying over this problem, see 'Pervoe', *PZM*, 1929, No. 5, pp. 132–133. For a wrangle over Semkovskii's sincerity, see *ibid.*, pp. 137, 149–151, *et pass.* Note on p. 164 Deborin's effort to call off his young followers, who were hounding Semkovskii for more convincing proofs of his sincerity in endorsing the Conference's resolutions.

[63] See Deborin, *Sovremennye*, p. 107. Cf. Karev, 'Marksistskaia', *Bol'shevik*, 1924, No. 10, p. 53, for Karev speaking of the dictatorship of Marxism already in 1924. Cf. also Bukharin, *Ataka*, p. 133, for an exchange with the eminent psychologist Chelpanov concerning the 'ideological dictatorship of Marxism'.

[64] Deborin, *Sovremennye*, pp. 106–107. This policy did not cease with the

Conference of April, 1929. See, e.g., the back cover of Shmidt, *Zadachi*, for Rubanovskii, *Problemy materii*, a mechanist work, advertised on the cover of the proceedings of the Conference that condemned mechanism.

[65] For Luppol on the 'ideological dictatorship', see Luppol, ed., *Obshchestvennye nauki*, pp. 6–7. For his distinction between Communists and non-Communist specialists, see Deborin, *Sovremennye*, pp. 119–120.

16. THE GREAT BREAK FOR NATURAL SCIENTISTS

[1] Stalin, *Sochineniia*, XII, 118. The article was originally published in *Pravda*, Nov. 7, 1929.

[2] 'Spinal year' is the expression of M. Dobb, *Soviet Economic Development*, pp. 228–229. The standard translation is simply, 'the great change'. See J. Stalin, *Problems of Leninism*, p. 289.

[3] E. Kol'man, 'Ob obostrenii', *Kommunisticheskaia revoliutsiia*, 1929, No. 1, pp. 18–19.

[4] See Ostrovitianov, 'O perspektivakh', *Nauchnyi rabotnik*, 1929, No. 7–8, p. 10, and also A. Vyshinskii, 'O nashikh kadrakh', *Nauchnyi rabotnik*, 1930, No. 1, pp. 33–36. Vyshinskii seems to have been in charge of the 're-elections'.

[5] See the reports in *Nauchnyi rabotnik*, 1929, No. 1, pp. 90–91; No. 2, pp. 88–92; No. 3, pp. 81–83; No. 4, p. 109. The report last cited gives the impression that Luzin was elected *qua* mathematician, but see 'Vybory novykh akademikov', *Izvestiia*, Jan, 13, 1929, for the original report.

[6] *Nauchnyi rabotnik*, 1929, No. 3, pp. 82–83.

[7] *Ibid.*, p. 85.

[8] See *VKA*, 1929, No. 35–36, p. 383; and Romanova, 'Nekotorye vyvody', *Izvestiia*, 1929, 10 Sept.

[9] *Nauchnyi rabotnik*, 1930, No. 1, pp. 96–97. See also 'Vrediteli v Akademii Nauk', *Izvestiia*, 1929, 16 Nov. The number of people fired is uncertain because of an ambiguity in the report.

[10] For the last reference before 1929 to secret arrests and secret trials of scholars, as far as the present writer knows, see 'K vysylke kontrrevoliutsionnernoi professury', *Pravda*, 1922, August 31. The scholars in question were in the humanities and social sciences; they were comparatively few; their punishment was to be banished from Soviet Russia.

[11] The definition is from Ushakov, *Tolkovyi slovar' russkogo iazyka* (M., 1935). The initial trial, still remembered in Soviet accounts as proof of the terrible danger necessitating the terror, was the Shakhty affair. See, e.g., Kim, *Sorok let sovetskoi kul'tury*, pp. 193–194.

[12] See *Nauchnyi rabotnik*, 1929, No. 12, pp. 88–89, and 1930, No. 1, pp. 96–97, and *VARNITSO*, 1930, No. 2, pp. 73–76. See also *Izvestiia*, 1929, 16 Nov.

[13] Lepeshinskii, 'Problema intelligentsii', *VKA*, 1931, No. 1, p. 83.

[14] *Nauchnyi rabotnik*, 1930, No. 2, pp. 115–116.

[15] E. Kol'man, 'Politika, ekonomika i matematika', in Kol'man, ed., *Na bor'bu*, pp. 53–54. This was originally a speech to the Society of Dialectical Materialist Mathematicians in November, 1930.

[16] Quoted from the preface to *ibid.*, p. 8.

[17] See A. Lunacharsky, 'Proletarskaia revoliutsiia i nauka', in Luppol, ed., *Kul'turnaia revoliutsiia*, pp. 28–29, for the 1928 speech. Compare Lunacharsky, 'Intelligentsiia i sotsializm', *VARNITSO*, 1930, No. 2, pp. 10–13.

[18] Radek, 'Po tu ili druguiu storonu barrikady', *VARNITSO*, 1930, No. 7–8, p. 5.

[19] *Ibid.*

[20] 'K XIII godovshchinu OPGU', *Nauchnyi rabotnik*, 1930, No. 11–12, p. 168. The anniversaries were counted from the founding of the OGPU's predecessor, the Cheka, on December 20, 1917.

[21] M. Gorky, 'Somov i drugie', in Gorky, *Sobranie sochinenii*, XVIII, 7–78. The play was left unfinished. For Gorky's irritation with the Russian intelligentsia, see especially his *Zhizn' Klima Samgina* [*The Life of Clim Samgin*], various editions, and in English translation. This enormous novel (four volumes) was written in the period from 1925 to Gorky's death in 1936, and ends symbolically with the feckless *intelligent* who is the chief protagonist being trampled to death by a revolutionary crowd. Gorky, who was after all an *intelligent* himself, could also admire the Russian intelligentsia. See, e.g., his letter to Ol'denburg on the 200th anniversary of the Academy of Sciences, in *Sobranie sochinenii*, XXIX, 440–441.

[22] *Ibid.*, XVIII, 53–54.

[23] *Ibid.*, p. 55.

[24] *Ibid.*, XXV, 81–82. This was originally published in *Izvestiia*, Dec. 12–13, 1929, that is, in the immediate aftermath of the purge in the Academy of Sciences. There is of course a very sharp contrast between this article and the letter of 1925 cited in note 21.

[25] Afinogenov, 'Strakh', in his *P'esy* (Moscow, 1956), pp. 83–158. For an English translation, see Lyons, *Six Soviet Plays*, pp. 389–469.

[26] See the review in *Pravda*, Jan, 7, 1932, of the Moscow première. (The play had originated in Leningrad.)

[27] See *Nauchnyi rabotnik*, 1930, No. 1, pp. 109–110, and cf. Kim, *Sorok let*, p. 198.

[28] O. Iu. Shmidt and B. Ia. Shmulevich, *Nauchnye kadry*, pp. 34–35. Cf. the editorial in *VARNITSO*, 1930, No. 2, pp. 4–5 for a proposal to liquidate the separate Unions of Engineers, Scientific Workers, and Educators, in order to end the specialist's feeling of aloofness from ordinary workers. For the tripling of the student body in 1928–1932, see N. I. Boldyrev, ed., *Direktivy VKP* (*b*), II, 77. For the approximate doubling of the teaching staff in 1929–1931 (admittedly not quite the same period), see N. DeWitt, *Soviet Professional Manpower*, p. 175.

[29] O. Iu. Shmidt, 'Rol' matematiki', p. 31.

[30] The decree may be found in KPSS, II, 512–522. That theoretical subjects were pushed into the background is evident from the Decree of September 19, 1932, which restored traditional curricula. For that Decree, see Boldyrev, ed., *Direktivy VKP* (*b*), pp. 77–89.

[31] For the abolition of courses in physics and chemistry, see *ibid.*, p. 79. For a report of professorial protest, see Ostrovitianov, 'K predstoiashchemu plenumu', *Nauchnyi rabotnik*, 1930, No. 1, p. 4.

[32] For Lysenko's own account of his initial failure to impress a meeting of biologists in January, 1929, and of his subsequent success with peasant farmers, see his *Agrobiologiia*, (3rd ed.; Moscow, 1948), pp. 16–17. The quotation is from a speech of Commissar Iakovlev to the All-Union Conference on Struggle with Drought, which met in October, 1931. See *Biulleten' iarovizatsii*, 1932, No. 1, p. 3.

[33] *Kazanskii meditsinskii zhurnal*, 1931, No. 4–5, pp. 520–521.

[34] See, e.g., Shmidt, 'Rol' matematiki', *Trudy*, p. 30. For a review of the sharp disputes over planning, see A. Kaufman, 'The Origin of "The Political Economy of Socialism"', *Soviet Studies*, January, 1953, pp. 243–272. Soviet references to the leading 'geneticists', as those planners were called who 'treated the plan as a plan-prognosis, the basic positions of which should be determined by the method of extrapolation', suggest that they were condemned as 'enemies of the people'. The quoted characterization of their position is from the volume *SSSR* of *BSE* (2nd ed.), pp. 535–536.

[35] A. Ia. Boiarskii, B. S. Iastremskii, and V. I. Khotimskii, *Teoriia matematicheskoi statistiki* (Moscow, 1930). The insurance specialist was Iastremskii, for whose biography see *BSE* (1st ed.), LXV, 801–802. Cf. the biography of Khotimskii in LX, 111–112.

[36] Kol'man, ed., *Na bor'bu*, p. 5.

[37] The young complainant was S. A. Ianovskaia. See Kommunisticheskaia Akademiia. *Za povorot*, p. 39. That the new theory of statistics was already moribund by 1936 is suggested by the absence of the paper Boiarskii gave at a convention of 1930 from the convention's *Proceedings*, which were not published until 1936. See *Trudy pervogo*, especially p. 370. For proof of the theory's complete death, see A. G. Kurosh, ed., *Matematika v SSSR za tridtsat' let*: 1917–1944 (Moscow, 1948).

[38] See, e.g., *Zhurnal russkogo botanicheskogo obshchestva*, which terminated in 1931, and was succeeded by *Botanicheskii zhurnal* in 1932. Even earlier V. L. Komarov, one of the principal agents in the Bolshevization of the Academy of Sciences, had become chief editor of the journal. But the first sign of partyness (*partiinost'*) in the journal was an editorial in the 1933, No. 5 issue: 'K 16-letiiu Oktiabria.' It should also be noted that the *Kazan Medical Journal*, which carried the pistol-waving review, was otherwise the usual staid academic journal. Even the journal of popular science (*Priroda*) published by the Academy of Sciences, was very long in giving up articles written from a non-Marxist viewpoint, although it suffered great changes in its editorial staff.

[39] *VARNITSO*, 1930, No. 7–8.

[40] For an example of the earlier headshaking, see M. Pistrak, 'Puti ideologicheskogo zavoevaniia', *Kommunisticheskaia revoliutsiia*, 1928, No. 14, p. 73. The drive to capture the scientific societies can be traced to the Third Congress of the Union of Scientific Workers, which met in February, 1929. See the report of the Union's Secretary, K. V. Ostrovitianov,

'Itogi III s'ezda nauchnykh rabotnikov', *Nauchnyi rabotnik*, 1929, No. 4, and especially the excerpts from the resolutions adopted, pp. 25–26.

41 See, e.g., B. M. Zavadovskii, 'Itogi', *EIM*, 1930, No. 2–3, pp. 142–160; S. Vail', 'II vsesoiuznyi s'ezd', *EIM*, 1930, No. 2–3, pp. 160–165.

42 See Zavadovskii, 'Itogi', *EIM*, 1930, No. 2–3.

43 See Vail', "II vsesoiuznyi s'ezd', *EIM*, 1930, No. 2–3.

44 See the report in *Priroda*, 1930, No. 10, pp. 1044–1045. For evidence of Ioffe's Soviet patriotism, see *Pravda*, Feb, 2, 1932, for his speech to the Seventeenth Conference of the Communist Party on the service of physics in the construction of socialism.

45 The evidence for this statement is the complete absence of any allegations of such a connection in the reports of the turmoil among the mathematicians. It is also significant that the group of five who tried to reorganize the Moscow Mathematical Society at the end of 1930 did not include supporters of Iastremskii's 'revolutionary' theory of statistics.

46 The most detailed account of the Egorov affair is in L. A. Liusternik, L. G. Shpirel'man, A. Gel'fond, L. Pontriashii, and Nekrasov, 'Deklaratsiia', *Nauchnyi rabotnik*, 1930, No. 11–12, pp. 67–71.

47 See I. G. Dolin, 'Na vstreche aspirantov', *Nauchnyi rabotnik*, 1930, No. 1, p. 18.

48 For the Congress of Mathematicians see *Trudy pervogo s'ezda*. See E. Kol'man, 'Vreditel'stvo v nauke', *Bol'shevik*, 1931, No. 2, pp. 78–79, for the news that the Congress refused to send greetings to the Party Congress. To be sure, the unsigned report of the Mathematicians' Congress in *Nauchnyi rabotnik*, 1930, No. 8–9, pp. 106–107 quotes a fervent pledge of co-operation in the construction of socialism, which was allegedly sent to the Party Congress. Apparently this was premature reporting. Note that the resolutions of the Mathematicians' Congress, as printed in the *Trudy*, do not contain such a greeting.

49 Reported in *Bol'shevik*, 1931, No. 2, pp. 78–79.

50 *Ibid.*

51 For the news of Egorov's arrest see *VARNITSO*, 1930, No. 11–12, p. 73. The 'Initiating Group's' Declaration stated, without any supporting evidence, that 'participation in a counter-revolutionary organization' was the reason for Egorov's arrest. *Nauchnyi rabotnik*, 1930, No. 11–12, p. 68. For the report of the Society's business meeting following the arrest, see *ibid.*, p. 70. Note that the member of the Communist Youth was expelled for his participation, and the older mathematician who chaired the meeting was deprived of an academic post. Since then both have been rehabilitated.

52 *Ibid.*, p. 68.

53 *Ibid.*

54 See 'Ot redaktsii', *Matematicheskii sbornik*, XXXVIII, No. 1–2, inside front cover, for the declaration of the 'Initiating Group' to the effect that they were in control of this famous old publication of the Moscow Mathematical Society. But see the appeal in the following, No. 3–4, issue: 'Sovetskie matematiki, podderzhivaite svoi zhurnal!' ['Soviet Mathematicians, Support Your Journal!'], which contains the tacit ad-

mission that the journal had been boycotted. Cf. also P. S. Aleksandrov, 'Moskovskoe matematicheskoe obshchestvo', *Uspekhi matematicheskikh nauk, Novaia seriia*, 1946, I, No. 1 (11), p. 236, for a brief statement of the events described above, in which the cessation of the Society's activities during 1931 is noted. Egorov's date of death is given in current Soviet references to him. See, e.g., his biography in *BSE* (2nd ed.), XV, 468–469.

[55] Kommunisticheskaia Akademiia, *Za povorot*, pp. 31–32. The speaker was P. P. Bondarenko.

[56] 'Postanovlenie TsK VKP (b) ot 15 marta 1931 g. po dokladu presidiuma Kommunisticheskoi Akademii', *VKA*, 1931, No. 2–3, p. 5.

[57] Stalin, *Sochineniia*, XIII, 29–42.

[58] *Ibid.*, pp. 38–39.

[59] See the reports of their establishing 'methodological control', in *VKA*, 1931, No. 2–3, pp. 42–43; No. 5–6, pp. 118–119; No. 8–9, pp. 114–115; No. 12, pp. 69–71.

[60] 'Nauchnye obshchestva pered sudom sovetskoi obshchestvennosti', *Varnitso*, 1931, No. 3, p. 59.

[61] Bukharin, 'Doklad', *Pravda*, 1931, 9 April.

[62] *Pravda*, 1931, 14 April.

[63] Stalin, *Sochineniia*, XIII, 72. Stalin's implicit admission that this had been Bolshevik policy for the past two years was hardly cancelled by the startlingly incongruous assertion immediately following: 'We have always considered "*spetseedstvo*" [literally: "eating specialists", i.e., hostility to them] a harmful and shameful thing, and we still do'.

[64] Bukharin, *et al.*, 'Podgotovka', *Pravda*, 1931, 24 August.

[65] Kol'man, 'Bor'ba za ovladenie naukoi', *Pravda*, 1931, 4 October.

[66] A. Stetskii, 'Ob uproshchenstve i uproshchentsakh', *Pravda*, June 4, 1932. The examples cited in the text above are Stetskii's.

[67] For his speech of the year before in a quite different spirit, see Stetskii, 'O Komakademii i nauchnoi rabote', *VKA*, 1931, No. 2–3, pp. 6–17.

[68] 'Ob uchebnykh programmakh i rezhime v vysshei shkole i tekhnikumakh', in Boldyrev, ed., *Direktivy VKP (b)*, II, 77–89.

[69] The order for a general Party purge, dated 12 January 1933, may be found in *KPSS*, II, 741. For reports of the purge in some institutions of higher learning, see Nemilov, 'O partiinoi chistke', *FNIT*, 1933, No. 12, pp. 115–119; and S. Paramonov, 'O partiinoi chistke', *ibid.*, pp. 120–122. For a speech to the Party organization at the Academy of Sciences warning against 'left excesses' in the treatment of the non-Party specialist, see V. P. Volgin, *Akademiia nauk SSSR*, pp. 85–91.

[70] See Nemilov, 'O partiinoi chistke', *FNIT*, 1933, No. 12, pp. 117–118.

[71] There is statistical evidence that the majority of professors and senior researchers were not men of twenty-five, as Shmidt had suggested they would be. See De Witt, *Soviet Manpower*, pp. 178–179. The median age among professors *in 1947* was 53.3, which indicates that, *seventeen years after Shmidt's comment*, more than half of the Soviet Union's professors were

people who had received much of their higher education before the Revolution. That this older generation of scientists was still felt to be ideologically alien after the great break is evident in many comments. See, e.g., E. Kol'man, 'Triumf marksizma—nauki proletariata', in Institut Marksa-Engel'sa-Lenina, *Marksizm i estestvoznanie*, pp. 64 *et passim*. For an example of the indomitable few who persisted even after the great break in expressions of disapproved ideologies, see V. I. Vernadskii, 'Problema vremeni v sovremennoi nauke', *Izvestiia Akademii Nauk, otdelenie matematicheskikh i estestvennykh nauk*, 1932, No. 4.

[72] This has been the main stress of Soviet commentators on the great break. See, e.g., E. Iaroslavskii, *O roli intelligentsii v SSSR* (Moscow, 1939), pp. 25 ff.

[73] In Russia, as in the rest of Europe, Christianity was established by royal edict. Vladimir, the Primary Chronicle tells us, ordered the chief heathen idol flogged through the streets and thrown in the Dnieper. 'And then Vladimir sent through the whole town, saying: "If anyone does not come to the river tomorrow morning [for baptism], be he rich or poor, or beggar or slave, he will be my enemy"'. D. S. Likhachev, ed., *Povest' vremennykh let* (M., 1950), I, 80.

17. THE GREAT BREAK FOR PHILOSOPHERS

[1] In 1924 Stalin organized a seminar on Leninism at the Academy. See *VKA*, 1924, Kn. 8, p. 379. For his speech in December, 1929, see Stalin, *Sochineniia*, XII, 141–172.

[2] For evidence of this effort to forget, note the absence of an article on the Communist Academy in *BSE* (2nd ed.). Cf. *Istoriia SSSR*, 1958, No. 6, pp. 159–162, for a recent effort to stimulate historical research on the first phase of higher learning in the Soviet Union. Note especially the recognition, on p. 160, that the cardinal importance of the Institute of Red Professorship 'has been hushed up in our historical literature until recently'. This hush has been even more marked in the case of the Communist Academy.

[3] Stalin, *Sochineniia*, XIII, 107.

[4] See *Trudy pervoi vsesoiuznoi konferentsii agrarnikov-marksistov*, I, 11–60.

[5] *Ibid., passim.*

[6] Adoratskii, 'I. V. Stalin', *Izvestiia*, 21 December 1929.

[7] Stalin, *Sochineniia*, XII, 142.

[8] *Ibid.*

[9] See *Nauchnyi rabotnik*, 1930, No. 3, p. 104. Cf. also 'Krasnoi professury, instituty', *BSE* (1st ed.), XXXIV, 600–601.

[10] See again, *Nauchnyi rabotnik*, 1930, No. 3, p. 104.

[11] Ral'tsevich, *Na dva fronta*, p. 5.

[12] For Iudin's biography, see *BSE* (2nd ed), XLIX, 359–360.

[13] Ral'tsevich, *Na dva fronta*, p. 5.

[14] This account has been pieced together from fragmentary, retrospective, somewhat inconsistent reports and comments, See Iudin's account in *Za marksistsko-leninskoe uchenie*, pp. 7–8. For a more revealing

account, see Ral'tsevich, *Na dva fronta*, pp. 5–6. Cf. also Ral'tsevich's account in *Raznoglasiia*, p. 170.

[15] See, e.g., Ral'tsevich, 'Protiv mistitsizma', *RIK*, 1930, 28 February, pp. 7–15. Cf. the entirely favourable review of Luppol's *Lenin i filosofiia*, in *Molodaia gvardiia*, 1930. No. 6 (March). Tarasenkov, 'Filosofiia', *Iunyi kommunist*, 1930, No. 7 (April), p. 58, recalls Lenin's criticism of Deborin, but immediately preceding is an entirely Deborinite article, Matusov, 'Zadachi', *Iunyi kommunist*, 1930, No. 7. It will be noted that Tarasenkov's article appeared in April, after Iaroslavskii's criticism of the Deborinites. For the Deborinites still speaking as the officially approved philosophers early in 1930, see, e.g., Karev, 'K voprosu', *Bol'shevik*, 1930, No. 2 (31 Jan.).

[16] M. Furshchik, 'O liberal'nom i marksistskom ponimanii etiki', *Bol'shevik*, 1930, No. 6 (31 March).

[17] See the exchange of letters in *Antireligioznik*, 1930, No. 5 (May), pp. 66–71, and No. 6 (June), pp. 70–78, and No. 7 (July), pp, 83–86. Cf. also Ral'tsevich, *Na dva fronta*, pp. 6–7.

[18] Reprinted in *Antireligioznik*, 1930, No. 6, pp. 73–78.

[19] For Mitin's biography, see *BSE* (2nd ed.), XXVII, 598, and *World Biography* (N.Y., 1948). For evidence of his adherence to the Deborinite bloc before 1930, see Mitin, 'Na filosofskom fronte', *Kommunisticheskaia revoliutsiia*, 1929, No. 8 (April), and his speech to the First All-Union Conference of the Society of Militant Materialist Dialecticians, reprinted in *PZM*, 1929, No. 5, pp. 134–136.

[20] See Deborin, 'Itogi', *PZM*, 1930, No. 6, pp. 1–19. For the resolution adopted, see 'Ob itogakh', *PZM*, 1930, No. 4, p. 4.

[21] Reported by Iudin in *Raznoglasiia*, p. 219.

[22] *Ibid.*, pp. 219–220. Cf. also Ral'tsevich, *Na dva fronta*, pp. 6–7.

[23] Reported by Iudin, in *Raznoglasiia*, pp. 219–220.

[24] Cited in *PZM*, 1930, No. 5, p. 141.

[25] Mitin, Ral'tsevich, Iudin, 'O novykh zadachakh', *Pravda*, 1930, 7 June.

[26] *Ibid.*

[27] See Sarab'ianov's letter in *Antireligioznik*, 1930, No. 7, p. 86. Note that Sarab'ianov blames this method on the Deborinites. But they never matched the Bolshevizers in this respect. See, for a most egregious example, the treatment of Karev in Vesna, 'O melkoburzhuaznykh shataniiakh', *Pravda*, 1930, 31 July. Cf. Karev's letter to the editor in *Pravda*, 1930, 10 August.

[28] Deborin, *Lenin kak myslitel'* (3rd ed., 1929), p. 26. The first edition appeared in 1924.

[29] *Ibid.*, p. 59.

[30] Mitin, *Boevye voprosy*, p. 7. For Deborin's anti-Leninist essay of 1908, see above, p. 34.

[31] Perhaps the last time this characterization was used was at the Conference of October, 1930. See Deborin ridiculing it, in *Raznoglasiia*, p.16.

[32] The Resolution of the Party Organization of the Institute of Red Professorship, which was adopted on the eve of the Conference of October,

1930, contains this formulation. See *Bol'shevik*, 1930, No. 19–20, p. 106, *et passim.*

33 Mitin, *Boevye voprosy*, p. 25.

34 See 'Dialektiku—v massy', *RIK*, 1930, No. 19–20, pp. 21–22.

35 Cf. Mitin, *Boevye voprosy*, pp. 44–59. Cf. also Vandek and Timosko, 'Vstupitel'naia stat'ia', in Plekhanov, *Protiv filosofskogo revizionizma* (Moscow, 1935); and V. A. Fomina, 'Rol' G. V. Plekhanova', in Shchipanov, ed., *Iz istorii russkoi filosofii* (Moscow, 1951), pp. 629–703. Since 1954 Soviet authors have been modifying this view of Plekhanov.

36 In their first major article the Bolshevizers wrote: 'We must now concretize the position . . .' that Lenin took on partyness. See Mitin, Ral'tsevich, Iudin, 'O novykh zadachakh', *Pravda*, 1930, 7 June.

37 Mitin, 'K voprosu', *RIK*, 1930, No. 19–20 (31 October), pp. 36–37.

38 'Dialektika—v massy', *RIK*, 1930, No. 19–20, p. 25.

39 Mitin, 'K voprosu', *RIK*, 1930, No, 19–20, p. 37.

40 See the report of the Communist Academy's plenary meeting of 17–18 June 1930, in *VKA*, 1930, Kn. 39, especially pp. 25–26, and 42, for Pokrovsky and Miliutin criticizing the work of the Institute of Philosophy. They were less severe toward it than toward other institutes, and they made no reference to the Bolshevizers' recent article in *Pravda*. The same session elected Karev and Gessen to membership in the Academy. See *Ibid.*, p. 88. Cf. the report of this meeting in *Pravda*, 22 June, 1930, in which R(al'tsevich) makes it seem that the Institute of Philosophy was the problem child of the Academy. For another report, which makes clear the tendentious nature of R(al'tsevich)'s reporting, see *Nauchnyi rabotnik*, 1930, No. 8–9, pp. 102–105.

41 The most notable, after Mitin, were V. N. Ral'tsevich, Ernst Kol'man, V. P. Egorshin, and Konstantin Milonov.

42 See Maksimov's letter to the editor. *Pravda*, 1930, 17 August; and cf. Deborin's reply in 24 August.

43 See especially Deborin, Luppol, Sten, Karev, *et al.*, 'O bor'be', *PZM*, 1930, No. 5, pp. 139–149, a joint reply of ten Deborinites to the Bolshevizers' original article in *Pravda*. Cf. also Karev, 'Zametka o vrede putanitsy', *PZM*, 1930, No 7–8; Novikov, 'Voinstvuiushchaia putanitsa', *loc. cit.*; and Podvolotskii, 'XVI s'ezd', *loc. cit.*

44 Deborin *et al.*, 'O bor'be', *PZM*, 1930, No. 5, p. 145.

45 *Ibid.*

46 For a stenographic report of the Conference, see *Raznoglasiia* or *VKA*, 1930, *kn.* 40–41 and 42.

47 For Miliutin's membership in the Control Commission, see his biography in *MSE* (1st ed.), 218. For his speech, see *Raznoglasiia*.

48 *Raznoglasiia*, pp. 15–16.

49 *Ibid.*, pp. 132–146.

50 *Ibid.*, p. 98.

51 See above, p. 39.

52 See *Raznoglasiia*, pp. 231–236.

53 *Ibid.*, p. 84. The Deborinite was Novikov.

54 See Mitin, *Boevye voprosy*, for a reprint of his speech. The quotation occurs on pp. 17–18.

55 *Raznoglasiia*, p. 63. The speaker was Milonov.

56 *Ibid.*, p. 67.

57 *Ibid.*, p. 159.

58 *Ibid.*, p. 125.

59 *Ibid.*, pp. 256–268.

60 *RIK*, 1930, No. 19–20, p. 30.

61 The 'Rezoliutsiia' in *Raznoglasiia*, pp. 276–281, is an altered version of the resolution originally adopted on November 2, 1930. The alteration, which was made on January 11, 1931 following Stalin's intervention, is pointed out in the version printed in *VKA*, 1931, *kn.* 1, pp, 18–19. For a Bolshevizer complaining that the Resolution of the Communist Academy repeated the Resolution of the Institute of Red Professorship 'in a more "delicate" form', see Ral'tsevich, *Na dva fronta*, p. 8.

62 See *RIK*, 1930, No. 19–20 (31 October). Ral'tsevich was the new editor. Previously, the magazine had an editorial 'college', including Deborin, Luppol, and Sten.

63 For Sten's declaration, see *Raznoglasiia*, p. 121. For the expulsion, see 'Pochemy molchit tov. Sten?' *Pravda*, 1930, 23 November. Cf. also Iudin, 'K filosofskoi diskusii', *Pravda*, 1930, 18 November.

64 *Ibid.* On Lominadze, see Deutscher, *Stalin*, p. 333.

65 Iudin, 'K filosofskoi diskussii', *Pravda*, 1930, 18 November.

66 *PZM*, 1930, No. 9, came out in September or October. *PZM*, 1930, No. 10–12, appeared in February, 1931, after the Central Committee's Decree of 25 January 1931.

67 The conversation took place on 9 December 1930. See Stalin, *Sochineniia*, XIII, 401. For Mitin's report of the conversation (at a meeting in the Communist Academy on 1 January 1931), see Mitin, *Boevye voprosy*, pp. 43–44. For his more extensive report several years later, see Mitin, 'Nekotorye itogi', *PZM*, 1936, pp. 21–26.

68 Mitin, *Boevye voprosy*, p. 44.

69 See *VKA*, 1931, No. 1, pp. 18–19.

70 'Postanovlenie Ts. K. V.K.P. (b) o zhurnale Pod znamenem marksizma', *Pravda*, 1931, 26 January.

71 See 'Men'shevistvuiushchii idealizm', *BSE* (1st ed.), XXXVIII, 827–830. They were condemned *before* the major trials of the late 'thirties. See Mitin, 'Nekotorye', *PZM*, 1936, No. 1, pp. 26–27. Sten seems to have been condemned already in 1933. See Kommunisticheskaia Akademiia, *Materialy nauchnoi sessii*, p. 103; and Kol'man, 'Triumf marksizma', in Institut Marksa-Engel'sa-Lenina, *Marksizm i estestvoznanie*, p. 69.

72 See Deborin, ed., *Na boevom postu.*

73 For an unverified report of his further punishment, see *Sotsialisticheskii vestnik*, 1942, No. 19–20, p. 236. He is supposed to have been exiled to a town on the Volga, where he died in 1942. I was told much the same story in Leningrad in the summer of 1958.

74 Iaroslavskii, *Tret'ia sila*, p. 37. See *passim* for Iaroslavskii's charges against Riazanov.

[75] It appears that Riazanov clashed with the Bolshevizers on this issue already in 1930. See *Za marksistsko-leninskoe uchenie*, pp. 14–15.

[76] See *Bol'shevik*, 1931, No. 5, p. 76, for the charge that Riazanov printed an abridged version of Marx's letter to Jenny Longuet ***without giving notice of the abridgment.*** Cf. Riazanov, 'Novye dannye', *VSA*, 1923, Kn. 6, pp. 371–372, for Riazanov's history of this letter, whose criticisms of Karl Kautsky had been omitted from the version published in 1899. The original had fallen into the possession of Menshevik émigrés, who had offered Riazanov an abridged copy, which he refused. Cf. *VM*, 1925, No. 4, pp. 177–181, where the 1899 version is reprinted ***with explicit and detailed notice of the abridgment.*** (On p. 157 the editors promised an article by Riazanov in a future issue; unfortunately, I did not have access to later issues of this journal.)

Bol'shevik, 1931, No. 5, pp. 75–76, prints a 1925 letter to Riazanov from a Menshevik émigré, promising him the original of the letter. There is no ground for accusing Riazanov of dissimulation or dishonesty. The 1899 version of Marx's letter, clearly labelled as an abridgment, was republished in *VM* before he obtained the original. The possibility remains that he obtained the original on terms restricting the time of its publication, and that he honoured these terms.

[77] See above, pp. 200 ff. For Adoratskii's switch to the Bolshevizers, see *Raznoglasiia*, pp. 184–187. For his paean to Stalin as the supreme theorist, see *Izvestiia*, 1929, 21 December.

[78] Iudin, 'God raboty', *PZM*, 1932, No. 1–2, p. 123.

[79] See 'Postananovlenie Ts. K. V.K.P. (b)', *VKA*, 1931, No. 2–3, pp. 3–5.

[80] Mitin, 'Nekotorye itogi i zadachi', *PZM*, 1936, No. 1, p. 38.

[81] See again Iudin's biography in *BSE* (2nd ed.), XLIX, 359–360. It is encouraging to note that this biography does not call him a philosopher. (Until recently he was Soviet ambassador to China.) His chief effort in philosophy was *Kratkii filosofskii slovar'* (M. Rozental', co-author), which Leopold Infeld has called 'a monument of shame of the past period'. E. Stillman, ed., *Bitter Harvest* (N.Y., 1959), p. 240.

[82] Stalin, 'O nekotorykh voprosakh istorii bol'shevizma', *Sochineniia*, XIII. The reference to 'archive rats' occurs on p. 96.

[83] Dobrynin, 'Za leninskuiu pereotsenku nasledstva Plekhanova', *RAPP*, 1931, No. 3. The offensive passage is on pp. 50–51. Cf. Iudin's attack in 'Lenin i filosofskaia diskussiia', *PZM*, 1931, No. 9–10, pp. 22–23. And cf. the letter of apology from the editors of RAPP, in *Pravda*, 1931, 19 Dec.

[84] Mints, 'V redaktsiiu "Bol'shevika"', *Bol'shevik*, 1931, No. 23–24, p. 135.

[85] *Ibid.*, pp. 135–138. For Iaroslavskii's confession, and disapproval of Mints', see *Bol'shevik*, 1931, No. 21, pp. 83–86.

[86] See Iudin, 'God raboty', *PZM*, 1932, No. 1–2, especially p. 119. Note that Iudin omits mention of the Conference of October, 1930. This omission became standard in Soviet accounts of the controversy. See, e.g., 'Men'shevistvuiushchii idealizm', *BSE* (1st ed.), XXXVIII, 827–830.

87 The present writer was unable to find the letter, but repeated references to it establish its existence. See, e.g., Mitin, 'Nekotorye itogi', *PZM*, 1936, No. 1, pp. 26–27.

88 Kommunisticheskaia Akademiia, *Materialy nauchnoi sessii*, p. 137.

89 *Ibid.*, pp. 154–158, 160, *et pass.*

90 *Ibid.*, pp. 191–197.

91 *Ibid.*, p. 203.

92 *Ibid.*

93 See, e.g., Iudin, 'God raboty', *PZM*, 1932, No. 1–2, p. 124; and Mitin, in Kommunisticheskaia Akademiia, *Materialy nauchnoi sessii*, pp. 46–51. Cf. also Mitin, 'Nekotorye itogi', *PZM*, 1936, No. 1, p. 41: 'There is no daring for great monographic works'.

94 See Timiriazev's speech in *Raznoglasiia*, pp. 231–236. Cf. also Timiriazev, Perov, and Varjas, 'S bol'noi golovy na zdorovuiu', *PZM*, 1931, No. 1, pp. 77–82.

95 See *Raznoglasiia*, p. 214. The speaker was Kerzhentsev. Cf. also the speeches of Iaroslavskii and O. Iu. Shmidt.

96 See Amelin and Cheremnykh, 'Advokaty teoreticheskoi bazy pravogo opportunizma', *PZM*, 1931, No. 1–2, pp. 83–112.

97 See Kommunisticheskaia Akademiia, *Za povorot*, *passim*, for representative speeches of the Bolshevizers in the field of natural science.

98 See, e.g., Iu. P. Shein, 'Opredelenie skorosti', *Priroda*, 1931, No. 9, pp. 877–886; and Shein, 'Metafizika fiziki', *Problemy marksizma*, 1931, No. 2, pp. 94–102. For a number of examples from minor authors, see Stetskii, 'Ob uproshchenstve', *Pravda*, 1932, 4 June.

99 See, e.g., Ianovskaia's speech in Kommunisticheskaia Akademiia, *Za povorot*, p. 39.

100 Stetskii, 'Ob uproshchenstve', *Pravda*, 1932, 4 June.

101 See A. A. Maksimov, 'Filosofiia i estestvoznanie', *PZM*, 1936, No. 1, pp. 56–57. Note also that courses in the philosophy of science gave way to courses in the history of individual sciences.

102 See, e.g., V. Egorshin, 'Kak I. E. Tamm kritikuet marksistov', *PZM*, 1933, No. 2, pp. 232–260. Cf. also Kommunisticheskaia Akademiia, *Materialy nauchnoi sessii*', pp. 303–391, for Maksimov's speech on 'Marxism and Natural Science' to a meeting in March, 1933. Both he and those who spoke in the discussion were remarkably restrained.

18. THE 'CRISIS' IN PHYSICS

1 L. S. Feuer, 'Dialectical', *Philosophy of Science*, April, 1949, p. 116; and Lazar Volin, 'Science', in Christman, ed., *Soviet Science*, p. 89. Cf. Wetter, *Dialectical Materialism*, pp. 416, 423, 424; and Deutscher, *Prophet Unarmed*, p. 180.

2 For a list of Fridman's works and an account of his life, see *Geofizicheskii sbornik*, 1927, Vol. V, *vypusk* I, pp. 7–63. For a list of Fok's works and an account of his life, see Akademiia Nauk, *Materialy k biobibliografii uchenykh SSSR, Seriia fiziki*, vypusk 7.

3 From an unsigned editorial in *Uspekhi fizicheskikh nauk*, 1931, No. 1, p. 2.

[4] See his biography in *BSE* (1st ed.), XXIX. Cf. also Ioffe, *Moia zhizn'*.

[5] See their biographies in *BSE* (1st ed.), LVIII and LVIX.

[6] Frank's paper, 'Uber die Begriffe und Sätze der klassischen Mechanik, die in der Quantenmechanik von Bedeutung sind', is noted in *Shestyi s'ezd russkikh fizikov*, p. 59. His articles are 'Gidromekhanika', *BSE* (1st ed.), XVI, and 'Volny', *ibid.*, XII.

[7] See 'Priroda', *Elektrichestvo*, 1930, No. 3, pp. 133–134; No. 8, p. 349.

[8] See the account in *Sorena*, 1933, No. 9, pp. 155–165.

[9] *Elektrichestvo*, 1930, No. 3, p. 136.

[10] See the review of literature concerning relativity in *PZM*, 1922, No. 9–10, especially p. 208, for the author's estimate that three-fourths of this literature 'are devoted to the propaganda of idealism and metaphysics, and sometimes scientific charlatanry as well. . . .'

[11] One of the last outlets for explicitly non-Marxist appraisals of the new physics was *Priroda*, a journal of popular science, published by the Academy of Sciences. See, e.g., N. V. Belov, 'Printsip prichinnosti', *Priroda*, 1929, No. 12. Cf. also Belov's introduction to a translation of Dirac on quantum mechanics in *Priroda*, 1931, No. 8, where Belov has already ceased to speak openly as a follower of Reichenbach in philosophical matters.

[12] See above, p. 72.

[13] Khvol'son, *Kharakteristika*, pp. 205–206.

[14] Fridman and Frederiks, *Osnovy teorii otnositel'nosti* I, 26–27.

[15] *Uspekhi fizicheskikh nauk*, 1931, No. 1, p. 6.

[16] The astronomer P. K. Shternberg was a Bolshevik before Timiriazev, but he died in 1920. See his biography in *BSE* (1st ed.), LXII.

[17] For his earliest criticism of relativity, see A. K. Timiriazev, 'Periodicheskaia', *KN*, 1921, No. 1; 'Printsip', *KN*, 1921, No. 2; and 'Obzor', *KN*, 1921, No. 4.

[18] See, e.g., Timiriazev, 'Oprovergaet', *PZM*, 1922, No. 4; 'Neskol'ko', *PZM*, 1923, No. 6–7; and 'Po povodu', *PZM*, 1926, No. 11. It should be noted that Timiriazev believed in the possibility of exploiting atomic energy, even though he denied the universal applicability of Einstein's famous equation.

[19] See, e.g., Timiriazev, 'Teoriia otnositel'nosti Einshteina i makhizm', *VKA*, 1924, *kn.* 7.

[20] Lenin, *Sochineniia*, XXXIII, 206–207.

[21] Nevskii, 'Restavratsiia', *PZM*, 1922, No. 7–8, p. 124; 'Sovetskaia nauka', *PIR*, 1923, No. 3, pp. 114–115; 'Marksizm', *PZM*, 1923, No. 12, pp. 210–211.

[22] Trotsky, *Literature and Revolution*, p. 219.

[23] Deborin, 'Lenin', *PZM*, 1924, No. 2, pp. 14–15. This essay was reprinted in 1925 as an introduction to Lenin's *Materialism and Empiriocriticism*.

[24] Reported in *VKA*, 1924, Kn. 7, p. 366.

[25] Stukov, 'V plenu', *PZM*, 1923, No. 4–5, and 'V chem', *PZM*, 1923, No. 6–7; I. E. Orlov, 'Klassicheskaia', *PZM*, 1924, No. 3, and

'Sushchestvuet', *PZM*, 1924, No. 1, and 'Novye', *VM*, 1924, No. 2. Rubanovskii's *Problemy materii* was not available to the present writer, but references to it indicate that it followed Timiriazev's line. For Kharazov's work, see 'Matematicheskaia', *VKA*, 1925, Kn. 10, and 'Malyi', *ibid.*, Kn. 12, 14.

[26] Maksimov, 'Populiarno-nauchnaia', *PZM*, 1922, No. 7–8.

[27] Maksimov, 'O printsipe', *PZM*, 1922, No. 9–10.

[28] Maksimov, 'Teoriia', *PZM*, 1923, No. 4–5, p. 144.

[29] Tseitlin, 'Teoriia . . .' in *Teoriia otnositel'nosti i materializm* (Leningrad, 1925), pp. 188–189.

[30] Marx-Engels, *Correspondence*, p. 530.

[31] See reference in note 29. Cf. also Tseitlin, 'Problema', *DVP*, 1928, No. 3, pp. 146 *et pass.*

[32] A possible exception is Tseitlin's 'Problema vsemirnogo tiagoteniia' *VKA*, 1925, Kn. 13, which the present writer is not competent to evaluate.

[33] See the preface, signed 'July, 1913', to Mach's posthumous *Die Prinzipien der physikalischen Optik* (Leipzig, 1921).

[34] Einstein, 'Ernst Mach', *Physikalische Zeitschrift*, 1916, No. 7. In his later years Einstein changed his views on Mach's philosophy. See Schilpp, ed., *Albert Einstein*, p. 21.

[35] V. Bazarov, 'Obzor', *VKA*, 1923, Kn, 3, p. 340. The original is in A. V. Vasil'ev, *Prostranstvo*, p. 148.

[36] Bazarov, 'Obzor', *VKA*, 1923, Kn. 3, p. 343. Bazarov himself was condemned as a 'wrecker' in 1931. See Popov, *Outline*, p. 440, and Shcheglov, *Bor'ba*, p. 55.

[37] See especially *Teoriia otnositel'nosti i ee filosofskoe istolkovanie* (Moscow, 1923), a collection of essays by Moritz Schlick, Bazarov, Bogdanov, Iushkevich. Cf. their participation in the discussion of Timiriazev's speech to the Communist Academy, in *VKA*, 1924, Kn. 7, pp. 357–365. Bazarov was expelled from the Communist Academy in June, 1930. See *Nauchnyi rabotnik*, 1930, No. 8–9, p. 105. Iushkevich continued intellectual work, as a translator, into the 'thirties. Bogdanov died in 1928. For a convenient summary of their interpretation of relativity, see V. Iurinetz, 'Die Relativitätstheorie und die russische marxistische Literatur', *Unter dem Banner des Marxismus*, 1925, No. 1, pp. 172–175.

[38] See above, p. 116.

[39] See A. Thalheimer, 'O nekotorykh', *PZM*, 1925, No. 1–2. There is a German version in *Unter dem Banner des Marxismus*, 1925, No. 2, pp. 302–338.

[40] S. Semkovskii, *Dialekticheskii materializm i printsip otnositel'nosti*, p. 11.

[41] See the discussion of Einstein's methodology in Mercier and Kervaire, eds., *Jubilee of Relativity Theory* (1956), especially the papers by Born and Pauli. Cf. also Schilpp, ed., *Albert Einstein, Philosopher-Scientist.*

[42] Ukrains'ka Akademiia Nauk, Istoriychno-filologichnyi viddil, *Zapysky*, 1926, Kn. 9, pp. 361–364. The reviewer was Khvol'son.

[43] See above, pp. 185 ff.

[44] B. M. Gessen, *Osnovnye idei teorii otnositel'nosti* (1928). This book was

unavailable to the present writer, who formed his impression of it from the many references to it in other works.

45 Gessen, 'Teoretiko-veroiatnostnoe obosnovanie', *Uspekhi fizicheskikh nauk*, 1929, No. 5, p. 629.

46 Gessen's earliest statement of this view occurs in his introduction to a translation of an article by Smoluchowski. See *PZM*, 1927, No. 4, p. 164. Cf. also Gessen, 'Statisticheskii metod v fizike', *EIM*, 1929, No. 1. In his preface to a translation of an article by Mises late in 1930, Gessen attacked Mises sharply for his 'bourgeois' philosophy. Presumably this was a result of the Bolshevizers' attacks on Gessen. See *Uspekhi fizicheskikh nauk*, 1930, No. 4, pp. 437–439.

47 Gessen, 'Idealisticheskie', *Molodaia gvardiia*, 1929, No. 3.

48 See *VKA*, 1924, Kn. 7, pp. 368–371 and *VKA*, 1929, Kn. 31/1/, p. 244. The date of his dismissal from the editorship of the State Publishing House's Scientific Section, 1930, is given in *BSE* (1st ed.), XXX, 514.

49 See *VKA*, 1926, Kn. 15, pp. 327, 337; and *EIM*, 1929, No. 1, p. 180.

50 S. I. Vavilov, *Eksperimental'nye* (1928). Timiriazev wrote a hostile review in *PIR*, 1928, No. 4, pp. 183–185, calling Vavilov a 'Machist', but Vavilov did not reply. Cf. also Vavilov's reviews in *Uspekhi fizicheskikh nauk*, 1930, No. 1, p. 160, No. 3, pp. 433–434, and No. 5–6, pp. 788–790. These were reviews of books full of philosophical interest to Marxists, yet Vavilov completely ignored their philosophical aspects.

51 For his reply to Timiriazev, see *VKA*, 1924, Kn. 7, pp. 365–368. For the Statute's call for a discussion of relativity, see *VKA*, 1925, Kn. 12, p. 391.

52 Other cases of truly competent specialists who were sympathetic to Marxism, yet did not participate extensively in the Marxist discussions of the new physics, were V. K. Semenchenko and Ia. Shatunovskii. The case of the eminent astronomer V. G. Fesenkov may be regarded as an exception, because of his articles, 'Astronomicheskie dokazatel'stva printsipa otnositel'nosti', *VKA*, 1925, Kn. 13, and 'Razvitie vzgliadov na stroenie vselennoi', *Molodaia gvardiia*, 1926, No. 1. But in neither of these articles does he discuss the philosophical issues or take a stand for or against the various positions of Soviet Marxists who had written on the same topics.

53 For biographical material on Ioffe, see references in note 4. Cf. also V. M. Dukel'skii, 'Akademik A. F. Ioffe', in P. I. Lukirskii, ed., *Sbornik posviashchennyi*, pp. 5–30.

54 Ioffe, *Moia zhizn'*, p. 21.

55 For a report of the exchange between Ioffe and Timiriazev at the Congress of Physicists, see *PZM*, 1927, No. 2–3, p. 181. For Ioffe's article, see 'Chto', *Pravda*, 1927, No. 1.

56 *Ibid.*

57 For Timiriazev's reply, see *Pravda*, 26 Feb. 1927. For the Deborinite defence of Ioffe, see Gessen and Egorshin, 'Ob otnoshenii', *PZM*, 1927, No. 2–3.

58 The stenographic record of the debate, 'Priroda elektricheskogo

toka', is in *Elektrichestvo*, 1930, Nos, 3, 8, 10. The fact that students suggested the debate emerges in No. 3, p. 137.

59 See, e.g., *Elektrichestvo*, No. 8, p. 350.

60 See *Elektrichestvo*, 1930, No. 3, pp. 137–138, for the comments of Gornshtein, and No. 8, pp. 348–349, for the comments of Shirvindt. A third Communist philosopher spoke (see No. 10, p. 431), but his comments are not included.

61 For Mitkevich's distress, see *Elektrichestvo*, 1930, No. 8, p. 346. For Mitkevich's claim that the philosophers had 'basically' taken his side, see No. 10, p. 435.

62 For the rare exceptions, see *Elektrichestvo*, No. 3, p. 138, where Ioffe interjects a comment in Gornshtein's speech; and No. 8, p. 350, where Frenkel says that Mitkevich 'requires, not the "philosophical matter" filling space, which satisfies Comrade Shirvindt; he requires physical matter [filling space]'.

63 Mitkevich (or Mitkewich), 'The Work of Faraday', in *Science at the Crossroads*, p. 3. Gessen, having publicly apologized for his Deborinite errors (Kommunisticheskaia Akademiia, *Za povorot*, pp. 55–56), was also a member of the Soviet delegation to the Congress, but he gave a sociological analysis of Newton's *Principia* (see again *Science at the Crossroads*) rather than a philosophical analysis of recent physics.

64 See, e.g., Mitkevich, 'K voprosu', *Sorena*, 1932, No. 3.

65 *Ibid.* There is a convenient reprint in Mitkevich, *Osnovnye* (2nd ed.), pp. 85–103.

66 For the comments of Shirvindt and Shein, and the reply of Mitkevich, see *Trudy leningradskogo elektromekhanicheskogo instituta*, 1933, No. 2, pp. 9–11, and 1934, No. 1, p. 3.

67 See A. A. Maksimov, 'Sodoklad', in Kommunisticheskaia Akademiia, *Za povorot*; and Maksimov, 'Lenin', *PZM*, 1931, No. 1–2.

68 Timiriazev himself was selected to write the philosophical section of a long encyclopedia article on the theory of relativity that appeared in 1931; but it is significant that the technical sections of the article were done by eminent physicists who gave a straightforward, uncritical account. See 'Otnositel'nosti, teoriia', *Tekhnicheskaia entsiklopediia*, XV.

69 Egorshin, 'Lenin', *RIK*, 1930, No. 19–20, p. 103. For other examples of the 'Bolshevization of physics', see Iu. Shein, 'Opredelenie skorosti, *Priroda*, 1931, No. 9; Shein, 'Metafizika fiziki', *Problemy marksizma*, 1931, No. 2; V. G. Fridman, 'Printsipa ekvivalentnosti', *Priroda*, 1931, No. 9; E. Kol'man (or Colman), 'Dynamic and Statistical Regularity', in *Science at the Crossroads*, an adaptation of his 'K voprosu', *PZM*, 1931, No. 1–2; Kol'man, 'Khod zadom', *Nauchnoe slovo*, 1931, No. 1; Kol'man, 'Boevy voprosy', *PZM*, 1931, No. 3.

70 L'vov, 'Novoe', *Izvestiia*, 22 Oct. 1929.

71 L'vov, 'Al'bert Einshtein', *Novyi mir*, 1931, Kn. 10.

72 *Ibid.*, p. 195.

73 See above, pp. 269–270.

74 Tamm, 'O rabote', *PZM*, 1933, No. 2. For his earlier challenge to

Mitkevich, see Tamm, 'Rukovodiashchie idei', *Uspekhi fizicheskikh nauk*, 1932, No. 1.

[75] See Lenin, *Materialism and Empirio-criticism* (M., 1947), p. 267.

[76] Semkovskii gave the dialectical materialist speech at the Sixth Mendeleev Congress in December, 1932. See his 'Fizika', in Institut Marksa-Engel'sa-Lenina, *Marksizm i estestvoznanie.* Gessen was selected to do the following articles for the *Great Soviet Encyclopedia* in 1933: 'Einshtein, filosofskie vzgliady', 'Energiia', and 'Entropiia'. See *BSE* (1st ed.), LXIII and LXIV.

[77] A. A. Maksimov, 'Marksizm i estestvoznanie', in Kommunisticheskaia Akademiia, *Materialy*, pp. 302–333.

[78] *Ibid.*, pp. 317–326.

[79] *Ibid.*, pp. 328, 329, 320.

[80] A. A. Maksimov, 'O filosofskikh vozzreniiakh', *PZM*, 1937, No. 7.

[81] For Ioffe's praise, see Ioffe, 'O polozhenii', *PZM*, 1937, No. 11–12, p. 143. For the Bolshevizer's disclaimer, see Kol'man, 'Pis'mo', *PZM*, 1937, No. 11–12, pp. 232–233. For news of Semkovskii's fate, see Shelkoplias, 'K itogam', *PZM*, 1936, No. 1, p. 73.

[82] In a debate in 1936, Mitkevich said: 'In my many years of struggle I have felt myself almost completely isolated. None of the physicists has so far supported me. . . .' Mitkevich, *Osnovnye* (2nd ed.), p. 161. He went on to claim that a Professor V. I. Romanov had now come to his defence. It should also be noted that Timiriazev got the physicist N. P. Kasterin to support his war on relativity in the 'thirties.

[83] Cited in Akademiia Nauk, otdelenie matematicheskikh i estestvennykh nauk, *Matematika i estestvoznanie v SSSR* (1938), p. 87.

19. THE CRISIS IN BIOLOGY

[1] See above, pp. 22–23.

[2] Berg, *Nomogenez.* The English translation is Berg, *Nomogenesis* (London, 1926). Another eminent non-Marxist scientist who propounded a similar theory of evolution was D. N. Sobolev. See his 'Evoliutsiia kak organicheskii rost', *Priroda*, 1929, No. 5. See also V. I. Vernadskii, *Biosfera*, and the same author's 'Evoliutsiia vidov i zhivoe veshchestvo', *Priroda*, 1928, No. 3, for another eminent non-Marxist scientist expounding a grand theory of evolution that evoked an anxious ambivalence among Soviet Marxists.

[3] See K. A. Timiriazev, *Nauka i demokratiia*, and cf. Lenin's 'enraptured' reaction in *Lenin, Sochineniia*, XXXV, 380.

[4] See Riazanov's long comment in Plekhanov, *Fundamental Problems of Marxism* (N.Y., 1929), pp. 138–141. The Russian original of this edition, annotated by Riazanov, was published in 1923. Cf. also D. Gul'be, 'Darvinizm i teoriia mutatsii', *PZM*, 1924, No. 7–8, for an extended argument in support of Riazanov's view. Note the editors' expression of neutrality and invitation to opposing views. For another Marxist agreeing with Riazanov, see I. I. Stepanov's comment in *PZM*, 1925, No. 3, p. 213.

[5] Sarab'ianov's defence of Berg originated as a paper read to the Dis-

cussion Club of the Communist Party's Moscow Committee early in 1923. See Sarab'ianov, 'Nazrevshii vopros', *SK*, 1923, No. 20. He later reprinted the essay in his *Istoricheskii materializm* (Moscow, 1925). Cf. also Bukharin, *Historical Materialism* (N.Y. 1925), p. 81, for this well-known Marxist using mutations as evidence of dialectical development in nature. The Russian original was published in 1921.

[6] See *PIR*, 1922, No. 7, pp. 265–266, for the original. For Sarab'ianov quoting M. M. Zavadovskii, see Sarab'ianov, *Istoricheskii materializm* (Moscow, 1925), pp. 149–150.

[7] See C. D. Darlington, 'Purpose and Particles in the Study of Heredity', in *Science, Medicine, and History: Essays on the Evolution of Scientific Thought and Medical Practice Written in Honour of Charles Singer* (London, 1953), II, pp. 472–481, for the historical background of the problems that vexed Riazanov and Sarab'ianov. Note especially pp. 474–475, for the simultaneous attraction that many thinkers have felt toward both points of view that Darlington contrasts: the Epicurean and the Aristotelian. See also J. Needham, *History Is on Our Side* (London, 1946), p. 134, for a West European Marxist expressing a sympathetic interest in Berg's nomogenesis. Needham was apparently unaware that Berg's theory has been ceaselessly denounced in Soviet Marxist writings, even though Berg himself (d. 1950) has been very favourably regarded since the 'forties.

[8] Most of the publications of the Timiriazev Institute are unavailable outside the Soviet Union, but it is possible to get a good idea of their content from the frequent reviews in *PIR* and *Knigonosha*. The transcripts of the discussions in the Communist Academy appeared in *VKA*, or in *EIM*, as did also the meagre reports of laboratory research in the Academy and the Timiriazev Institute.

[9] At the end of 1927 Levit said that there were about three hundred members of the Societies of Materialist Doctors and Biologists (*PZM*, 1928, No. 1, p. 222). On the other hand, a formal report of the Society of Materialist Doctors mentioned forty-one 'actual' or 'full' (*deistvitel'nye*) members for the end of 1928 (*EIM*, 1929, No. 1, p. 178), and Pokrovsky gave a figure of 883 for the people involved in *all* the institutes, sections, associations and societies of the Communist Academy in 1929 (*VKA*, 1929, Kn. 32/2/, p. 212). Apparently Levit was straining the term 'members' to include the audiences at the Societies' lectures and discussions. The present writer established a file of those who *participated* in the published biological discussions, and determined their ages and specialties by consulting *NR*.

[10] See the concluding pages of E. Nordenskiold, *The History of Biology* (N.Y., 1928), for evidence suggesting that Soviet Marxist biologists, in making this issue their favourite, were following the example of non-Marxist biologists abroad, as well as continuing and intensifying the heritage of Marxist interest in biology.

[11] See, e.g., B. M. Zavadovskii *et al.*, 'Darvinizm i marksizm', *VKA*, 1926, Kn. 14, pp. 226–274. Even the Timiriazev Institute, which was predominantly 'Lamarckist' from the start, included a defence of 'Morganism' by F. G. Dobrzhanskii (Th. Dobzhansky, now at Columbia

University) in the symposium, *Preformizm ili epigenezis?* See the reviews in *Knigonosha*, 1926, No. 31–32, p. 49, and in *PZM*, 1926, No. 11, pp. 250-253.

[12] 'Morganist' will be used here in preference to the other terms, in view of such remarks as the following: 'Contemporary genetics . . . owes its progress to the methodological and experimental line, on the whole and in the main a correct one, that it has succeeded in grasping mainly with the hands of T. H. Morgan's American school.' Dubinin, 'Priroda i stroenie gena', *EIM*, 1929, No. 1, p. 61.

[13] E. A. Bogdanov's report of his experiments was not accessible to the present writer, who relied on the review in *EIM*, 1929, No. 1, pp. 169–172, and on other comments and references. In trying to understand how the Timiriazev Institute became the chief centre of 'Lamarckism' among Soviet Marxists of the 'twenties, one should bear in mind that Academician S. G. Navashin, the eminent botanist who was the Director of the Institute, was probably sympathetic to 'Lamarckism'. See the review of his pamphlet, *Neomendelizm*, in *Knigonosha*, 1926, No. 28, p. 35. For Navashin's cryptically brief account of the Timiriazev Institute's work on 'experimental evolution', see his 'Avtobiografiia', *Zhurnal russkogo botanicheskogo obshchestva*, 1928, No. 1–2, p. 14.

[14] The most informative account of this episode is by N. A. Il'in, 'Zhizn' i deiatel'nost' Paulia Kammerera', *Nauchnyi rabotnik*, 1926, No. 10, pp. 3–19. For Kammerer's letter in English, see *Science*, Nov. 19, 1926, No. 64, pp. 493–494. Cf. also the obituaries in *Nature*, 1926, No. 118, pp. 555 and 635–636, and the spirited defence of Kammerer by E. W. MacBride in *Nature*, June 18, 1932, p. 901. Concerning the Soviet authorities' motives in offering Kammerer a laboratory, see the statement by O. Iu. Shmidt, the mathematician in charge of the Communist Academy's Section of Natural and Exact Sciences, in Shmidt, *Zadachi*, pp. 121–122.

[15] Agol, 'Pamiati professora P. Kammerera', *Antireligioznik*, 1926, No. 11, pp. 29–30.

[16] The characterization of Tennessee is in Agol, *ibid.*, p. 26. For another Soviet author on the illiberalism of capitalist countries, see *Knigonosha*, 1926, No. 31–32, p. 49.

[17] See his speech in Kommunisticheskaia Akademiia, *Za povorot*, pp. 60–63. Cf. *VKA*, 1926, Kn. I, pp. 258–260, for Serebrovskii's part in one of the earliest biological discussions at the Communist Academy: he attacked the concessions that B. M. Zavadovskii was then willing to make to the 'Lamarckist' position. In January, 1926, Serebrovskii gave his first major paper in defence of 'Morganism' and its compatibility with Marxism: 'Teoriia nasledsvennosti Morgana i Mendelia i marksisty', *PZM*, 1926, No. 3, pp. 98–117.

[18] See references in note 17.

[19] The biographical data have been obtained from *NR*.

[20] Their fate is suggested by the exclusion of their names from encyclopedias, bibliographies, and other works where one expects to find them. The condemnation of Agol and Levit is clearly indicated in the news dispatches cited by Zirkle, *Death of a Science in Russia*, pp. 3–4.

Serebrovskii, it should be added, has been posthumously rehabilitated in the last few years.

[21] *EIM*, 1929, No. 3, p. 135.

[22] Shmidt, *Zadachi*, p. 80. The graduate student was Iu. M. Vermel'. The same source contains the denunciations of 'Lamarckism' by Agol, Levit, Levin, E. A. Finkel'shtein, and B. M. Zavadovskii.

[23] See *ibid*., pp. 121–122, for the Director of the Communist Academy's Section of Natural and Exact Sciences rebuffing Agol's demand and stating that 'Lamarckist' work would be allowed in the Academy. Cf. pp. 212–213 above, for the decree that put an end to the 'Lamarckists'' chief centre, the Timiriazev Institute, but guaranteed the continuation of their laboratory within the Communist Academy.

[24] M. M. Zavadovskii, 'Geny i ikh rol' v osushchestvlenii priznaka', *EIM*, 1929, No. 3. Note the comments of Levit (pp. 123–125), Dubinin (pp. 131–135), and especially Zavadovskii's reply on 138–140.

[25] See especially Deborin, *Sovremennye problemy*, p. 150, for Levin assigning biologists to the rival camps of 'Lamarckism' and 'Morganism' in accordance with their specialities. See also *EIM*, 1929, No. 2, p. 53, for Serebrovskii doing the same. The present writer's file of participants in the discussions tends to confirm Levin's classification.

[26] See, e.g., the writings of the leading 'Lamarckist' in Soviet Marxist circles, E. S. Smirnov, who took over the laboratory in the Communist Academy that was originally intended for Kammerer. See Smirnov, 'Novye dannye', *VKA*, 1928, Kn. 25/1/; and note the rebuttal by Serebrovskii (pp. 197–198), Levin (199–205), and Levit (210). Cf. also Smirnov, 'Problemy', *EIM*, 1929, No. 2; and Dubinin's rebuttal, 'Genetika i neolamarkizm', *EIM*, 1929, No. 4. For a curious instance of non-adaptive variations being used in support of 'Lamarckism', see P. V. Serebrovskii, 'Darvinizm', *PZM*, 1929, No. 4, and 'Dialektika', *PZM*, 1930, No. 10–12. Cf. the rebuttal by A. S. Serebrovskii, 'Ortogenez P. V. Serbrovskogo', *EIM*, 1930, No. 1.

[27] A. S. Serebrovskii, 'Chetyre stranitsy', *Pravda*, Sept. 11, 1927, p. 5.

[28] See Agol, 'Zadachi marksistov-lenintsev v biologii', *PZM*, 1930 No. 5.

[29] Engels, *Dialectics of Nature* (N.Y., 1940), p. 7.

[30] It should be noted that A. S. Serebrovskii made a modest effort to break away from this kind of thinking. See, e.g., his 'Opyt', *EIM*, 1929, No. 2, especially pp. 68–71. But cf. A. S. Serebrovskii, 'Teoriia', *PZM*, 1926, No. 3, especially pp. 109–112, for clear evidence that Serebrovskii originally held the views described in the text above. Moreover, his renunciation of reductionism in 1929 strikes the present writer as a superficial alteration of terminology. Cf. also A. S. Serebrovskii, 'Problema svedeniia v evoliutsionnom uchenii', *Nauchnoe slovo*, 1930, No. 9, pp. 29–48, which was not accessible to the present writer.

[31] For a possible exception to this generalization see the references in note 30.

[32] Quoted in Mestergazi, 'Epigenezis i genetika', *VKA*, 1927, Kn. 19, p. 226.

[33] *Ibid.*, *passim.*

[34] See Deborin, *Sovremennye problemy*, p. 88. Cf. also Serebrovskii's disdain for the film *Salamander*, based on the Kammerer incident, in which 'Comrade Lunacharsky . . . arranges class elements about this problem: for the inheritance of acquired characters are the revolutionary intelligentsia, the People's Commissariat of Education [of which Lunacharsky was the head], and others, while against it are clerics, bankers, fascists, and counterfeiters'. *EIM*, 1929, No. 2, p. 53.

[35] Compare the speeches of the 'Morganists' and 'Lamarckists' at the Conference of Marxist-Leninist Institutions in April, 1929, as printed in Deborin, *Sovremennye*, and in Shmidt, *Zadachi.* The paradoxical boast of the 'Morganists' that they reflected the views of a *minority* among biologists was especially revealing; the majority of biologists were regarded as 'bourgeois specialists' who were to be converted or replaced by 'red specialists'.

[36] For the closing of the Timiriazev Institute but the continuation of its biological laboratory within the Communist Academy, see pp. 212–213 above. The editorial policy of *EIM* and *VKA* in 1929 and 1930 indicated the preference for 'Morganism': 'Lamarckist' articles were accompanied by editorial expressions of disagreement, while 'Morganist' articles were given considerably more space and printed without comment.

[37] See, e.g., Dubinin, 'Genetika i neolamarkizm', *EIM*, 1929, No. 4, especially pp. 88–89. Cf. *EIM*, 1930, No. 1 (5), p. 197, for a report of a discussion in the Communist Academy on the subject. 'The Role of Genetics in Reconstructing Cattle Breeding in the USSR'. A. S. Serebrovskii gave the main speech. Cf. also the articles by Agol and Levit in *PZM*, 1930, No. 5, for the two leading 'Morganists' who were Party members speaking as formulators of the Party line in biology. Serebrovskii, who was not a Party member, was much less assertive in this respect.

[38] See, e.g., A. S. Serebrovskii, 'Ortogenez P. V. Serbrovskogo', *EIM*, 1930, No. 1 (5), pp. 30–31, where he contrasts the alleged uselessness of his opponent's work with the significance of Vavilov's both for theory and for practice.

[39] See N. I. Vavilov, 'Geograficheskaia lokalizatsiia', *PZM*, 1929, No. 6, pp. 146–149.

[40] A. S. Serebrovskii, 'Antropogenetika', *Mediko-biologicheskii zhurnal*, 1929, No. 5, pp. 3–19.

[41] *Ibid.*, p. 12.

[42] *Ibid.*, p. 15.

[43] *Ibid.*, p. 16.

[44] *Ibid.*, p. 18.

[45] *Ibid.*

[46] D. Bednyi, 'Evgenika', in his *Sochineniia*, XVIII, 132–133. I wish to thank Professor Harry J. Marks for his genuinely rhyming version, which I have perversely altered.

[47] *Ibid.*, especially p. 127.

[48] *Ibid.*, especially pp. 138–139.

[49] See Agol, 'Zadachi', *PZM*, 1930, No. 5.

[50] On a few occasions A. S. Serebrovskii dropped remarks that implied acceptance of Agol's equation. See *EIM*, 1930, No. 1 (5), p. 28, for the most noteworthy. But Serebrovskii did not speak or write explicitly and at length in support of the Deborinite philosophy.

[51] See, e.g., *EIM*, 1929, No. 2, p. 220, for Dubinin wringing his hands over a paper by Iu. A. Filipchenko, in which this outstanding popularizer of genetics mocked at Marxist philosophy.

[52] See Kommunisticheskaia akademiia, *Za povorot*, for the transcript of the conference and for the resolution adopted.

[53] The most extensive statement of the new position was by I. A. Vaisberg, 'Protiv mekhanisticheskoi genetiki', *Problemy marksizma*, 1931, No. 2, pp. 103–121, and No. 4, pp. 154–172.

[54] *Ibid.* Cf. V. Brand, Eklekticheskaia putanitsa i evoliutsionnaia teoriia', *RIK*, October, 1930, No. 19–20, pp. 123–126, for a more popular statement.

[55] Biologicheskii Institut Komakademii, *Marks, Engel's, Lenin o biologii*, p. 147. This collective work was virtually a formal statement by the Institute.

[56] *Ibid.*, pp. 147 and 150.

[57] See, e.g., A. N. Bakharev, *Selektsionno-geneticheskaia stantsiia I. V. Michurina; istoricheskii ocherk* (Moscow, 1933).

[58] See Lysenko, 'Osnovnye rezul'taty', *Biulleten' iarovizatsii*, 1932, No. 4, p. 3: 'Plants of a winter variety grown from appropriately treated seed stock, without changing hereditarily, behaved like spring plants. Such spring plants with a winter heredity have been called vernalized, and the method of preparing the seed stock of winter plants is called vernalization.'

[59] The first Marxist discussion at which Lysenko's work was mentioned, as far as I know, was transcribed as 'Osnovnye ustanovki', *Sovetskaia botanika*, 1934, No. 3, pp. 3–68. Lysenko was mentioned in the speeches of Keller, Zavadskii, and Prezent. For a clear indication of this discussion's twin origin in the scientific changeover from 'bourgeois' to red specialists, and in 'the struggle for the harvest', see B. A. Keller, 'Botanicheskii institut', *Sovetskaia botanika*, 1934, No. 1, pp. 3–13.

[60] See 'Osnovnye ustanovki', *Sovetskaia botanika*, 1934, No. 3, pp. 3–68.

BIBLIOGRAPHY

THE following bibliography is limited to books, journals, and articles that are cited in the text and notes. Even so it has been necessary to make further limitations, lest the bibliography get entirely out of hand. Items that are cited in the notes without mention of the authors' names—e.g., reports of the activities of various institutions—are not repeated here. Decrees of the Soviet government and the Communist Party that are cited in the notes are repeated here only if they are cited from other sources than *Sobranie uzakonenii*, *Sobranie zakonov*, or *KPSS*.

The following abbreviations have been used in citing journals: *PZM* for *Pod znamenem marksizma* (Under the Banner of Marxism) and *VSA* or *VKA* for *Vestnik sotsialisticheskoi akademii* or *Vestnik kommunisticheskoi akademii* (Herald of the Socialist or Communist Academy).

ACTON, H. B. *The Illusion of the Epoch; Marxism-Leninism as a Philosophical Creed.* London, 1955.

ADORATSKII, V. V. 'I. V. Stalin, kak teoretik leninizma' (J. V. Stalin as a Theorist of Leninism), *Izvestiia*, December 21, 1929.

—— 'Ideologiia (Pravovaiia)' (Ideology, Legal), *Entsiklopediia gosudarstva i prava* (Encyclopedia of State and Law), II, columns 13–16. Moscow, 1926.

—— *K voprosu o nauchnoi biografii Lenina* (On the Problem of a Scholarly Biography of Lenin) Moscow, 1933.

—— 'Metod dialekticheskogo materializma' (The Method of Dialectical Materialism), *Pis'ma Marksa i Engel'sa* (Letters of Marx and Engels). Moscow, 1923. pp. xvii–xlvii.

—— 'O rabotakh Lenina po filosofii' (Concerning Lenin's Works on Philosophy), *VKA*, 1929, Kn. 32/2/.

—— 'Ob ideologii' (Concerning Ideology), *PZM*, 1922, No. 11–12.

AGOL, I. I. 'Metafizika i dialektika v biologii' (Metaphysics and Dialectics in Biology), *PZM*, 1926, No. 3.

'Pamiati professora Kammerera' (In Memory of Professor Kammerer), *Antireligioznik*, 1926, No. 11.

—— *Vitalizm, mekhanicheskii materializm i marksizm* (Vitalism, Mechanical Materialism and Marxism). Moscow-Leningrad, 1928.

—— 'Zadachi marksistov-lenintsev v biologii' (The Tasks of Marxist-Leninists in Biology), *PZM*, 1930, No. 5.

AKSEL'ROD, L. I. 'Deistvennost' i dialektika v filosofii K Marksa' (Practicability and Dialectics in the Philosophy of K. Marx), *VSA*, 1923, Kn. 4.

—— 'Dve techeniia' (Two Tendencies), *Na rubezhe; k kharakteristike sovremennykh iskanii; kriticheskii sbornik* (On the Border; Toward a Characterization of Contemporary Searchings; a Critical Collection). S. Petersburg, 1909.

—— 'Estetika N. G. Chernyshevskogo' (N. G. Chernyshevskii's Esthetics), *VKA*, 1929, Kn. 34/4/.

—— *Etiudy i vospominaniia* (Studies and Reminiscences). Leningrad, 1925.

—— *Filosofskie ocherki; otvet filosofskim kritikam istoricheskogo materializma* (Philosophical Essays; Reply to the Critics of Historical Materialism). 3rd ed.; Moscow, 1924.

—— *Idealisticheskaia dialektika Gegelia i materialisticheskaia dialektika Marksa* (The Idealist Dialectics of Hegel and the Materialist Dialectics of Marx). Moscow-Leningrad, 1934.

—— 'Iz proshlogo' (From the Past), *Zapiski Instituta Lenina* (Memoranda of the Institute of Lenin), I (1929).

—— *Marks, kak filosof* (Marx as a Philosopher). Kharkov, 1924.

—— 'Metodologicheskie voprosy iskusstva' (Methodological Problems of Art), *Krasnaia nov'* (Red Virgin Soil), 1926, Nos. 6, 7, 12.

—— 'Moi otvet' (My Answer), *Dialektika v prirode* (Dialectics in Nature), 1927, No. 2.

—— 'Nadoelo!' (Enough!), *Krasnaia nov'* (Red Virgin Soil), 1927, No. 3.

—— 'Novaia raznovidnost' revizionizma' (A New Variety of Revisionism), *Iskra* (The Spark), November 5 (18), 1904.

—— 'Otvet na "Nashi raznoglasiia" A. Deborina' (Reply to A. Deborin's 'Our Disagreements'), *Krasnaia nov'* (Red Virgin Soil), 1927, No. 5.

—— *Protiv idealizma* (Against Idealism). 3rd ed.; Moscow, 1933.

—— 'Spinoza i materializm' (Spinoza and Materialism), *Krasnaia nov'* (Red Virgin Soil), 1925, No. 7. Translated in Kline, *Spinoza in Soviet Philosophy*.

—— (or Axelrod, Esther Luba). *Tolstois Weltanschauung und ihre Entwicklung*. Bern, 1901.

—— *V zashchitu dialekticheskogo materializma: protiv skholastiki* (In Defence of Dialectical Materialism: Against Scholasticism). Moscow-Leningrad, 1928.

ALEKSANDROV, P. S. 'Moskovskoe matematicheskoe obshchestvo' (The Moscow Mathematical Society), *Uspekhi matematicheskikh nauk* (The Progress of the Mathematical Sciences), *novaia seriia* (New Series), 1946, I, No. 1 (11).

AMELIN, K., and P. CHEREMNYKH. 'Advokaty teoreticheskoi bazy pravogo opportunizma' (Advocates of the Theoretical Basis of Right Opportunism), *PZM*, 1931, No. 1–2.

ASMUS, V. F. 'Dialektika v sisteme Dekarta' (Dialectics in Descartes' System), *VKA*, 1928, Kn. 25/1/.

—— *Ocherki istorii dialektiki v novoi filosofii* (Essays on the History of Dialectics in Modern Philosophy). Moscow-Leningrad, 1929.

BABAKHAN, N. 'V zashchitu leninizma' (In Defence of Leninism), *PZM*, 1923, No. 4–5.

BABB, H. W. (translator) and J. N. HAZARD (ed.). *Soviet Legal Philosophy*. Cambridge, 1951.

BALDWIN, J. M. (ed.). *Dictionary of Philosophy and Psychology*. 2 vols. New York, 1928.

BAMMEL', G. K. (ed.). *Demokrit v ego fragmentakh i svidetel' stvakh drevnosti*

(Demokritos in His Fragments and in the Testimonials of Antiquity). Moscow, 1935.

—— '"Dialektika prirody" Engel'sa i filosofskiia bluzhdaniia estestvoznaniia' (Engels' *Dialectics of Nature* and the Philosophical Confusions of Natural Science), *VKA*, 1926, Kn. 15.

—— 'Istochniki izucheniia Demokrita' (Sources for the Study of Demokritos), *PZM*, 1923, Nos. 6–7, 8–9, 12.

—— 'Literatura o leninizme' (The Literature on Leninism), *Pechat' i revoliutsiia* (The Press and the Revolution), 1924, Nos. 4, 6.

—— 'Materialisticheskaia sistema Demokrita' (Demokritos' Materialist System), *VKA*, 1926, Kn. 14.

—— 'O nashem filosofskom razvitii za desiat' let revoliutsii' (Concerning our Philosophical Development during Ten Years of Revolution), *PZM*, 1927, No. 10–11, 12.

BARYKINA, V. E. *Dialekticheskii metod Gegelia i Marksa* (The Dialectical Method of Hegel and Marx). Orel, 1930.

BAUER, E. S. 'Osnovnye oshibki biologii' (Fundamental Errors of Biology) *Dialektika v prirode* (*Dialectics in Nature*), 1927, No. 2.

BAUER, R. A. *The New Man in Soviet Psychology*. Cambridge, 1952.

BAZAROV, V. A. *Na dva fronta* (On Two Fronts). St. Petersburg, 1910.

—— 'Obzor nauchno-populiarnoi literatury po teorii otnositel'nosti' (Review of the Popular Science Literature on the Theory of Relativity), *VSA*, 1923, Kn. 3.

—— *et al. Ocherki po filosofii marksizma* (Essays in the Philosophy of Marxism). St. Petersburg, 1908.

BEDNYI, D. 'Evgenika' (Eugenics), *Sochineniia*, XVIII (1932).

BELOV, N. V. 'Printsip prichinnosti v sovremennoi fizike' (The Principle of Causality in Contemporary Physics), *Priroda* (Nature), 1929, No. 12.

BERDIAEV (or BERDYAEV) N. A. 'The "General Line" of Soviet Philosophy', *American Review*, Oct., 1933.

—— *Sub'ektivizm i individualizm v obshchestvennoi filosofii* (Subjectivism and Individualism in Social Philosophy). St. Petersburg, 1901.

BERG, L. S. *Nomogenez* (Nomogenesis). Leningrad, 1922.

—— *Nomogenesis*. London, 1926.

Biologicheskii Institut Komakademii. *Marks, Engel's, Lenin o biologii* (Marx, Engels, Lenin on Biology). Moscow, 1933.

BOCHENSKI, I. M. *Der sowjetrussische dialektische Materialismus*. Bern, 1950.

BOGDANOV, A. A. 'Avtobiografiia' (Autobiography), in *Entsiklopedicheskii slovar' Granat* (Granat Encyclopedic Dictionary), XLI, *chast'* 1, columns 29–33, Moscow, 1924.

—— *Empiriomonizm; stat'i po filosofii* (Empirio-monism; Articles on Philosophy). 3 vols. Moscow, 1904–7.

—— 'Ernst Mach und die Revolution', *Neue Zeit*, Feb. 14, 1908.

—— *Filosofiia zhivogo opyta* (The Philosophy of Living Experience). Moscow, 1920.

—— 'Ob'ektivnoe ponimanie printsipa otnositel'nosti' (An Objective Understanding of the Principle of Relativity), *VKA*, 1924, Kn. 8.

—— *Osnovnye elementy istoricheskogo vzgliada na prirodu* (Basic Elements of the Historical View of Nature). St. Petersburg, 1899.

—— 'Otkrytoe pis'mo Plekhanovu' (An Open Letter to Plekhanov), *Vestnik zhizni* (The Herald of Life), 1907, No. 7.

—— *Poznanie s istoricheskoi tochki zreniia* (Cognition from an Historical Point of View). St. Petersburg, 1901.

—— 'Predely nauchnosti rassuzhdeniia' (The Limits of the Scientific Quality of Reasoning), *VKA*, 1927, Kn. 21.

—— *Tektologiia: vseobshchaia organizatsionnaia nauka* (Tectology: The Universal Science of Organization). 3 vols. Moscow, 1925–1928.

—— 'Uchenie o refleksakh i zagadki pervobytnogo myshleniia' (The Doctrine of Reflexes and Enigmas of Primitive Thought), *VKA*, 1925, Kn. 10.

—— 'Uchenie ob analogiiakh' (The Doctrine of Analogies), *VSA*, 1923, Kn. 2.

BOIARSKII, A. IA., B. S. IASTREMSKII, and V. I. KHOTIMSKII. *Teoriia matematicheskoi statistiki* (The Theory of Mathematical Statistics). Moscow, 1930.

BOLDYREV, N. I. (ed.). *Direktivy VKP (b) i postanovleniia sovetskogo pravitel'stva o narodnom obrazovanii* (Directives of the CPSU and Decrees of the Soviet Government Concerning Education). 2 vols. Moscow, 1947.

BORICHEVSKII, I. A. 'Dogmaticheskoe bogoslovie pod pokrovom filosofii' (Dogmatic Theology Under Cover of Philosophy), *Kniga i revoliutsiia* (The Book and the Revolution), 1922, No. 3 (15).

—— *Drevniaia i sovremennaia filosofiia v ee predel'nykh poniatiiakh* (Ancient and Contemporary Philosophy in its Limiting Concepts). Part I. Moscow, 1925.

—— 'Idealisticheskaia legenda o Kante' (The Idealist Legend about Kant), *VSA*, 1923, Kn. 4.

—— *Kratkii ocherk istorii drevnego materializma* (A Brief Sketch of the History of Ancient Materialism). Leningrad, 1930.

—— 'Neskol'ko slov o tak nazyvaemoi "russkoi filosofii"' (Some Words about So-called 'Russian Philosophy'), *Kniga i revoliutsiia* (The Book and the Revolution), 1922, No. 3 (15).

—— *Vvedenie v filosofiiu nauki* (Introduction to the Philosophy of Science). Petrograd, 1922.

BOSSE, G. G. (ed.). *Mekhanisticheskoe estestvoznanie i dialekticheskii materializm* (Mechanistic Natural Science and Dialectical Materialism). Vologda, 1925.

BRAND, V. 'Eklekticheskaia putanitsa i evoliutsionnaia teoriia' (Eclectic Confusion and Evolutionary Theory), *Revoliutsiia i kul'tura*, 1930, No. 19–20.

BREHIER, E. *Histoire de la philosophie*. 2 vols. Paris, 1950.

BROWN, E. J. *The Proletarian Episode in Russian Literature*, 1928–1932. New York, 1953.

BRUSHLINSKII, V. K. 'O kategorii mery u Gegelia' (On the Category of Measure in Hegel), *VKA*, 1929, Kn. 35–36.

BUKHARIN, N. I. 'A. A. Bogdanov', *Pravda*, 1928, 8 April.

—— *Ataka* (Attack). 2nd ed.; Moscow, 1924.

—— 'Avtobiografiia' (Autobiography), *Entiklopedicheskii slovar' Granat*

(The Granat Encyclopedic Dictionary), XL 1, chast' I, appendix, pp. 52–56.
—— 'Doklad' (Speech), *Pravda*, 1931, 9 April.
—— *Ekonomika perekhodnogo perioda* (Economics of the Transitional Period). Moscow, 1920.
—— *Historical Materialism; a System of Sociology*. New York, 1925.
—— 'K postanovke problem teorii istoricheskogo materializma' (Toward a Formulation of the Problems of the Theory of Historical Materialism), *VSA*, 1923, Kn. 3.
—— 'Lenin kak marksist' (Lenin as a Marxist), *VKA*, 1924, Kn. 7.
—— 'Po skuchnoi doroge: otvet moim kritikam' (On a Dreary Road: Reply to My Critics), *Krasnaia nov'* (Red Virgin Soil), 1923, No. 1.
—— *Teoriia istoricheskogo materializma; populiarnyi uchebnik marksistskoi sotsiologii* (The Theory of Historical Materialism; a Popular Textbook of Marxist Sociology). Moscow, 1921.
BUTIAGIN, A. S., and IU. A. SALTANOV. *Universitetskoe obrazovanie v SSSR* (University Education in the USSR). Moscow, 1957.
CARR, E. H. *The Interregnum*, 1923–1924. London, 1954.
CHAMBRE, H. *Le marxisme en union soviétique; idéologie et institutions; leur évolution de 1917 à nos jours*. Paris, 1955.
CHEREMNYKH, P. and K. AMELIN. (Advokaty teoreticheskoi bazy pravogo opportunizma' 'Advocates of the Theoretical Basis of Right Opportunism), *PZM*, 1931, No. 1–2.
CHICHIKALOV, A. 'Itogi filosofskoi diskussii' (The Results of the Philosophical Discussion), *Bol'shevik*, 1929, No. 19.
CHIZHEVSKII, A. L. *Fizicheskie faktory istoricheskogo protsessa* (Physical Factors of the Historical Process). Kaluga, 1924.
CHRISTMAN, R. C. (ed.). *Soviet Science*. Washington, D. C., 1952.
CHWOLSON, O. D. *Hegel, Haeckel, Kossouth und das zwölfte Gebot*. Brunswick, 1906.
COLE, G. D. H. *A History of Socialist Thought*. 3 vols. London, 1952–1956.
CURTISS, J. S. *The Russian Church and the Soviet State*. Boston, 1953.
DANZAS, J. 'La pensée philosophique en U.R.S.S.', *La vie intellectuelle*, 1936, No. 41.
DARLINGTON, C. D. 'Purpose and Particles in the Study of Heredity', in *Science, Medicine, and History: Essays on the Evolution of Scientific Thought and Medical Practice Written in Honour of Charles Singer*. London, 1953.
DEBORIN, A. M. *Dialektika i estestvoznanie* (Dialectics and Natural Science). Moscow-Leningrad, 1929.
——'Engel's i dialekticheskoe ponimanie prirody' (Engels and the Dialectical Conception of Nature), *PZM*, 1925, No. 10–11.
—— 'Engel's i dialektika v biologii' (Engels and Dialectics in Biology), *PZM*, 1926, Nos. 1–2, 3, 9–10.
—— *Filosofiia i marksizm* (Philosophy and Marxism). Moscow, 1930.
—— 'Filosofiia Makha i russkaia revoliutsiia' (Mach's Philosophy and the Russian Revolution), *Golos sotsial-demokrata* (The Voice of the Social Democrat), 1908, No. 4–5.

—— 'Freidizm i sotsiologiia' (Freudism and Sociology), *Voinstvuiushchii materialist* (The Militant Materialist), 1925, No. 4.

—— 'Gibel' Evropy ili torzhestvo imperializma' (The Ruin of Europe or the Triumph of Imperialism), *PZM*, 1922, No. 1–2.

—— 'Itogi i zadachi na filosofskom fronte' (Results and Tasks on the Philosophical Front), *PZM*, 1930, No. 6.

—— 'K piatiletiiu zhurnala *PZM*' (On the Fifth Anniversary of the Journal *PZM*), *PZM*, 1926, No. 12.

—— 'Legkomyslennyi kritik' (A Thoughtless Critic), *VKA*, 1924, Kn. 7.

—— *Lenin i krizis noveishei fiziki* (Lenin and the Crisis of the Newest Physics). Leningrad, 1930.

—— *Lenin kak myslitel'* (Lenin as a Thinker). Moscow, 1924, 1925.

—— 'Lenin—voinstvuiushchii materialist' (Lenin, a Militant Materialist), *PZM*, 1924, No. 1.

—— 'Lenin—voinstvuiushchii materialist' (Lenin, a Militant Materialist), in Lenin, *Materializm i empirio-krititsizm* (Materialism and Empirio-criticism). Leningrad, 1925.

—— 'Mekhanisty v bor'be s dialektikoi' (The Mechanists in Struggle with Dialectics), *VKA*, 1927, Kn. 19.

—— (ed.), *Na boevom postu; sbornik k shestidesiatiletiiu D. B. Riazanova* (At Battle Station; Collection in Honour of the Sixtieth Anniversary of D. B. Riazanov). Moscow, 1930.

—— 'Nashi raznoglasiia' (Our Disagreements), *Letopisi marksizma* (Chronicles of Marxism), 1926, No. 2.

—— *et al.* 'O bor'be na dva fronta v filosofii' (Concerning the Struggle on Two Fronts in Philosophy), *PZM*, 1930, No. 5.

—— *Ocherki po istorii materializma XVII-XVIII vv.* (Essays in the History of Materialism in the 17th to 18th Centuries). Moscow, 1929.

—— 'Oktiabr' i marksistsko-leninskaia dialektika' (October and Marxist-Leninist Dialectics), *Pravda*, 1929, 10 November.

—— 'Poslednee slovo revizionizma' (The Latest Word in Revisionism), *Voinstvuiushchii materialist* (The Militant Materialist), 1924, No. 1.

—— (ed.). *Sovremennye problemy filosofii marksizma* (Contemporary Problems of the Philosophy of Materialism). Moscow, 1929.

—— *Vvedenie v filosofiiu dialekticheskogo materializma* (Introduction to the Philosophy of Dialectical Materialism). Petrograd, 1916.

DEUTSCHER, I. *The Prophet Unarmed; Trotsky: 1921–1929*. London, 1959.

DEWITT, N. *Soviet Professional Manpower; its Education, Training, and Supply.* Washington, D.C., 1955.

Dialektika v prirode (Dialectics in Nature). 5 vols. 1926–1929.

DMITRIEV, G. 'Logika Gegelia i logika Marksa' (Hegel's Logic and Marx's Logic), *VKA*, 1929, Kn. 35–36.

DOBB, M. *Soviet Economic Development Since 1917*. London, 1948.

DOBRYNIN. 'Za leninskuiu pereotsenku nasledstva Plekhanova' (For a Leninist Revaluation of Plekhanov's Heritage), *RAPP*, 1931, No. 3.

DOLIN, I. G. 'Na vstreche aspirantov' (Meeting the Graduate Students), *Na chnyi rabotnik'* 1930, No. 1.

DUBININ, N. P. 'Genetika i neolamarkizm' (Genetics and neo-Lamarckism), *Estestvoznanie i marksizm*, 1929, No. 4.

—— 'Priroda i stroenie gena' (The Nature and Structure of the Gene), *Estestvoznanie i marksizm*, 1929, No. 1.

DUCHINSKII, F. F. 'Darvinizm, lamarkizm i neodarvinizm' (Darwinism, Lamarckism, and Neo-Darwinism), *PZM*, 1926, No. 7–8.

—— 'Neodarvinizm i problema evoliutsii cheloveka' (Neo-Darwinism and the Problem of Man's Evolution), *PZM*, 1930, No. 2–3.

—— 'Osnovnye problemy biologii v svete dialektiki' (Fundamental Problems of Biology in Light of Dialectics), *Dialektika v prirode* (Dialectics in Nature), 1929, No. 4.

DYNNIK, M. A. 'Uchenie Gegelia o sluchainosti' (Hegel's Doctrine of Contingency), *VKA*, 1929, Kn. 34/4/.

EGORSHIN, V. P. 'Estestvoznanie i klassovaia bor'ba' (Natural Science and Class Struggle), *PZM*, 1926, No. 6.

—— 'K voprosu o politike marksizma v oblasti estestvoznaniia' (On the Problem of the Policy of Marxism in the Field of Natural Science), *PZM*, 1926, No. 7–8.

—— 'Kak I. E. Tamm kritikuet marksistov' (How I. E. Tamm Criticizes the Marxists), *PZM*, 1933, No. 2.

—— 'Lenin i krizis fiziki XX veka' (Lenin and the Crisis of Physics in the Twentieth Century), *Revoliutsiia i kul'tura*, 1930, No. 19–20.

—— and B. Gessen, 'Ob otnoshenii tov. Timiriazeva k sovremennoi nauke' (Concerning Comrade Timiriazev's Relationship to Contemporary Science), *PZM*, 1927, No. 2–3.

—— 'Sovremennoe estestvoznanie i marksizm' (Contemporary Natural Science and Marxism), *Molodaia gvardiia* (The Young Guard), 1926, No. 5.

EINSTEIN, A. 'Ernst Mach', *Physikalische Zeitschrift*, 1916, No. 7.

ENCHMEN, E. *Teoriia novoi biologii i marksizm* (The Theory of the New Biology and Marxism), *vypusk* I. Petrograd, 1923.

—— *Vosemnadtsadt' tezisov o 'Teorii novoi biologii'* (Eighteen Theses on the 'Theory of the New Biology'). Piatigorsk, 1920.

ENGELS, F. *Anti-Dühring; Herr Eugen Dühring's Revolution in Science.* New York, 1939.

—— and P. and L. LAFARGUE. *Correspondance.* Paris, 1956.

—— *Dialectics of Nature.* New York, 1940.

—— 'Dialektika prirody' (Dialectics of Nature), in *Arkhiv K. Marksa i F. Engel'sa* (Archives of K. Marx and F. Engels), 1925, No. 2.

—— *Dialektika prirody* (Dialectics of Nature). Moscow, 1930.

—— *Dialektika prirody* (Dialectics of Nature). Moscow, 1952.

—— *Ludwig Feuerbach and the Outcome of German Classical Philosophy.* New York, 1941.

Entsiklopediia gosudarstva i prava (Encyclopedia of the State and Law). 3 vols. Moscow, 1925–1927.

Entsiklopedicheskii slovar' Granat (The Granat Encyclopedic Dictionary). 52 vols. Moscow, 1911–1934.

FESENKOV, V. G. 'Astronomicheskie dokazatel'stva printsipa otnositel'nosti' (Astronomical Proofs of the Principle of Relativity), *VKA*, 1925, Kn. 13.

—— 'Razvitie vzgliadov na stroenie vselennoi' (The Development of Views on the Structure of the Universe), *Molodaia gvardiia*, 1926, No. 1.

FEUER, L. S. 'Dialectical Materialism and Soviet Science', *Philosophy of Science*, 1949, April.

FLORENSKII, P. A. *Mnimosti v geometrii* (Imaginaries in Geometry). Moscow, 1922.

FOMINA, V. A. 'Rol' G. V. Plekhanova v rasprostranenii marksistskoi filosofii v Rossii' (G. V. Plekhanov's Role in the Dissemination of Marxist Philosophy in Russia), in Shchipanov, I. Ia. *Iz istorii russkoi filosofii* (From the History of Russian Philosophy). Moscow, 1951.

FRANK, PHILIPP. *Einstein; His Life and Times*. New York, 1947.

—— 'Gidromekhanika' (Hydromechanics), in *Bol'shaia sovetskaia entsiklopediia* (Great Soviet Encyclopedia), 1st edition.

—— *Le principe de causalité et ses limites*. Paris, 1937.

—— 'Volny' (Waves), in *Bol'shaia sovetskaia entsiklopediia* (Great Soviet Encyclopedia), 1st edition.

FRANK, S. L. 'Filosofskie otkliki. Novaia kniga Berdiaeva. Filosofskaia raspria v marksizme' (Philosophical Comments. Berdiaev's New Book. The Philosophical Quarrel in Marxism), *Russkaia mysl'*, 1910, No. 4.

FRANKFURT, IU. V. 'Mekhanisticheskii "dukh" revizuet psikhologicheskie vzgliady osnovopolozhnikov marksizma' (A Mechanistic 'Spirit' Revises the Psychological Views of the Founders of Marxism), *VKA*, 1928, Kn. 26/2/.

FREDERIKS, V. K., and A. A. FRIDMAN. *Osnovy teorii otnositel'nosti. vypusk I: tenzorial'noe ischislenie* (Foundations of the Theory of Relativity. Part I: Tensor Calculus). Leningrad, 1924.

FRENKEL, IA. I. *Teoriia otnositel'nosti* (The Theory of Relativity). Petrograd, 1923.

FRIDMAN, A. A. and V. K. FREDERIKS. *Osnovy teorii otnositel'nosti. vypusk I: tenzorial'noe ischislenie* (Foundations of the Theory of Relativity. Part I: Tensor Calculus). Leningrad, 1924.

FRIDMAN, V. G. 'Printsip ekvivalentnosti Einshteina i uchenie N'iutona o masse i tiagoteniia' (Einstein's Principle of Equivalence and Newton's Doctrine of Mass and Gravitation), *Priroda*, 1931, No. 1.

FROLICH, PAUL. *Rosa Luxemburg; Her Life and Work*. London, 1940.

FURSHCHIK, M. M. 'O liberal'nom i marksistskom ponimanii etiki' (Concerning the Liberal and the Marxist Conception of Ethics), *Bol'shevik*, 1930, No. 6.

GAY, PETER. *The Dilemma of Democratic Socialism; Eduard Bernstein's Challenge to Marx*. New York, 1952.

GESSEN, B. M., and I. PODVOLOTSKII. 'Filosofskie korni pravogo opportunizma' (The Philosophical Roots of Right Opportunism), *PZM*, 1929, No. 9.

—— 'Idealisticheskie techeniia v sovremennoi fizike i bor'ba s nimi' (Idealistic Tendencies in Contemporary Physics and the Struggle with Them), *Molodaia gvardiia* (The Young Guard), 1929, No. 3.

—— and I. K. Luppol. 'O kruzhkakh po izucheniiu dialekticheskogo materializma sredi molodykh nauchnykh rabotnikov' (Concerning Circles for the Study of Dialectical Materialism among Young Scientific Workers), *Kommunisticheskaia revoliutsiia* (Communist Revolution), 1928, No. 14.

—— and Egorshin. 'Ob otnoshenii tov. Timiriazeva k sovremennoi nauke' (Concerning the Relationship of Comrade Timiriazev to Contemporary Science), *PZM*, 1927, No. 2–3.

—— *Osnovnye idei teorii otnositel'nosti* (Fundamental Ideas of the Theory of Relativity). Moscow, 1928.

—— and V. P. Egorshin. 'Piatyi s'ezd russkikh fizikov' (The Fifth Congress of Russian Physicists), *PZM*, 1927, No. 1.

—— (or Hessen, B.). 'The Social and Economic Roots of Newton's "Principia"' in *Science at the Crossroads*. London, 1931. Also a separate printing: Sydney, 1946.

—— 'Statisticheskii metod v fizike i novoe obosnovanie teorii veroiatnostei R. Mizesa' (The Statistical Method in Physics, and R. Mises' New Foundation for the Theory of Probability), *Estestvoznanie i marksizm*, 1929, No. 1.

—— 'Teoretiko-veroiatnostnoe obosnovanie ergodicheskoi gipotezy' (The Foundation of the Ergodic Hypothesis in Theory of Probability), *Uspekhi fizicheskikh nauk*, 1929, No. 5.

GIRINIS, S. V. (ed.). *Ocherednoe izvrashchenie marksizma* (The Current Perversion of Marxism). Moscow, 1924.

GOL'DSHTEIN, E. 'Kruzhok po izucheniiu dialekticheskogo materializma pri iacheike fiziko-matematicheskogo fakul'teta I moskovskogo Gosudarstvennogo Universiteta' (The Circle for the Study of Dialectical Materialism in the Party Cell of the Physico-Mathematical Faculty of the First Moscow State University), *PZM*, 1923, No. 4–5.

GOL'TSMAN, A. 'Einshtein i materializm' (Einstein and Materialism), *PZM*, 1924, No. 1.

GONIKMAN, S. L. 'Dialektika tov. Bukharina' (Comrade Bukharin's Dialectics), *PZM*, 1922, No. 3.

—— 'Protiv filosofskogo likvidatorstva' (Against Philosophical Liquidationism), *Pravda*, 1924, 10 January.

—— 'Revizor' (The Inspector General), *PZM*, 1924, No. 10–11.

—— 'Uchenie Gegelia o "deistvitel'nosti"' (Hegel's Doctrine on 'Reality'), *PZM*, 1924, No. 1, 3.

—— 'Voinstvuiushchii i "terpimyi" marksizma' (Militant and 'Tolerant' Marxism), *Bil'shovik* (The Ukrainian Bolshevik), 1925, No. 2–3.

GORDON, G. O., and LUPPOL and VOLGIN (editors). *Obshchestvennye nauki SSSR. 1917–1927* (The Social Sciences in the USSR, 1917–1927). Moscow, 1928.

GORKY, M. *Autobiography*, New York, 1949.

—— *Culture and the People*, New York, 1939.

—— and Lenin, V. I. *Pis'ma, vospominaniia, dokumenty* (Letters, Reminiscences, Documents). Moscow, 1958.

—— *Sobranie sochinenii* (Collected Works). 30 vols. Moscow, 1949–55.

GORTER, G. *Istoricheskii materializm* (Historical Materialism). Moscow, 1919.

GREBENEV, K. V. 'Kriticheskaia zametka' (Critical Note), *VKA*, 1929, Kn. 34/4/.

GREDESKUL, N. A. 'Byt' li estestvoznanie mekhanicheskim ili stat' dialekticheskim?' (Is Natural Science to be Mechanical or Is It to Become Dialectical?), *PZM*, 1928, No. 1.

GRIB, V. 'Dialektika i logika kak nauchnaia metodologiia' (Dialectics and Logic as a Scientific Methodology), *PZM*, 1928, No. 6.

GUBANOV, N. 'Obshchestvo voinstuiushchikh materialistov-dialektikov i bor'ba za ateizm' (The Society of Militant Materialist Dialecticians and the Struggle for Atheism), *Antireligioznik* (The Antireligionist), 1929, No. 6.

GUL'BE, D. 'Darvinizm i teoriia mutatsii s tochki zreniia dialekticheskogo materializma' (Darwinism and the Theory of Mutations from the Point of View of Dialectical Materialism), *PZM*, 1924, No. 7–8.

HAZARD, J. N. (ed.), and H. W. BABB (translator). *Soviet Legal Philosophy*. Cambridge, 1951.

HECKER, JULIUS F. *Moscow Dialogues; Discussions on Red Philosophy*. London, 1933.

HESSEN, B. (or GESSEN, B. M.). 'The Social and Economic Roots of Newton's "Principia"' in *Science at the Crossroads*. London, 1931. Also a separate printing: Sydney, 1946.

IANKOVSKII, R., and V. RAL'TSEVICH (eds.). '*Materializm i empiriokrititsizm' V. I. Lenina* (V. I. Lenin's 'Materialism and Empirio-criticism'). Moscow, 1935.

IANOVSKAIA, S. IA. 'Kategoriia kolichestva u Gegelia i sushchnost' matematiki' (The Category of Quantity in Hegel and the Essence of Mathematics), *PZM*, 1928, No. 3.

IAROSLAVSKII, E. *O roli intelligentsii* (Concerning the Role of the Intelligentsia). Moscow, 1939.

—— *Tret'ia sila* (The Third Force). Moscow, 1932.

IASTREMSKII, B. S., A. IA. BOIARSKII and V. I. KHOTIMSKII. *Teoriia matematicheskoi statistiki* (The Theory of Mathematical Statistics). Moscow, 1930.

IL'IN, N. A. 'Zhizn' i deiatel'nost' Paulia Kammerera' (The Life and Activity of Paul Kammerer), *Nauchnyi rabotnik*, 1926, No. 10.

Institut Marksa-Engel'sa-Lenina. *Marksizm i estestvoznanie* (Marxism and Natural Science). Moscow, 1933.

IOFFE, A. F. 'Chto govoriat opyty o teorii otnositel'nosti' (What Experiments Say about the Theory of Relativity), *Pravda*, 1927, No. 1.

—— *Moia zhizn' i rabota; avtobiograficheskii ocherk* (My Life and Work; Autobiographical Sketch). Moscow, 1933.

—— 'O polozhenii na filosofskom fronte sovetskoi fiziki' (Concerning the Situation on the Philosophical Front of Soviet Physics), *PZM*, 1937, No. 11–12.

IUDIN, P. F. 'God raboty' (A Year's Work), *PZM*, 1932, No. 1–2.

—— 'K filosofskoi diskussii' (On the Philosophical Discussion), *Pravda*, 1930, 18 November.

—— and M. Rosental'. *Kratkii filosofskii slovar'* (Brief Philosophical Dictionary). 1st–4th eds.; Moscow, 1939–1954.

—— 'Lenin i filosofskaia diskussia 1908–1910 g.' (Lenin and the Philosophical Discussion of 1908–1910), *PZM*, 1931, No. 9–10.

IURENETS, V. A. (or JURINETZ, V.). 'Die Relativitätstheorie und die russische marxistische Literatur' *Unter dem Banner des Marxismus*, 1925, No. 1.

IUSHKEVICH, P. S. *Stolpy filosofskoi ortodoksii* (Pillars of Philosophical Orthodoxy). St. Petersburg, 1910.

IVANOV, D. D., and N. A. FIGUROVSKII (eds.). *Istoriia estestvoznaniia; literatura, opublikovannaia v SSSR (1917–1947)* (The History of Natural Science; Literature Published in the USSR 1917–1947). Moscow-Leningrad, 1949.

IVANOVSKII, V. N. *Metodologischeskoe vvedednie v nauku i filosofiiu* (Methodological Introduction to Science and Philosophy). Minsk, 1923.

JEANS, JAMES. 'The Physics of the Universe', *Nature*, 1929, pp. 689–700; translated into Russian in *Priroda* (Nature), 1929, No. 1.

'K vysylke kontrrevoliutsionnernoi professury' (On the Banishment of the Counter-revolutionary Professors), *Pravda*, August 31, 1922.

KAGAN, V. F. 'Aksioma' (Axiom), *Bol'shaia sovetskaia entsiklopediia* (Great Soviet Encyclopedia), 1st edition.

KALININ, M. I. 'Rol' intelligentsii v nashem stroitel'stve' (The Role of the Intelligentsiia in Our Construction), *Izvestiia*, 1925, 12 December.

KAMENEV, L. 'Ne po doroge' (Not Our Way), *Proletarii*, February 12 (25), 1909.

KAMMERER, P. 'Pis'mo' (Letter), *VKA*, 1926, Kn. 18; translated in *Science*, 1926, No. 64.

KANATCHIKOV, S. 'Mysli o partiino-vospitatel'noi rabote' (Thoughts on Party Educational Work), *Pravda*, 1922, 16 and 18 August.

KAREV, N. A. 'K itogam i perspektivam sporov s mekhanistami' (The Results and Perspectives of the Debates with the Mechanists), *PZM*, 1928, No. 4.

—— 'K voprosu o teoreticheskikh korniakh pravogo uklona' (On the Question of the Theoretical Roots of the Right Deviation), *Bol'shevik*, 1930, No. 2.

—— 'Marksistskaia kritika i kritika marksizma' (Marxist Criticism and the Criticism of Marxism), *Bol'shevik*, 1924, No. 10.

—— 'Na putiakh izucheniia marksistskoi filosofii' (On the Path of Study of Marxist Philosophy), *PZM*, 1923, No. 2–3.

—— 'Neskol'ko zamechanii po povodu stat'i professora Samoilova' (Several Remarks on Professor Samoilov's Article), *PZM*, 1926, No. 4–5.

—— 'O deistvitel'nom i nedeistvitel'nom izucheniia Gegelia' (Concerning Real and Unreal Study of Hegel), *PZM*, 1924, No. 4–5.

—— 'O nashikh filosofskikh sporakh' (Concerning Our Philosophical Disputes), *Molodaia gvardiia* (The Young Guard), 1929, No. 1.

—— 'Spinoza i materializm' (Spinoza and Materialism), *Krasnaia nov'* (Red Virgin Soil), 1927, No. 6.

—— 'Zametka o vrede putanitsy i popovskoi propovedi i ob iskusstve zametat' sledy' (A Note on the Harm of Confusion and Popish Sermonizing, and on the Art of Covering One's Tracks), *PZM*, 1930, No. 7–8.

—— 'Zasedanie Obshchestva Voinstvuiushchikh Materialistov' (The Meeting of the Society of Militant Materialists), *PZM*, 1927, No. 4.

KAUFMAN, A. 'The Origin of "The Political Economy of Socialism"', *Soviet Studies*, January, 1953.

KAUTSKY, K. 'Ein Brief über Marx und Mach', *Der Kampf*, 1909, No. 10, pp. 451–452.

—— *Der Einfluss der Volksvermehrung auf den Fortschritt der Gesellschaft.* Wien, 1880.

—— *Ethics and the Materialist Conception of History.* Chicago, 1907.

—— *The Social Revolution.* Chicago, 1910.

—— *Vermehrung und Entwicklung in Natur und Gesellschaft.* Stuttgart, 1910.

KELLER, B. A. 'Botanicheskii institut Akademii Nauk k XVII s'ezdu kommunisticheskoi partii' (The Botanical Institute of the Academy of Sciences and the Seventeenth Congress of the Communist Party), *Sovetskaia botanika*, 1934, No. 1.

KHARAZOV, G. A. 'Malyi printsip otnositel'nosti' (The Small Principle of Relativity), *VKA*, 1925, Kn. 12, 14.

—— 'Matematicheskaia kritika teorii otnositel'nosti' (Mathematical Criticism of the Theory of Relativity), *VKA*, 1925, Kn. 10.

KHOTIMSKII, V. I., A. IA. BOIARSKII, and B. S. IASTREMSKII. *Teoriia matematicheskoi statistiki* (The Theory of Mathematical Statistics), Moscow, 1930.

KHVOL'SON, O. D. 'Chto dal Oktiabr' russkoi fizike?' (What Did October Give to Russian Physics?), *Nauchnyi rabotnik* (The Scientific Worker), 1927, No. 12.

—— *Kharakteristika razvitiia fiziki za poslednee 50 let* (A Characterization of the Development of Physics for the Past Fifty Years). Leningrad, 1924.

—— *Teoriia otnositel'nosti A. Einshteina i novoe miroponimanie* (A. Einstein's Theory of Relativity and the New Conception of the World). Petrograd, 1922, and Leningrad, 1925.

—— *Znanie i vera v fizike* (Knowledge and Faith in Physics). Petrograd, 1916.

KIM, M. P. *Sorok let sovetskoi kul'tury* (Forty Years of Soviet Culture). Moscow, 1957.

KLINE, G. L. *Spinoza in Soviet Philosophy.* London, 1952.

KOL'MAN, E. G. 'Boevye voprosy estestvoznaniia i tekhniki v rekonstruktivnyi period' (Controversial Questions of Natural Science and Technology in the Period of Reconstruction), *PZM*, 1931, No. 3.

—— 'Bor'ba za vladenie nauki v novykh usloviiakh' (The Struggle for the Mastery of Science in the New Circumstances), *Pravda*, Oct, 4, 1931.

—— 'Ideologicheskie i organizatsionnye problemy v oblasti estestvoznaniia i tekhniki' (Ideological and Organizational Problems in the Field of Natural Science and Technique), *Kommunisticheskaia revoliutsiia* (Communist Revolution), 1928, No. 17–18.

—— 'K voprosu o dinamicheskoi i statisticheskoi zakonomernosti' (On the Question of Dynamic and Statistical Regularity), *PZM*, 1931, No. 1–2. (An English version is in *Science at the Crossroads*.)

—— 'K voprosu o sluchainosti' (On the Problem of Contingency), *PZM*, 1926, No. 6.

—— 'Khod zadom filosofii Einshteina' (Einstein's Philosophy Takes a Step Backward), *Nauchnoe slovo*, 1931, No. 1.

—— (ed.). *Na bor'bu za materialisticheskuiu dialektiku v matematike* (In Struggle for Materialist Dialectics in Mathematics). Moscow-Leningrad, 1931.

—— 'Ob obostrenii ideologicheskoi bor'by v oblasti nauki i o probleme sozdaniia novykh nauchnykh kadrov' (On the Sharpening of Ideological Struggle in the Field of Science and on the Problem of Creating New Scientific Cadres), *Kommunisticheskaia revoliutsiia* (Communist Revolution), 1929, No. 1.

—— 'Politika, ekonomika i matematika' (Politics, Economics and Mathematics), in Kol'man (ed.), *Na bor'bu za materialisticheskuiu dialektiku a matematike*. Moscow-Leningrad, 1931.

—— 'Triumf marksizma—nauki proletariata' (The Triumph of Marxism, Science of the Proletariat), in Institut Marksa-Engel'sa-Lenina, *Marksizm i estestvoznanie* (Marxism and Natural Science). Moscow, 1933.

—— 'Vreditel'stvo v nauke' (Wrecking in Science), *Bol'shevik*, 1931, No. 2.

Kommunisticheskaia Akademiia. *Dvadtsat' piat' let 'Materializma i empiriokrititsizma'* (Twenty-five Years of 'Materialism and Empiriocriticism'). Moscow, 1934.

—— *Materialy nauchnoi sessii; k piatidesiatiletiiu so dnia smerti Marksa* (Materials of the Scholarly Session; on the Fiftieth Anniversary of Marx's Death). Moscow, 1934.

—— *Za povorot na fronte estestvoznaniia; diskussiia na zasedaniiakh prezidiuma Komakademii 23–XII–1930–6–I–1931* (For a Turn on the Front of Natural Science; The Discussion at Meetings of the Communist Academy's Presidium, 23–XII–1930 to 6–I–1931). Moscow, 1931.

Kommunisticheskaia Partiia Sovetskogo Soiuza v rezoliutsiiakh i resheniiakh s'ezdov, konferentsii i plenumov Ts K (The Communist Party of the Soviet Union in the Resolutions and Decisions of Congresses, Conferences, and Plenary Meetings of the CC), 2 vols. 7th ed.; Moscow, 1953.

KON, A. 'Kritika "kritikov"' (A Critique of the 'Critics'), *PZM*, 1922, No. 5–6.

'Korennye voprosy dialekticheskogo materializma' (Basic Questions of

Dialectical Materialism), *PZM*, 1928, No. 1; an account of a debate in the Meierhold Theater on 19 December 1927.

KORNILOV, K. N. 'Mekhanicheskii materializm v sovremennoi psikhologii' (Mechanical Materialism in Contemporary Psychology), *PZM*, 1926, No. 4–5.

—— 'Vozzreniia sovremennykh mekhanistov na zakon sokhranenii energii i psikhiku' (The Views of Contemporary Mechanists on the Law of the Conservation of Energy and on the Psyche), *Psikhologiia* (Psychology), II (1929), *vypusk* 1.

KOROLEV, F. F. *Ocherki po istorii sovetskoi shkoly i pedagogiki, 1917–1920* (Essays in the History of the Soviet School and Pedagogy, 1917–1920). Moscow, 1958.

KORSCH, K. (Or Korsh in Russian transliteration). *Marxismus und Philosophie.* Berlin, 1923. The Russian translation is *Marksizm i filosofiia.* Moscow, 1924.

KOSHTOIANTS, Kh. 'Aleksandr Filipovich Samoilov', in Samoilov, *Izbrannye stat'i i rechi* (Selected Articles and Speeches). Moscow-Leningrad, 1946.

KOZO-POLIANSKII, B. M. 'K vyiasneniiu nekotorykh nashikh pozitsii v darvinizme' (Toward a Clarification of Some of our Positions in Darwinism), *Dialektika v prirode* (Dialectics in Nature), 1927, No. 2.

KRIKORIAN, Y. (ed.). *Naturalism and the Human Spirit.* New York, 1944.

'Kritika teoreticheskikh osnov bukharinskoi kontseptsii istoricheskogo materializma' (Critique of the Theoretical Foundations of the Bukharinite Conception of Historical Materialism), *VKA*, 1929, Kn. 35–36; speeches and discussion in the Communist Academy's Institute of Philosophy.

KRIVTSOV, S. S. 'A. A. Bogdanov', *PZM*, 1928, No. 4. 'I. I. Skvortsov-Stepanov', *VKA*, 1928, Kn. 30/6/.

KRUPSKAIA, N. 'Obstanovka, v kotoroi pisalas' stat'ia Lenina "O znachenii voinstvuiushchego materializma"' (The Circumstances in which Lenin Wrote his Article 'Concerning Militant Materialism'), *PZM*, 1933, No. 1.

KUROSH, A. G. (ed.). *Matematika v SSSR za tridtsat' let: 1917–1944* (Mathematics in the USSR for Thirty Years: 1917–1944). Moscow, 1948.

LAFARGUE, P. *Social and Philosophical Studies.* Chicago, 1906.

LAIDOV, M. N. (or LIADOV). 'Functions of a Communist University', *Labour Monthly*, 1926, July.

LAZAREV, P. P. *Ocherki istorii russkoi nauki* (Essays in the History of Russian Science). Moscow-Leningrad, 1950.

LENIN, V. I. *Filosofskie tetradi* (Philosophical Notebooks). Moscow, 1933.

—— *Materialism and Empirio-criticism.* Moscow, 1947.

—— *Materializm i empiriokrititsizm* (Materialism and Empirio-criticism). Leningrad, 1925.

—— and Gorky, M. *Pis'ma, vospominaniia, dokumenty* (Letters, Reminiscences, Documents), Moscow, 1958.

—— *Sochineniia* (Works), 3rd and 4th editions (cf. notice at head of footnotes).

LEPESHINSKII, P. N. 'Problema intelligentsii v rekonstruktivnyi period i sovetskie spetsialisty' (The Problem of the Intelligentsia in the Period of Reconstruction and Soviet Specialists), *VKA*, 1931, No. 1.

LIADOV, M. N. (or LAIDOV). 'Functions of a Communist University', *Labour Monthly*, 1926, July.

—— 'O studenchestve, akademizme i demokratii' (Concerning the Student Body, Academism, and Democracy), *Pravda*, 1924, No. 6.

—— 'Razvernutaia "rabochaia demokratiia" i akademizm' (Fully Developed 'Working Class Democracy' and Academism), *Pravda*, 1923, 18 December.

—— 'Spasaet li filosofiia ot opportunizma?' (Will Philosophy Rescue One from Opportunism?), *Pravda*, 1924, 13 January.

LIPSHITS, S. IU. *Russkie botaniki; biografo-bibliograficheskii slovar'* (Russian Botanists; Biographical and Bibliographical Dictionary). 4 vols. to date. Moscow, 1947.

LIUSTERNIK, L. A., *et al.* 'Deklaratsiia initsiativnoi gruppy po reorganizatsii matematicheskogo obshchestva' (Declaration of the Initiating Group for the Reorganization of the Mathematical Society), *Nauchnyi rabotnik*, 1930, No. 11–12.

LOSSKII, N. O. *Dialekticheskii materializm v SSSR* (Dialectical Materialism in the USSR). Paris, 1934.

LUKIRSKII, P. I. (ed.). *Sbornik posviashchennyi semidesiatiletiiu Akademika A. F. Ioffe* (Festschrift on the Seventieth Anniversary of Academician A. F. Ioffe). Moscow, 1950.

LUNACHARSKY, A. V. *Etiudy; sbornik statei* (Sketches; A collection of Articles). Moscow, 1922.

—— 'Intelligentsiia i sotsializm' (The Intelligentsia and Socialism), *VARNITSO*, 1930, No. 2.

—— *Ot Spinozy do Marksa* (From Spinoza to Marx). Moscow, 1925.

—— 'Proletarskaia revoliutsiia i nauka' (The Proletarian Revolution and Science), in Luppol (ed.), *Kul'turnaia revoliutsiia i naychnye rabotniki* (The Cultural Revolution and Scientific Workers). Moscow, 1928.

—— *Protiv idealizma; etiudy polemicheskie* (Against Idealism; Polemical Sketches). Moscow, 1924.

—— *Revoliutsionnye siluety* (Revolutionary Silhouettes). 2nd ed.; Kharkov, 1924.

LUPPOL, I. K. 'Antireligioznaia propaganda i nauchnye rabotniki' (Antireligious Propaganda and Scientific Workers), *Antireligioznik* (The Anti-religionist), 1929, No. 6.

—— *Deni Didro* (Denis Diderot). Moscow, 1924. Translated as *Diderot, ses idées philosophiques*. Paris, 1936.

—— 'Filosofiia' (Philosophy), in Luppol (ed.), *Obshchestvennye nauki SSSR, 1917–1927* (The Social Sciences in the USSR, 1917–1927). Moscow, 1928.

—— *Istoriko-filosofskie etiudy* (Historico-Philosophical Studies). Moscow, 1935.

—— 'K vyboram v Akademiiu Nauk SSSR' (Toward the Elections to the Academy of Sciences of the USSR), *Nauchnyi rabotnik* (The Scientific Worker), 1928, No. 11.

—— (ed.). *Kul'turnaia revoliutsiia i nauchnye rabotniki* (The Cultural Revolution and Scientific Workers). Moscow, 1928.

—— *Lenin i filosofiia* (Lenin and Philosophy). 1st ed.; Moscow, 1927. 2nd ed.; Moscow, 1929.

—— '"Materializm i empiriokritisizm" v otsenke vstretivshei ego kritiki' ('Materialism and Empirio-criticism' as the Critics Received It), *PZM*, 1927, No. 1, pp. 19–38.

—— *Na dva fronta* (On Two Fronts). Moscow-Leningrad, 1930.

—— 'Neskol'ko slov ob irratsional'nom metode v istorii filosofii' (Some Words on an Irrational Method in the History of Philosophy), *Voinstvuiushchii materialist* (The Militant Materialist), 1925, No. 3.

—— and Gessen. 'O kruzhkakh po izuchenii dialekticheskogo materializma sredi molodykh nauchnykh rabotnikov' (Concerning the Circles for the Study of Dialectical Materialism among Young Scientific Workers), *Kommunisticheskaia revoliutsiia* (Communist Revolution), 1928, No. 14.

—— 'O novom uchebnike po istoricheskomu materializmu' (On a New Textbook of Historical Materialism), *PZM*, 1924, No. 12.

—— and Volgin and Gordon (eds.). *Obshchestvennye nauki SSSR, 1917–1927* (The Social Sciences in the USSR, 1917–1927). Moscow, 1928.

—— 'Perspektivy vysshego filosofskogo obrazovaniia v RSFSR' (The Perspectives of Higher Philosophical Education in the RSFSR), *Nauchnyi rabotnik* (The Scientific Worker), 1927, No. 5–6.

L'VOV, V. E. 'Al'bert Einshtein v soiuze s religiei' (Albert Einstein in Union with Religion), *Novyi mir*, 1931, No. 10.

—— 'Novoe v fizike: krizis zakona sokhraneniia energii' (A New Development in Physics: a Crisis of the Law of the Conservation of Energy), *Izvestiia*, October 22, 1929.

LYONS, E. *Six Soviet Plays*. Boston, 1934.

LYSENKO, T. D. *Agrobiologiia; raboty po voprosam genetiki, selektsii i semenovodstva* (Agrobiology; Works on Problems of Genetics, Selection, and Seed Production). 3rd ed.; Moscow, 1948.

—— 'Osnovnye rezul'taty rabot po iarovizatsii sel'sko-khoziaistvennykh rastenii' (The Basic Results of Work on the Vernalization of Agricultural Plants), *Biulleten' iarovizatsii*, 1932, No. 4.

—— *et al. The Situation in Biological Science*. New York, 1949.

MACH, E. *Die Prinzipien der physikalischen Optik*. Leipzig, 1921.

MAKSIMOV, A. A. 'Filosofiia i estestvoznanie za piat' let' (Philosophy and Natural Science for Five Years), *PZM*, 1936, No. 1.

—— 'K voprosu o dialektike v istorii estestvoznaniia' (On the Problem of Dialectics in the History of Natural Science), *PZM*, 1924, No. 4–5.

—— 'Lenin i krizis estestvoznaniia epokhi imperializma' (Lenin and the Crisis of Natural Science in the Epoch of Imperialism), *PZM*, 1931, No. 1–2.

—— 'Metodologiia izmereniia i dialekticheskii materializm' (The

Methodology of Measurement and Dialectical Materialism), *PZM*, 1929, No. 7–8.
—— 'N'iuton i filosofiia' (Newton and Philosophy), *PZM*, 1927, No. 4.
—— 'O filosofskikh vozzreniiakh Akademika V. F. Mitkevicha i o putiakh razvitiia sovetskoi fiziki' (On the Philosophical Views of Academician V. F. Mitkevich, and on the Paths of Development of Soviet Physics), *PZM*, 1937, No. 7.
—— 'O printsipe otnositel'nosti A. Einshteina' (Concerning A. Einstein's Principle of Relativity), *PZM*, 1922, No. 9–10.
—— 'O zadachakh marksizma v oblasti estestvoznaniia' (Concerning the Tasks of Marxism in the Field of Natural Science), *Molodaia gvardiia* (The Young Guard), 1926, No. 9.
—— 'Ob istochnikakh i rezul'tatakh uproshchenstva v estestvoznanii' (On the Sources and Results of Vulgarization in Natural Science), *PZM*, 1926, No. 1–2.
—— 'Populiarno-nauchnaia literatura o printsipe otnositel'nosti' (Popular Science Literature Concerning the Principle of Relativity), *PZM*, 1922, No. 7–8, 11–12.
—— 'Sovremennoe estestvoznanie i "Materializm i empiriokrititsizm" Lenina' (Contemporary Natural Science and Lenin's *Materialism and Empirio-criticism*), *PZM*, 1927, No. 10–11.
—— 'Teoriia otnositel'nosti i materializm' (The Theory of Relativity and Materialism), *PZM*, 1923, No. 4–5.
MAMAI, N. *Kommunisticheskaia partiia v bor'be za ideino-politicheskoe vospitanie mass v pervye gody Nepa* (The Communist Party in Struggle for the Ideological and Political Education of the Masses in the First Years of NEP). Moscow, 1954.
MARCUSE, H. *Reason and Revolution; Hegel and the Rise of Social Theory*. New York, 1954.
—— *Soviet Marxism; a Critical Analysis*. New York, 1958.
'Marksistskoe mirovozzrenie i industrializatsiia strany' (The Marxist World View and the Industrialization of the Country), *Dialektika v prirode* (Dialectics in Nature), 1928, No. 3.
Marksizm i darvinizm (Marxism and Darwinism). Moscow, 1927.
MARTEL, R. *Le mouvement antireligieux en U.R.S.S. (1917–1932)*. Paris, 1933.
MARX, K., and ENGELS, F. *Briefwechsel*. 4 vols. Berlin, 1949.
—— *Capital*. New York, 1906. The 'Kerr' edition.
—— and Engels, F. *Correspondence, 1846–1895*. New York, 1934.
—— *The Eighteenth Brumaire of Louis Bonaparte*. London, 1926.
—— *The German Ideology*. New York, 1947.
—— and Engels, F. *Historisch-kritische Gesamtausgabe*. 6 vols. Frankfurt, 1927–1932.
—— *Letters to Dr. Kugelmann*. London, n.d.
—— and Engels, F. *Pis'ma* (Letters). Moscow, 1923.
—— *Sochineniia* (Works). 2nd ed.; Moscow, 1955–.
MASARYK, T. G. *The Spirit of Russia*. 2 vols. London, 1919.
M(ATUSOV), N. 'Zadachi i soderzhanie filosofii' (The Tasks and the

Content of Philosophy), *Iunyi kommunist* (The Young Communist), 1930, No. 7.

MAYER, G. *Friedrich Engels; a Biography*. New York, 1936.

MAYNARD, J. *Russia in Flux*. New York, 1951.

MEHRING, F. *Karl Marx; the Story of His Life*. New York, 1935.

MERCIER, A. (ed.). *Jubilee of Relativity Theory*. Basel, 1956.

MESTERGAZI, M. M. 'Epigenezis i genetika' (Epigenesis and Genetics), *VKA*, 1927, Kn. 19.

MILIUTIN, V. P. 'O direktivakh po sostavleniiu plana rabot KA' (Concerning the Directives for the Establishment of a Plan of Works for the Communist Academy), *VKA*, 1929, Kn. 32/2/.

MILONOV, K. 'Dialektika i filosofiia' (Dialectics and Philosophy), *Molodaia gvardiia* (The Young Guard), 1926, No. 6.

—— 'Gegel' i materializm' (Hegel and Materialism), *VKA*, 1928, Kn. 25/1/.

—— 'Ob odnom noveishem perevorote' (Concerning a Very New Revolution), *Voinstvuiushchii materialist* (The Militant Materialist), 1924, No. 1.

—— 'Protiv mekhanicheskogo miroponimaniia' (Against the Mechanical World View), *VKA*, 1926, Kn. 16 and Kn. 18.

MININ, S. 'Filosofiiu za bort!' (Philosophy Overboard!), *PZM*, 1922, No. 5–6.

—— 'Kommunizm i filosofiia' (Philosophy and Communism), *PZM*, 1922, No. 11–12.

von MISES, R. *Positivism: a Study in Human Understanding*. Cambridge, 1951.

MITIN, M. B. *Boevye voprosy materialisticheskoi dialektiki* (Controversial Questions of Materialist Dialectics). Moscow, 1936.

—— *Dialekticheskii i istoricheskii materializm* (Dialectical and Historical Materialism). Moscow, 1933.

—— 'K voprosu o partiinosti filosofii' (On the Question of the Partyness of Philosophy), *Revoliutsiia i kul'tura* (Revolution and Culture), 1930, No. 19–20.

—— 'Na filosofskom fronte' (On the Philosophical Front), *Kommunisticheskaia revoliutsiia*, 1929, No. 8.

—— 'Nekotorye itogi i zadachi raboty na filosofskom fronte' (Some Results and Tasks of the Work on the Philosophical Front), *PZM*, 1936, No. 1.

—— and V. N. Ral'tsevich and P. F. Iudin. 'O novykh zadachakh marksistsko-leninskoi filosofii' (Concerning the New Tasks of Marxist-Leninist Philosophy), *Pravda*, June 7, 1930.

—— 'O pravoi "filosofii" klassovoi bor'by v rekonstruktivnyi period' (The Right 'Philosophy' of Class Struggle in the Period of Reconstruction), *Kommunisticheskaia revoliutsiia*, 1929, No. 10–11.

MITKEVICH, V. F. 'K voprosu o prirode elektricheskogo toka' (On the Question of the Nature of an Electrical Current), *Sorena*, 1932, No. 3.

—— *Osnovnye fizicheskie vozzreniia* (Basic Physical Views). 2nd ed.; Moscow, 1936.

—— (or Mitkewich, W. Th.). 'The Work of Faraday and Modern Developments in the Application of Electrical Energy', in *Science at the Crossroads*. London, 1931.

MOLOTOV, V. 'Doklad' (Speech), *Pravda*, April 14, 1931.

Na rubezhe; k kharakteristike sovremennykh iskanii; kriticheskii sbornik (On the Border; Toward a Characterization of Contemporary Searchings; a Critical Collection). St. Petersburg, 1909.

NAGEL, E. 'The Meaning of Reduction in the Natural Sciences', in Stauffer (ed.), *Science in Civilization*. Madison, Wisconsin, 1949.

'Nash sbornik No. 2' (Our Collection No. 2), *Dialektika v prirode* (Dialectics in Nature), 1928, No. 3.

'Nashi zadachi' (Our Tasks), *Estestvoznanie i marksizm* (Natural Science and Marxism), 1929, No. 1.

'Nauchnye obshchestva pered sudom sovetskoi obshchestvennosti' (The Scientific Societies before the Court of Soviet Public Opinion), *VARNITSO*, 1931, No. 3.

Nauka i nauchnye rabotniki SSSR (Science and Scientific Workers of the USSR). Moscow, 1925, for Moscow; 1928, for all parts of USSR but Moscow and Leningrad; Moscow, 1930, for Moscow; Moscow, 1934, for Leningrad.

NEEDHAM, J. *History Is on Our Side*. London, 1946.

'Nekotorye itogi i perspektivy' (Some Results and Perspectives), *Estestvoznanie i marksizm* (Natural Science and Marxism), 1929, No. 3.

NEMILOV, A. 'O partiinoi chistke vuzov i nauchno-issledovatel'skikh institutov Leningrada' (On the Party Purge in the Universities and Scientific Research Institutes of Leningrad), *Front nauki i tekhniki*, 1933, No. 12.

NEVSKII, V. I. 'Marksizm i estestvoznanie' (Marxism and Natural Science), *PZM*, 1923, No. 12.

—— 'Restavratsiia idealizma i bor'ba s "novoi" burzhuaziei' (The Restoration of Idealism and the Struggle with the 'New' Bourgeoisie), *PZM*, 1922, No. 7–8.

—— 'Sovetskaia nauka' (Soviet Science), *Pechat' i revoliutsiia*, 1923, No. 3.

NORDENSKIOLD, E. *The History of Biology*. New York, 1928.

NOVIKOV, S. 'Voinstvuiushchaia putanitsa' (Militant Confusion), *PZM*, 1930, No. 7–8.

NOVINSKII, I. 'Za marksistsko-leninskuiu liniiu v biologii' (For a Marxist-Leninist Line in Biology), *Revoliutsiia i kul'tura*, 1930, No. 19–20.

'O proletkul'takh; pis'mo Ts. K. R.K.P.' (On the Proletcult Organizations; a Letter from the Central Committee of the Russian Communist Party), *Pravda*, December 1, 1920.

ORLOV, I. E. 'Chto takoe materiia?' (What Is Matter?), *Krasnaia nov'* (Red Virgin Soil), 1924, July, Kn. 4.

—— 'Dialektika eksperimenta' (The Dialectics of Experiment), *VSA*, 1923, Kn. 6.

—— 'Ischislenie sovmestnosti predlozhenii' (The Calculus of the Compatibility of Propositions), *Matematicheskii sbornik* (Mathematical Collection), XXXV.

—— 'Klassicheskaia fizika i reliativizm' (Classical Physics and Relativism), *PZM*, 1924, No. 3.

—— *Logika estestvoznaniia* (The Logic of Natural Science). Moscow-Leningrad, 1925.

—— 'Mekhanika i dialektika v estestvoznanii' (Mechanics and Dialectics in Natural Science), *Dialektika v prirode* (Dialectics in Nature), 1927, No. 2.

—— 'Novye variatsii na staruiu temu' (New Variations on an Old Theme) *Voinstvuiushchii materialist* (The Militant Materialist), 1924, No. 2.

—— 'O dialekticheskoi taktike v estestvoznanii' (Concerning Dialectical Tactics in Natural Science), *Dialektika v prirode* (Dialectics in Nature), 1928, No. 3.

—— 'O printsipe nauchnogo ob'iasneniia iavlenii' (On the Principle of a Scientific Explanation of Phenom ena), *Voinstvuiushchii materialist* (The Militant Materialist), 1925, No. 3.

—— 'O ratsionalizatsii umstvennogo truda' (On the Rationalization of Mental Labour), *PZM*, 1926, No. 12.

—— 'O zakonakh sluchainikh iavlenii' (On the Laws of Accidental Phenomena), *PZM*, 1924, No. 7–8.

—— 'Ob induktivnom dokazatel'stve' (Concerning Inductive Proof), *Voprosy filosofii i psikhologii* (Problems of Philosophy and Psychology), 1916, Kn. 135 (v), pp. 356–388.

—— 'Realizm v estestvoznanii i induktivnyi metod' (Realism in Natural Science and the Inductive Method), *Voprosy filosofii i psikhologii* (Problems of Philosophy and Psychology), 1916, Kn. 131 (i), pp. 1–35.

—— 'Teoriia sluchainosti i dialektika' (The Theory of Contingency and Dialectics), *PZM*, 1926, No. 9–10.

—— 'Zadachi dialekticheskogo materializma v fizike' (The Tasks of Dialectical Materialism in Physics), in *Teoriia otnositel'nosti i materializm* (The Theory of Relativity and Materialism). Leningrad, 1925.

—— 'Zdravyi smysl i ego ideolog' (Common Sense and Its Ideologist), *PZM*, 1925, No. 1–2.

'Osnovnye ustanovki i puti razvitiia sovetskoi ekologii' (Fundamental Viewpoints and Paths of Development for Soviet Ecology), *Sovetskaia botanika*, 1934, No 3.

OSTROVITIANOV, K. V. 'Itogi III s'ezda nauchnykh rabotnikov' (Results of the Third Congress of Scientific Workers), *Nauchnyi rabotnik*, 1929, No. 4.

—— 'K predstoiashchemu plenumu tsentral'nogo soveta' (On the Forthcoming Plenum of the Central Council), *Nauchnyi rabotnik*, 1930, No. 1.

—— 'O perspektivakh i metodakh dal'neishei raboty sektsii nauchnykh rabotnikov' (Concerning the Perspectives and Methods of Further Work of the Section of Scientific Workers), *Nauchnyi rabotnik*, 1929, No. 7–8.

PARAMONOV, S. 'O partiinoi chistke v AN' (On the Party Purge in the Academy of Sciences), *Front nauki i tekhniki*, 1933, No. 12.

PASHUKANIS, E. 'General Theory of Law and Marxism', in Babb and Hazard, *Soviet Legal Philosophy*. Cambridge, 1951.

PAVLOV, I. P. *Lectures on Conditioned Reflexes*. New York, 1928.

—— 'Pamiati Aleksandrova Filippovicha Samoilova' (In Memory of Alexander Filippovich Samoilov), *Kazanskii meditsinskii zhurnal* (The Kazan Medical Journal), 1931, No. 4–5.

'Pered obnovleniem Akademii Nauk SSSR' (Before the Renovation o the Academy of Sciences of the USSR), *Izvestiia*, 1928, 21 July.

PEROV, S. S. 'Dialektika v biokhimii' (Dialectics in Biochemistry), *Dialektika v prirode* (Dialectics in Nature), 1929, No. 4.

—— 'Dialektika v dispersnoi khimii' (Dialectics in Dispersive Chemistry), *Dialektika v prirode* (Dialectics in Nature), 1928, No. 3.

—— and A. K. Timiriazev and S. Varjas. 'S bol'noi golovy na zdorovuiu' (Blaming Others for One's Own Misdeeds), *PZM*, 1931, No. 1.

—— and A. K. Timiriazev and S. Varjas. 'Zaiavlenie v redaktsiiu "Dialektiki v prirode"' (Declaration to the Editors of 'Dialectics in Nature'), *Dialektika v prirode* (Dialectics in Nature), 1928, No. 3.

'Pervaia vsesoiuznaia konferentsiia marksistsko-leninskikh nauchno-issledovatel'skikh uchrezhdenii' (The First All-Union Conference of Marxist-Leninist Scientific Research Institutions), *VKA*, 1928, Kn. 26/2/, and Kn. 27/3/.

PISTRAK, M. 'Puti ideologicheskogo zavoevaniia uchitel'stva povyshennoi shkoly' (Ways to the Ideological Winning of the Teaching Staff of the Higher School), *Kommunisticheskaia revoliutsiia* (Communist Revolution), 1928, No. 14.

PLEKHANOV, G. V. *Fundamental Problems of Marxism*. New York, 1929.

—— *God na rodine* (A Year in the Homeland). 2 vols. Paris, 1921.

—— *Izbrannye filosofskie proizvedeniia* (Selected Philosophical Works). 5 vols. Moscow, 1956–59.

—— *Protiv filosofskogo revizionizma* (Against Philosophical Revisionism). Moscow, 1935.

—— *Sochineniia* (Works). 26 vols. Moscow, 1922–27.

PLEKHANOVETS. 'Zametki chitatelia o knige tov. Bukharina "Teoriia istoricheskogo materializma"' (A Reader's Notes on Comrade Bukharin's Book, *The Theory of Historical Materialism*), *PZM*, 1922, No. 11–12.

'Podgotovka nauchnykh kadrov i Akademiia Nauk' (The Training of Scientific Cadres and the Academy of Sciences), *VKA*, 1929, Kn. 35–36.

PODVOLOTSKII, I. P. '"Dialektika v prirode"' (Dialectics in Nature), *VKA*, 1927, Kn. 20.

—— and Gessen. 'Filosofskie korni pravogo opportunizma' (The Philosophical Roots of Right Opportunism), *PZM*, 1929, No. 9.

—— *Marksistskaia teoriia prava* (The Marxist Theory of Law), 2nd ed.; Moscow, 1925.

—— 'Shest'nadtsatyi s'ezd VKP (b) i zadachi na filosofskom fronte'

(The Sixteenth Congress of the CPSU, and the Tasks on the Philosophical Front), *PZM*, 1930, No. 7–8.

POKROVSKY, M. N. 'A. A. Bogdanov', *VKA*, 1928, Kn. 26/2/.

—— 'O deiatel'nosti Kommunisticheskoi Akademii' (Concerning the Activity of the Communist Academy), *VKA*, 1927, Kn. 22.

—— 'O podgotovke nauchnoi smeny' (On the Preparation of the Scientific Change-over), *Kommunisticheskaia revoliutsiia* (Communist Revolution), 1929, No. 13.

—— 'Obshchestvennye nauki v SSSR za 10 let' (The Social Sciences in the USSR for Ten Years), *VKA*, 1928, Kn. 26/2/.

POPOV, N. *Outline History of the Communist Party of the Soviet Union.* 2 vols. New York, 1934.

'Postanovleni TsK VKP (b) o zhurnale Pod znamenem Marksizma' (Decree of the CC CPSU Concerning the Journal Under the Banner of Marxizm), *Pravda*, January 26, 1930.

'Postanovlenie TsK VKP (b) ot marta 1931 g. po dokladu prezidiuma Kommunisticheskoi Akademii' (Decree of the CC CPSU of March 15, 1931, in Accordance with the Report of the Presiduim of the Communist Academy), *VKA*, 1931, No. 2–3.

POZNER, V. M. 'Fenomenalizm i realizm v sovremennom estestvoznanii' (Phenomenalism and Realism in Contemporary Natural Science), *Sputnik kommunista* (The Communist's Companion), 1923, No. 18.

PRIMAKOVSKII, A. P. *Bibliografiia po logike* (Bibliography on Logic). Moscow, 1955.

'Priroda elektricheskogo toka' (The Nature of an Electrical Current), *Elektrichestvo*, 1930, No. 3, 8, 10.

PROKOF'EV, P. 'Krizis sovetskoi filosofii' (The Crisis of Soviet Philosophy), *Sovremennye zapiski* (Contemporary Notes), XL, 1930.

—— 'Sovetskaia filosofiia' (Soviet Philosophy), *Sovremennye zapiski* (Contemporary Notes), XXXIII, 1927.

Protokoly soveshchaniia rasshirennoi redaktsii 'Proletariia' Iiun' 1909 g. (Proceedings of the Expanded Editorial Board of 'The Proletarian' June, 1909). Moscow, 1934.

RADEK, K. 'Po tu ili druguiu storonu barrikady' (On One Side of the Barricade or the Other), *VARNITSO*, 1930, No. 7–8.

RAL'TSEVICH, V N , and R. IANKOVSKII (eds.). *'Materializm i empiriokrititsizm' V. I. Lenina* (V. I. Lenin's 'Materialism and Empiriocriticism'). Moscow, 1935.

—— *Na dva fronta* (On Two Fronts). Moscow-Leningrad, 1931.

—— 'Protiv mistitsizma v estestvoznanii' (Against Mysticism in Natural Science), *Revoliutsiia i kul'tura*, 1930, No. 4.

Raznoglasiia po filosofskom fronte (Disagreements on the Philosophical Front). Moscow, 1931.

RAZUMOVSKII, I. P. 'Filosofiia prava' (The Philosophy of Law), in *Entsiklopediia gosudarstva i prava* (The Encyclopedia of the State and Law), III. Moscow, 1927.

—— 'Filosofskaia reviziia i voprosy prava' (Philosophical Revision and the Problems of Law), *PZM*, 1926, No. 7–8.

—— *Kurs istoricheskogo materializma* (A Course in Historical Materialism). 1st ed.; Moscow, 1924. 2nd ed.; Moscow, 1927.

—— 'Nashi "zamvriplekhanovtsy"' (Our 'Zamvrid-Plekhanovians'), *PZM*, 1923, No. 12.

—— 'Noveishie otkroveniia Karla Kautskogo' (The Newest Discoveries of Karl Kautsky), *VKA*, 1928, Kn. 29/5/.

—— 'Sushchnost' ideologicheskogo vozzreniia' (The Essence of the Ideological Outlook), *VSA*, 1923, Kn. 4.

—— 'V debriakh mekhanisticheskogo kritiki' (In the Jungles of Mechanistic Criticism), *VKA*, 1929, Kn. 31/1/.

REISNER, M. A. 'Freid i ego shkola o religii' (Freud and his School on Religion), *Pechat' i revoliutsiia* (The Press and The Revolution), 1924, Nos. 1 and 3.

—— 'Ideologiia i politika' (Ideology and Politics), *VKA*, 1929, Kn. 33/3/.

—— 'Problema psikhologii v teorii istoricheskogo materializma' (The Problem of Psychology in the Theory of Historical Materialism), *VSA*. 1923, Kn. 3.

—— 'Refleksy Pavlova i ideologiia Marksa' (The Reflexes of Pavlov and the Ideology of Marx), *Oktiabr' mysli* (The October of Thought), 1924, No. 5–6.

'Rezoliutsiia o sovremennykh problemakh filosofii marksizma-leninizma (po dokladu t. A. M. Deborina)' (Resolution Concerning the Contemporary Problems of the Philosophy of Marxism-Leninism [in accordance with the Speech of Comrade A. M. Deborin]), *VKA*, 1929, Kn. 32/2/.

'Rezoliutsiia o zadachakh marksistov v oblasti estestvoznaniia (po dokladu t. O. Iu. Shmidta)' (Resolution Concerning the Tasks of Marxists in the Field of Natural Science [in accordance with the Speech of Comrade O. Iu. Shmidt]), *VKA*, 1929, 32/2/.

RIAZANOV, D. B. 'Marks i Engel's o dialektike prirody' (Marx and Engels on the Dialectics of Nature), in Engels, *Dialektika prirody* (Dialectics of Nature). Moscow-Leningrad, 1930.

—— 'Novye dannye o literaturnom nasledstve K. Marksa i F. Engel'sa' (New Data on the Literary Heritage of K. Marx and F. Engels), *VSA*, 1923, Kn. 6.

—— 'Predislovie redaktora ko vtoromu izdaniiu' (The Editor's Foreword to the Second Edition), in Engels, *Dialektika prirody* (The Dialectics of Nature). Moscow-Lenigrad, 1930.

ROMANOVA, S. 'Nekotorye vyvody iz chistki v Akademii Nauk' (Some Conclusions from the Purge in the Academy of Sciences), *Izvestiia*, September 10, 1929.

ROZANOV, IA. S. *Filosofsko-sotiologicheskaia literatura marksizma za pervoe desiatiletie sovetskoi vlasti* (Philosophico-Sociological Literature of Marxism for the First Decade of the Soviet Regime). Moscow, 1928.

ROZENTAL', M., and P. IUDIN. *Kratkii filosofskii slovar'* (Brief Philosophical Dictionary). 1st–4th eds.; Moscow, 1939–54.

RUDAS, L. (or RUDASH, V. in Russian transliteration) 'Ortodoksal'nyi marksizm' (Orthodox Marxism) *VKA*, 1924, Kn. 8, 9.

RUDNIANSKII, S. *Besedy po filosofii materializma* (Talks on the Philosophy of Materialism). Kharkov, 1924.

RUDNIK, D. '"Materializm i empiriokrititsizm" kak dokument partiinoi bor'by' ('Materialism and Empirio-criticism' as a Document of Party Struggle), in V. Ral'tsevich and R. Iankovskii, (eds.). '*Materializm i emp iriokrititsizm' V. I. Lenina*. Moscow, 1935.

RUMII, V. 'Otvet odnomu iz talmudistov' (Reply to One of the Talmudists), *PZM*, 1923, No. 8–9.

RUNES, D. D., editor. *The Dictionary of Philosophy*. New York, n.d.

SAMOILOV, A. *Detskaia bolezn' 'levizny' v materializme* (The Childhood Disease of 'Leftism' in Materialism). Leningrad, 1926.

SAMOILOV, ALEKSANDR FILIPPOVICH. 'Dialektika prirody i estestvoznaniie' (The Dialectics of Nature and Natural Science), *PZM*, 1926, No. 4–5.

—— *Izbrannye stat'i i rechi* (Selected Articles and Speeches). Moscow-Leningrad, 1946.

SAPIR, I. D. 'Protiv idealizma v biologii' (Against Idealism in Biology), *Sputnik kommunista* (The Communist's Companion), 1923, Nos. 24, 25.

SAPOZHNIKOV, P. F. 'Novaia pobeda dialekticheskogo materializma' (A New Triumph of Dialectical Materialism), *Bol'shevik*, 1925, No. 16.

SARAB'IANOV,*z* V. N. 'Beglye vospominaniia' (Hasty Reminiscences), *Antireligio nik* (The Anti-religionist), 1927, No. 10.

—— 'Dialektika i formal'naia logika' (Dialectics and Formal Logic), *PZM*, 1922, No. 3.

—— *Istoricheskii materializm* (Historical Materialism). 1st–8th eds.; Moscow, 1922–26.

—— 'Kak inye tovarishchi tvoriat revizionizm' (How Some Comrades Create Revisionism), *PZM*, 1926, No. 6.

—— 'Nauka' (Science), *Antireligioznik* (The Anti-religionist), 1926, No. 2.

—— 'Nazrevshii vopros' (A Gathering Problem), *Sputnik kommunista* (The Communist's Companion), 1923, No. 20.

—— 'O filosofskikh raznoglasiiakh' (Concerning the Philosophical Disagreements), *Iunyi kommunist* (The Young Communist), 1928, No. 2.

—— 'O nekotorykh spornykh problemakh dialektiki' (Concerning Several Controversial Problems of Dialectics), *PZM*, 1925, No. 12.

Sbornik obshchestva istoricheskikh, filosofskih i sotsial'nykh nauk pri Permskom Universitete (Collection of the Society of Historical, Philosophical, and Social Sciences at Perm University), 1927, *vypusk* 2, and 1929, *vypusk* 3.

SCHILPP, P. A. (ed.). *Albert Einstein: Philosopher-Scientist*. New York, 1951.

SCHLESINGER, R. 'Neue sowjetrussische Literatur zur Sozialforschung: Moderne Phyzik und Philosophie', *Zeitschrift für Sozialforschung*, 1939, No. 1/2.

SCHMIDT, HEINRICH. *Philosophisches Wörterbuch*. Leipzig, 1931.

Science at the Crossroads. London, 1931.

SÉE, HENRI. *Matérialisme historique et interprétation économique de l'histoire.* Paris, 1927.

SEMASHKO, N. A. 'Smert' A. A. Bogdanova' (The Death of A. A. Bogdanov), *Pravda*, 1928, 8 April.

SEMKOVSKII, S. IU. *Dialekticheskii materializm i printsip otnositel'nosti* (Dialectical Materialism and the Principle of Relativity). Moscow, 1926.

—— *Etiudy po filosofii marksizma* (Studies in the Philosophy of Marxism). Moscow, 1925.

—— 'Fizika i khimiia kak nauki v svete marksizma' (Physics and Chemistry as Sciences in the Light of Marxism), in Institut Marksa-Engel'sa-Lenina, *Marksizm i estestvoznanie.* Moscow, 1933.

—— *Istoricheskii materializm* (Historical Materialism). 3rd ed.; Moscow, 1922.

SEREBROVSKII, A. S. 'Antropogenetika i evgenika v sotsialisticheskom obshchestve' (Anthropo-genetics and Eugenics in Socialist Society), *Mediko-biologicheskii zhurnal*, 1929, No. 5.

—— 'Chetyre stranitsy, kotorye vzvolnovali uchenyi mir' (Four Pages that Shook the Scientific World), *Pravda*, Steptember 11, 1927.

—— 'Opyt kachestvennoi kharakteristiki protsessa organicheskoi evoliutsii' (Sketch of a Qualitative Characterization of the Process of Organic Evolution), *Estestvoznanie i marksizm*, 1929, No. 2.

—— 'Ortogenez P. V. Serebrovskogo' (The Orthogenesis of P. V. Serebrovskii), *Estestvoznanie i marksizm*, 1930, No. 1.

—— 'Problema svedeniia v evoliutsionnom uchenii' (The Problem of Reduction in Evolutionary Theory), *Nauchnoe slovo*, 1930, No. 9.

—— 'Teoriia nasledsvennosti Morgana i Mendelia i marksisty' (The Theory of Heredity of Mendel and Morgan, and the Marxists), *PZM*, 1926, No. 3.

SEREBROVSKII, P. V. 'Darvinizm i "darvinizm"' (Darwinism and 'Darwinism'), *PZM*, 1929, No. 4.

—— 'Dialektika v evoliutsionnom protsesse' (Dialectics in the Evolutionary Process), *PZM*, 1930, No. 10–12.

SESEMANN, V. 'Die bolschewistische Philosophie in Sowjet-Russland', in *Festschrift N. O. Losskij zum 60 Geburtstage.* Bonn, 1932.

SHCHEGLOV, A. B. *Bor'ba Lenina protiv bogdanovskoi revizii marksizma* (Lenin's Struggle against the Bogdanovian Revision of Marxism). Moscow, 1937.

SHCHIPANOV, I. Ia. (ed.). *Iz istorii russkoi filosofii* (From the History of Russian Philosophy). Moscow, 1951.

SHEIN, IU. P. 'Metafizika fiziki; noveishee razvitie kvantovoi teorii i dialekticheskii materializm' (The Metaphysics of Physics; the Newest Development of Quantum Theory and Dialectical Materialism), *Problemy marksizma*, 1931, No. 2.

—— 'Opredelenie skorosti' (The Definition of Velocity), *Priroda*, 1931, No. 9.

SHELKOPLIAS, N. 'K itogam bor'by na filosofskom fronte' (On the Results of the Struggle on the Philosophical Front), *PZM*, 1936, No. 1.

Shestyi s'ezd russkikh fizikov; 5–16 Avgusta 1928 (The Sixth Congress of Russian Physicists; 5–16 August 1928, Moscow, 1928.

SHMIDT, O. Iu., and B. Ia. SHMULEVICH (eds.). *Nauchnye kadry i nauchnoissledovatel'skie uchrezhdeniia SSSR* (Scientific Cadres and Scientific Research Institutions of the USSR). Moscow, 1930.

—— 'Rol' matematiki v stroitel'stve sotsializma' (The Role of Mathematics in the Construction of Socialism), *Trudy pervogo vsesoiuznogo s'ezda matematikov, Khar'kov, 1930 g.* (Proceedings of the First All-Union Congress of Mathematicians, Kharkov, 1930). Moscow, 1936.

—— 'Vysshie uchebnye zavedeniia v SSSR' (Institutions of Higher Learning in the U.S.S.R.), in *Bol'shaia sovetskaia entsiklopediia* (Great Soviet Encyclopedia) (1st ed.), XIV.

—— *et al. Zadachi marksistov v oblasti estestvoznaniia* (The Tasks of Marxists in the Field of Natural Science). Moscow, 1929.

SHPET, G. G. *Vnutrenniaia forma slova* (The Internal Form of a Word). Moscow, 1927.

—— *Vvedenie v etnicheskuiu psikhologiiu* (Introduction to Ethnic Psychology). Moscow, 1927.

SITKOVSKII, E. 'Lenin i Plekhanov kak kritiki makhizma' (Lenin and Plekhanov as Critics of Machism), Kommunisticheskaia Akademiia. *25 let 'Materializma i empirikrititsizma'* (Twenty-Five Years of Materialism and Empirio-criticism). Moscow, 1934.

The Situation in Biological Science. New York, 1949.

SLEPKOV, V. 'Vitalizm, mekhanizm i dialektika' (Vitalism, Mechanism, and Dialectics), *PZM*, 1926, No. 9–10.

SMIRNOV, E. S. 'Novye dannye o nasledstvennom vliianii sredy i sovremennyi lamarkizm' (New Data Concerning the Hereditary Influence of Environment, and Contemporary Lamarckism), *VKA*, 1928, Kn. 25/1/.

—— 'Problemy ucheniia o nasledstvennosti' (Problems of the Theory of Heredity), *Estestvoznanie i marksizm*, 1929, No. 2.

SMIT, M., and A. K. TIMIRIAZEV (eds.). *Statisticheskii metod v nauchnom issledovanii; opyt kollektivnoi internauchnoi raboty* (The Statistical Method in Scientific Research; an Effort at Collective Inter-disciplinary Work). Moscow, 1925.

SOBOLEV, D. N. 'Evoliutsiia kak organicheskii rost' (Evolution as Organic Growth), *Priroda*, 1929, No. 5.

Sobranie uzakonenii i rasporiazhenii rabochego i krest'ianskogo pravitel'stva R.S.F.S.R. (Collection of Laws and Ordinances of the Workers' and Peasants' Government of the R.S.F.S.R.).

Sobranie zakonov i rasporiazhenii raboche-krest'ianskogo pravitel'stva SSSR (Collection of Laws and Ordinances of the Worker-Peasant Government of the USSR).

SOMERVILLE, J. *Soviet Philosophy; a Study of Theory and Practice.* New York, 1946.

SORIN, V. V. 'O filosofskikh tetradkakh Lenina' (Concerning the philosophical notebooks of Lenin), *Pravda*, 1926, 1 August.

Spornye voprosy genetiki i selektsii; raboty IV sessii Akademii sel'skokhoziaistvennykh nauk, 19–27 dek. 1936 g. (Controversial Problems of Genetics and Selection; Transcript of the Fourth Session of the Academy of Agricultural Sciences, 19–27 Dec., 1936). Moscow, 1937.

STALIN, J. V. *Problems of Leninism.* Moscow, 1947.

—— *Sochineniia* (Works). 13 vols. to date. Moscow, 1946–.

STAUFFER, R. C. (ed.). *Science in Civilization.* Madison, Wis., 1949.

STEKLOV, V. A. *Galileo Galilei.* Berlin, 1923.

STEN, IA. E. 'O tom, kak tov. Stepanov zabluzhdilsia sredi neskol'kikh tsitat Marksa i Engel'sa' (How Comrade Stepanov Went Astray among Some Quotations from Marx and Engels), *Bol'shevik,* 1924, No. 15–16.

—— 'Ob oshibkakh Gortera i Stepanova' (Concerning the Errors of Gorter and Stepanov), *Bol'shevik,* 1924, No. 11.

—— 'Nasha partiia i voprosy teorii' (Our Party and Problems of Theory), *Bol'shevik,* 1924, No. 12–13.

STEPANOV, I. I. *Dialekticheskii materializm i deborinskaia shkola* (Dialectical Materialism and the Deborinite School). Moscow-Leningrad, 1928.

—— 'Dialekticheskoe ponimanie prirody—mekhanisticheskoe ponimanie' (The Dialectical Conception of Nature is the Mechanistic Conception), *PZM,* 1925, No. 3.

—— *Elektrifikatsiia RSFSR v sviazi s perekhodnoi fazoi mirovogo khoziaistva* (The Electrification of the RSFSR in Connection with the Transitional Phase of the World Economy). Moscow, 1923.

—— 'Engel's i mekhanisticheskoe ponimanie prirody' (Engels and the Mechanistic Conception of Nature), *PZM,* 1925, No. 8–9.

—— *Istoricheskii materializm i sovremennoe estestvoznanie* (Historical Materialism and Contemporary Natural Science). 2nd ed.; Moscow, 1925.

—— 'O moikh oshibkakh, "otkrytykh i ispravlennykh" tov. Stenom' (About My Errors, 'Discovered and Corrected' by Comrade Sten), *Bol'shevik,* 1924, No. 14.

—— 'Predislovie' (Foreword), in Gorter, G., *Istoricheskii materializm* (Historical Materialism). Moscow, 1919.

—— *Skvortsov-Stepanov, politicheskii boets* (Skvortsov-Stepanov, a Political Warrior). Moscow, 1930.

STETSKII, A. 'O Komakademii i nauchnoi rabote' (On the Communist Academy and Scientific Work), *VKA,* 1931, No. 2–3.

—— 'Ob uproshchenstve i uproshchentsakh' (On Vulgarization and Vulgarizers), *Pravda,* June 4, 1932.

STOLIAROV, A. K. *Dialekticheskii materializm i mekhanisty; nashi filosofskie raznoglasiia* (Dialectical Materialism and the Mechanists; Our Philosophical Disagreements). 1st–4th eds.; Leningrad, 1928–30.

—— 'Filosofiia "kachestva" i kachestvo filosofii nekotorykh mekhanistov' (The Philosophy of 'Quality' and the Quality of the Philosophy of Some Mechanists), *PZM,* 1926, No. 6.

STRUMINSKII, V. *Psikhologiia; opyt sistematicheskogo islozheniia osnovnykh voprosov nauchnoi psikhologii s tochki zreniia dialekticheskogo materializma* (Psychology; an Attempt at a Systematic Exposition of the Fundamental Problems of Scientific Psychology from the Point of View of Dialectical Materialism). Orenburg, 1923.

STRUVE, P. B. *Kriticheskie zametki k voprosu ob ekonomicheskom razvitii Rossii* (Critical Notes on the Question of the Economic Development of Russia). St. Petersburg, 1894.

STUKOV, I. N. 'V chem sut' moikh vozrazhenii?' (What Is the Essence of My Objection?), *PZM*, 1923, No. 6–7.

—— 'V plenu u reliativizma' (A Prisoner of Relativism), *PZM*, 1923, No. 4–5.

TAMM, I. E. 'O rabote filosofov-marksistov v oblasti fiziki' (On the Work of Marxist Philosophers in the Field of Physics), *PZM*, 1933, No. 2.

—— 'Rukovodiashchie idei v tvorchestve Faradeia' (Leading Ideas in Faraday's Creative Work), *Uspekhi fizicheskikh nauk*, 1932, No. 1.

TARASENKOV, A. 'Filosofskie tetradi Lenina' (Lenin's Philosophical Notebooks), *Iunyi kommunist*, 1930, No. 7.

Teoriia otnositel'nosti i ee filosofskoe istolkovanie (The Theory of Relativity and its Philosophical Explanation). Moscow, 1923.

Teoriia otnositel'nosti i materializm (*The Theory of Relativity and Materialism*). Leningrad, 1925.

THALHEIMER, A. 'O nekotorykh osnovnykh poniatiiakh teorii otnositel'nosti s tochki zreniia dialekticheskogo materializma' (Some Fundamental Concepts of the Theory of Relativity from the Point of View of Dialectical Materialism), *PZM*, 1925, No. 1–2; a German version in *Unter dem Banner des Marxismus*, 1925, Heft 2.

THOMSON, J. J. *The Atomic Theory*. Oxford, 1914.

TIMIRIAZEV, A. K. 'Dialektika i estestvoznanie' (Dialectics and Natural Science), *PZM*, 1923, No. 4–5.

—— 'Einshtein, materializm i tov. Gol'tsman' (Einstein, Materialism, and Comrade Gol'tsman), *PZM*, 1924, No. 1.

—— *Filosofiia nauki; estestvenno-nauchnye osnovy materializma* (The Philosophy of Science; Natural Scientific Foundations of Materialism). Part I, *Fizika* (Physics). Moscow, 1923.

—— 'Iz oblasti "nashikh raznoglasii" s t. Deborinym' (From the Field of 'Our Disagreements' with Comrade Deborin), *Dialektika v prirode* (Dialectics in Nature), 1928, No. 3.

—— 'Lenin i sovremennoe estestvoznanie' (Lenin and Contemporary Natural Science), *PZM*, 1924, No. 2.

—— 'Neskol'ko zamechanii po povodu nastupleniia na materializm tov. Gol'tsmana' (Some Observations on Comrade Gol'tsman's Attack on Materialism), *PZM*, 1923, No. 6–7.

—— 'Ob ideologicheskoi baze vysshei skholy' (Concerning the Ideological Base of the Higher School), *Nauchnyi rabotnik* (The Scientific Worker), 1925, No. 1.

—— 'Obzor populiarnoi literatury o printsipe otnositel'nosti' (Survey of Popular Literature on the Theory of Relativity), *Krasnaia nov'* (Red Virgin Soil), 1921, No. 4.

—— 'Oprovergaet li sovremennaia elektricheskaia teoriia materii materializm?' (Does the Contemporary Electrical Theory of Matter Overthrow Materialism?), *PZM*, 1922, No. 4.

—— 'Otnositel'nosti, teoriia, i filosofiia" (Relativity, Theory of, and

Philosophy), *Tekhnicheskaia entsiklopediia* (Technical Encyclopedia), XV, 1930.

—— 'Periodicheskaia sistema elementov Mendeleeva i sovremennaia fizika' (Periodical System of Elements of Mendeleev and Contemporary Physics) *Krasnaia nov'* (Red Virgin Soil), 1921, No. 1.

—— 'Po povodu diskussii ob opytakh Deiton-Millera na piatom s'ezde russkikh fizikov' (Concerning the Discussion of Dayton Miller's Experiments at the Fifth Congress of Russian Physicists), *PZM*, 1927, No. 2–3.

—— 'Pokhod sovremennoi burzhuaznoi nauki protiv materializma v oblasti estestvoznaniia' (The Campaign of Contemporary Bourgeois Science against Materialism in the Field of Natural Science), *Sputnik kommunista* (The Communist's Companion), 1922, No. 18.

—— 'Printsip otnositel'nosti' (The Principle of Relativity), *Krasnaia nov'* (Red Virgin Soil), 1921, No. 2.

—— and S. S. Perov and S. Varjas. 'S bol'noi golovy na zdorovuiu' (Blaming Others for One's Own Misdeeds), *PZM*, 1931, No. 1.

—— and M. Smit (eds.). *Statisticheskii metod v nauchnom issledovannii; opyt kollektivnoi internauchnoi raboty* (The Statistical Method in Scientific Research; an Effort at Collective, Inter-disciplinary Work). Moscow, 1925.

—— 'Teoriia otnositel'nosti Einshteina i makhizm' (Einstein's Theory of Relativity and Machism), *VKA*, 1924, Kn. 7.

—— 'Teoriia otnositel'nosti i dialekticheskii materializm' (The Theory of Relativity and Dialectical Materialism), *PZM*, 1924, No. 7–8 and 10–11.

—— 'Voskresaet li sovremennoe estestvoznanie mekhanicheskii materializm XVIII veka?' (Is Contemporary Natural Science Reviving the Mechanical Materialism of the 18th Century?), *VKA*, 1926, Kn. 17.

—— and PEROV and VARJAS. 'Zaiavlenie v redaktsiiu' (Declaration to the Editors), *Dialektika v prirode* (Dialectics in Nature), 1923, No. 3.

TIMIRIAZEV, K. A. *Nauka i demokratiia* (Science and Democracy). Moscow, 1920.

TORBEK, G. 'Deiatel'nost' Kommunisticheskoi Akademii: rabota plenuma, prezidiuma, sektsii i institutov' (The Work of the Communist Academy: Activity of the Plenum, Presidium, Sections and Institutes), *VKA*, 1929, Kn. 33/3/.

TROITSKII, A. 'Zazorno li bol'sheviku zanimat'sia filosofiei?' (It It Shameful for a Bolshevik to Study Philosophy?), *Bol'shevik*. 1925, Nos. 17–18, 19–20.

TROTSKY, L. 'Darvinizm i marksizm' (Darwinism and Marxism), *Pravda*, 1923, No. 139, 140.

—— 'L. D. Trotskii o religii' (L. D. Trotsky on Religion), *Antireligioznik* (Anti-religionist), 1926, No. 4.

—— *Literature and Revolution*. New York, 1957.

—— 'Pis'mo tov. L. D. Trotskogo' (A Letter from Comrade L. D. Trotsky), *PZM*, 1922, No. 1–2.

—— 'Radio, nauka, tekhnika i obshchestvo' (Radio, Science, Technology, and Society), *Krasnaia nov'* (Red Virgin Soil), 1927, No. 2.

—— *Sochineniia* (Works). XXI. Moscow-Leningrad, 1927.

—— *The Third International after Lenin*. New York, 1936.

Trudy Instituta Krasnoi Professury (Works of the Institute of Red Professorship), 1923, 1 vol. only.

Trudy pervogo vsesoiuznogo s'ezda matematikov, Khar'kov 1930 g. (Proceedings of the First All-Union Congress of Mathematicians, Kharkov, 1930.) Moscow, 1936.

Trudy pervoi vsesoiuznoi konferentsii agrarnikov-marksistov (Proceedings of the First All-Union Conference of Marxist Agrarian Economists). 2 vols. in 1. Moscow, 1930.

TSEITLIN, Z. A. 'Metod dokazatel'stva zakona vzaimodeistviia tiazhelykh i elektricheskikh mass Niutona-Kavendisha-Maksvella sravnitel'no s metodom issledovaniia K. Marksa i F. Engel'sa' (The Method of Proof of the Law of Interaction of Gravitational and Electrical Masses of Newton-Cavendish-Maxwell in Comparison with the Method of Research of K. Marx and F. Engels), *PZM*, 1924, No. 6–7.

—— *Nauka i gipoteza* (Science and Hypothesis). Moscow, 1926.

—— 'O matematicheskom metode v estestvoznanii i politicheskoi ekonomii' (Concerning the Mathematical Method in Natural Science and Political Economy), *VKA*, 1927, Kn. 23.

—— 'Problema real'nogo obosnovaniia evklidovoi geometrii' (The Problem of the Real Foundation of Euclidean Geometry), *Dialektika v prirode* (Dialectics in Nature), 1928, No. 3.

—— 'Problema vsemirnogo tiagoteniia' (The Problem of Universal Gravitation), *VKA*, 1925, Kn. 13.

—— 'Ratsional'nyi i formal'nyi dialekticheskii materializm' (Rational and Formal Dialectical Materialism), *Voinstvuiushchii materialist* (The Militant Materialist), 1924, No. 1.

UDAL'TSOV, A. 'K kritike teorii kassov u A. A. Bogdanova' (Toward a Critique of the Theory of Classes in A. A. Bogdanov), *PZM*, 1922, No. 7–8.

VAIL', S. 'Vtoroi vsesoiuznyi s'ezd patologov' (The Second All-Union Congress of Pathologists). *Estestvoznanie i marksizm*, 1930, No. 2–3.

VAISBERG, I. A. 'Protiv mekhanisticheskoi genetiki' (Against Mechanistic Genetics), *Problemy marksizma*, 1931, No. 2.

VALENTINOV, N. *Vstrechi s Leninym* (Meetings with Lenin). New York, 1953.

VANDEK, V., and V. TIMOSKO. 'Vstupitel'naia stat'ia' (Introductory Article), Plekhanov, G. V. *Protiv filosofskogo revizionizma* (Against Philosophical Revisionism). Moscow, 1935.

VARJAS, S. (or VAR'IASH, A. in Russian transliteration). 'Formal'naia i dialekticheskaia logika' (Formal and Dialectical Logic), *PZM*, 1923, No. 6–7.

—— 'Istoriia filosofii i marksistskaia filosofiia istorii' (The History of Philosophy and the Marxist Philosophy of History), *VKA*, 1924, Kn. 9.

—— *Istoriia novoi filosofii* (The History of Modern Philosophy). 2 vols. Moscow-Leningrad. 1925–26.

—— *Logika i dialektika* (Logic and Dialectics). Moscow-Leningrad, 1928.

—— 'O tom, kak ne nado pisat' kritiku' (How Criticism Should Not Be Written), *PZM*, 1925, No. 5–6.

—— 'Ob obshchikh zakonakh dialektiki v knige Engel'sa "Dialektika prirody"' (Concerning the General Laws of Dialectics in Engels' Book, *Dialectics of Nature*) *Dialektika v prirode* (Dialectics in Nature), 1928, No. 3.

—— and S. S. Perov and A. K. Timiriazev. 'S bol'noi golovy na zdorovuiu' (Blaming Others for One's Own Misdeeds), *PZM*, 1931, No. 1.

—— and Timiriazev and Perov. 'Zaiavlenie v redaktsiiu' (Declaration to the Editors), *Dialektika v prirode* (Dialectics in Nature), 1928, No. 3.

VASIL'EV, A. V. *Prostranstvo, vremia, dvizhenie* (Space, Time, Motion). Petrograd, 1923. English translation: London, 1925.

VAVILOV, N. I. 'Geograficheskaia lokalizatsiia genov pshenits na zemnom share' (The Geographical Localization of Wheat Genes on the Earth), *PZM*, 1929, No. 6.

VAVILOV, S. I. *Eksperimental'nye osnovaniia teorii otnositel'nosti* (The Experimental Foundations of the Theory of Relativity). Moscow-Leningrad, 1928.

VERGER, V. 'Desiatiletie professional'noi nauchnoi obshchestvennosti' (A Decade of Public Opinion among Scientists), *Front nauki i tekhniki* (The Front of Science and Technology), 1933, No. 10–11.

VERNADSKII, V. I. *Biosfera* (The Biosphere). Leningrad, 1926.

—— 'Evoliutsiia vidov i zhivoe veshchestvo' (The Evolution of Spcies and Living Matter), *Priroda*, 1928, No. 3.

—— *Nachalo i vechnost' zhizni* (The Beginning and the Eternality of Life). Petrograd, 1922.

—— *Ocherk i rechi* (Essays and Speeches). 2 vols. in one. Petrograd, 1922.

—— 'Problema vremeni v sovremennoi nauke' (The Problem of Time in Contemporary Science), *Izvestiia Akademii Nauk, otdelenie matematicheskikh i estestvennykh nauk* (News of the Academy of Sciences, Division of Mathematical and Natural Sciences), 1932, No. 4.

VISHNEVSKII, A. F. 'V zashchitu materialisticheskoi dialektiki' (In Defence of Materialist Dialectics), *PZM*, 1925, No. 8–9.

VOLGIN, V. P. *Akademiia nauk za chetyre goda, 1930–1933* (The Academy of Sciences for Four Years, 1930–1933). Leningrad, 1934.

VOLIN, L. 'Science and Intellectual Freedom in Russia', in Christman (ed.) *Soviet Science*. Washington, D.C., 1952.

VOSTOKOV, P. 'La philosophie russe durant la periode post-revolutionnaire', *Le monde slave*, 1932, Nos. 11, 12.

'Vrediteli v Akademii Nauk' (Wreckers in the Academy of Sciences), *Izvestiia*, November 16, 1929.

Vsesoiuznaia Kommunisticheskaia Partiia, Tsentral'nyi Komitet (VKP, Ts. K.). 'O meropriiatiiakh po ukrepleniiu nauchnoi raboty v sviazi s itogami vtoroi vsesoiuznoi konferentsii marksistsko-leninskikh nauchno-issledovatel'skikh uchrezhdenii' (Concerning Measures to

Strengthen Scientific Work in Connection with the results of the Second All-Union Conference of Marxist-Leninist Scientific Research Institutions), *VKA*, 1929, Kn. 33/3/.

—— 'O nauchnykh kadrakh' (Concerning Scientific Cadres), *VKA*, 1929, Kn. 34/4/.

—— 'O zhurnale Pod znamenem marksizma' (Concerning the Journal *Under the Banner of Marxism*), *Pravda*, 1931, 26 January.

'Vybory novykh akademikov', *Izvestiia*, January 13, 1929.

VYROPAEV, B. N. 'Eshche ob empirizme i sovremennom estestvoznanii (Once again on Empiricism and on Contemporary Natural Science), *PZM*, 1926, No. 11.

VYSHINSKY, A. 'O nashikh kadrakh' (Concerning Our Cadres), *Nauchnyi rabotnik*, 1930, No. 1.

—— 'O podgotovke novoi tekhnicheskoi intelligentsii' (On the Preparation of a New Technical Intelligentsia), *Kommunisticheskaia revoliutsiia* (Communist Revolution), 1928, No. 23–24.

WETTER, G. A. *Dialectical Materialism; A Historical and Systematic Survey of Philosophy in the Soviet Union.* London, 1958.

—— *Der dialektische Materialismus; seine Geschichte und sein System in der Sowjetunion.* Freiburg, 1952.

WOLFE, B. D. *Three Who Made a Revolution.* New York, 1948.

Za marksistsko-leninskoe uchenie o pechati (For a Marxist-Leninist Conception of the Press). Moscow, 1932.

ZAVADOVSKII, B. M. 'Darvinizm i lamarkizm i problema nasledovaniia priobretennykh priznakov' (Darwinism and Lamarckism and the Problem of the Inheritance of Acquired Characteristics), *PZM*, 1925, No. 10–11.

—— 'Darvinizm i marksizm' (Darwinism and Marxism), *VKA*, 1926, Kn. 14.

——'Itogi IV vsesoiuznogo s'ezda fiziologov' (Results of the Fourth All-Union Congress of Physiologists), *Estestvoznanie i marksizm*, 1930, No. 2–3.

ZAVADOVSKII, M. M. 'Etapy biologii (populiarnyi ocherk)' (Stages of Biology; a Popular Sketch), *Krasnaia nov'* (Red Virgin Soil), 1924, Kn. 2.

—— 'Geny i ikh rol' v osushchestvlenii priznaka' (Genes and their Role in the Realization of a Characteristic), *Estestvoznanie i marksizm*, 1929, No. 3.

ZINOV'EV, G. 'Leninizm i dialektika' (Leninism and Dialectics), *Bol'shevik* (The Bolshevik), 1925, No. 16.

ZIRKLE, C. *Death of a Science in Russia.* Philadelphia, 1949.

INDEX

INDEX

www.ingramcontent.com/pod-product-compliance
Lightning Source LLC
LaVergne TN
LVHW011104300926
849160LV00008B/276